Death of a Rat

Understandings and Appreciations of Science

William D. Stansfield, Ph.D.

Prometheus Books

59 John Glenn Drive
Amherst, New York 14228-2197

Published 2000 by Prometheus Books

Inquiries should be addressed to
Prometheus Books
59 John Glenn Drive
Amherst, New York 14228-2197
VOICE: 716-691-0133, ext. 207
FAX: 716-564-2711
WWW.PROMETHEUSBOOKS.COM

04 03 02 01 00   5 4 3 2 1

Library of Congress Cataloging-in-Publication Data

Stansfield, William D., 1930–
    Death of a rat : understandings and appreciations of science / William D.
Stansfield.
        p. cm.
    Includes bibliographical references and index.
    ISBN 1-57392-814-3 (cloth : alk. paper)
    1. Life sciences—Popular works. 2. Science—Popular works. I. Title.
QH309 .S78 2000
174'.957—dc21   00-020932
                                                                      CIP

Printed in the United States of America on acid-free paper

CONTENTS

PREFACE

In 1995, a laboratory rat died at a research center in California. Science came to a standstill there for six weeks. Three people resigned from the company.[1] How is it possible that the death of one rat could have such far-ranging consequences for the advancement of science? In the eyes of the general public, rats are considered to be vermin (animals such as insects and rats that are destructive, annoying, or injurious to health). Wild rats can indeed carry disease. For example, the bacterium *Yersenia pestis* that causes bubonic plague in humans is ubiquitous in rats and can be spread by the bite of some flea species. This germ was responsible for the Black Death that swept through Europe in the fourteenth century and killed a quarter of its population.[2] Rats also infest granaries, causing millions of dollars worth of damage each year. Many people are employed year round as exterminators of vermin. Why then should the death of one laboratory rat cause such a fuss?

An answer to this question (in the first chapter) requires an understanding of the extent to which ethical considerations for the welfare of experimental animals has permeated the conduct of modern research in the United States. This understanding begets an appreciation for the numerous kinds of impediments that prevent science from advancing as rapidly as we might wish it would. Each of these two major concerns about science (ethics and impediments) are addressed in separate chapters under the titles of "The Frankenstein Model" and "Stones in the Road," respectively.

The main purpose of *Death of a Rat* is indicated by its subtitle; i.e., to provide an audience of inquisitive minds (high school or adult level) in the general population (laity, nonspecialists) with an entertaining way to broaden its understanding and appreciation of science. This objective will be pursued through analyses of specific case studies from the history of science. A glance at the table of contents reveals the kinds of questions asked and the case studies selected for analyses. All readers may not agree with the author's interpretations. In any event, it is hoped that the reader will be stimulated to do independent thinking and perhaps further research in the same and/or different sources than those listed in the notes of this book. Direct quotations and many reference sources are well documented at the end of each chapter so that the reader can verify the authenticity of the factual base on which the author's interpretations were made. Specialized scientific jargon has been avoided wherever possible. Terms likely to be unfamiliar to many readers are defined whenever this was practical. The book reads more like a novel than most scientific discourse with one important distinction—the contents of each chapter can be comprehended without having read any other chapter(s). The final chapter reveals some of the intricate ways in which the other chapters are interconnected. No attempt was made to place case histories in chronological order, so the reader may find it exhilaratingly possible to freely jump from one century to another like a time traveler.

As the reader progresses through the chapters, she may be pleasantly surprised to find that many new and/or enhanced understandings and appreciations concerning a broad spectrum of scientific issues have been acquired either with or without any conscious effort toward these goals.

NOTES

1. Andrew Lawler, "Key NASA Lab under Fire for Animal Care Practices," *Science* 268 (23 June 1995): 1692.

2. Raúl J. Cano and Jaime S. Colomé, *Microbiology* (St. Paul, Minn.; West Publishing Company, 1986), p. 683.

CHAPTER 1

THE FRANKENSTEIN MODEL

Hollywood has made a lot of money making films based on Mary W. Shelley's 1818 novel about the mad scientist Dr. Frankenstein who put together a living monster from parts of dead people. This scenario has often been used as a model for the way that many people think scientists in general conduct their business, i.e., without regard for the moral implications of what they do. The ghoulish aspects of even attempting to construct a living creature from pieces of cadavers sends chills up the spines of most sensitive people. However, modern medicine has had considerable success with human transplants of hearts, kidneys, corneas, livers, and the like, from living or recently dead donors.

And yet, as wonderful as it might be to resurrect the dead, or better yet, to prevent death altogether, should science be encouraged or even allowed to do research that might eventually bring these miracles to pass? Should science be involved with attempts to extend the human life span? Isn't medical science doing just that already? When someone is ill, is it not ethical to try by all means to keep the patient alive in the hope that he might somehow be brought back to health? Even if there is no present way to restore dead brain tissue, is it not ethical to keep the body of a brain-dead patient alive on artificial life-support machines such as the heart-lung machine? Physicians certainly do not want to harm their patients. (The Hippocratic Oath is often erroneously cited as stating, "First do no harm." However, the closest Hippocrates (or any of the unknown authors of the Hippocratic corpus) came to expressing this admonition is in *Epidemics*, book 1, where a translation states, "As to diseases, make a habit of two things—to help, or at least to do no harm."[1]) Turning off a heart-lung machine to such a patient would surely do harm to the body even though the patient's brain was incapable of feeling anything. Medical science has many more possibilities open to it today than at any time in the past, and along with that new knowledge comes additional responsibilities and vexing ethical decisions.

According to my dictionary, "morality" is defined as "the quality of being moral; a set of ideas of right and wrong." Likewise, "ethics" is defined as "the study of the general nature of morals and of the specific moral choices to be made by the individual in his relationship with others; the rules or standards governing the conduct of the members of a profession." The adjective "ethical" means "in accordance with the accepted principles of right and wrong governing the conduct of a group." Some ethicists like to make a distinction between "ethics" as a set of ideal principles to which human behavior should conform (such as the Golden Rule or the Ten Commandments), and "morals" as the behavior that is actually exhibited by an individual or a society. One of the greatest problems with these definitions is that many of our most important ethical or moral problems do not have either all "good" (right) or all "bad" (wrong) solutions.

Consider what happens by extending our life expectancy. Unless health can be equally prolonged, the older we get, the more medical care may be required to keep us alive. If just staying alive is the all-important goal, that is one thing; but many of us want more out of life than that. Granted, there are people who, despite severe disabilities, continue to find life not only worthwhile, but also manage to continue to be productive. The scientist who immediately comes to mind in this regard is the theoretical physicist Stephen Hawking who became so debilitated by amyotrophic lateral sclerosis (motor neuron disease or Lou Gehrig's disease) that he could hardly move a finger or utter anything comprehensible (except to a few people such as his translator, Dennis Sciama). Nevertheless, Hawking has, for many years, occupied the chair of Lucasian Professor of Mathematics at Cambridge University—the same position once held by Sir Isaac Newton. Hawking not only continues to enliven cosmology and cosmogony with his theories, but also makes these scientific propositions intelligible to the general public in books such as *A Brief History of Time: From the Big Bang to Black Holes*, which was on the *New York Times*'s best-seller list for one hundred weeks.

Not everyone can deal as successfully with their infirmities as Hawking has done. Many people find old age so full of pain and devoid of human dignity that, to them, death seems to offer their only escape. Most old folks hate being a burden to others, be it their children, other relatives, friends, or even society. They want to be self-sufficient, both physically and financially. They hate to see their life savings depleted by medical bills, leaving no inheritance for their children; or even worse, causing their children to use up their own savings for medical payments and geriatric homecare. Even being dependent on welfare is detestable. So here is a moral dilemma: Is it wise to extend life as long as possible, even though the quality of life becomes a burden both to the patient and to those who love him?

Science itself is amoral, i.e., the facts and theories of scientific knowledge exist apart from questions of morality. However, scientific knowledge must be gained by people (scientists) using the methods of science. Both the motivations of scientists and the methods they use to gather and analyze their data are likely to have ethical implications for the scientific community and beyond. The uses to which scientific knowledge is put by individuals, businesses, political groups, or whatever are almost sure to involve questions of right versus wrong.

For example, was it morally right for the United States to develop the atomic bomb during World War II, knowing that its use could cause the death of thousands of people, and that many of the survivors were likely to suffer long-term health problems such as leukemia or other forms of cancer? One could argue that the bomb could have been used only as a threat to bring the war with Japan to a speedy conclusion. Alternatively, films of the first test (on July 16, 1945, near Alamogordo, New Mexico) could have been shown to the enemy or the bomb could have been dropped on some uninhabited spot near Japan to demonstrate its destructive potential. If Japan had surrendered under these conditions, it might be argued that the explosive power of the bomb could be used to benefit mankind in giant earth-moving projects such as the building of new dams. The moral argument that actually was used, however, was based on an estimate of the number of American lives that would be lost if Japan had to be invaded by Allied troops. At that time, the sentiment among most Americans was that no price was too great to pay if it would save the lives of American GIs. At the time the bomb was dropped, Japan's air strength was poor and could not defend against U.S. bombers flying at high altitudes above thirty thousand feet. Unlike the United States, Russia had not declared war on Japan. Japan wanted to arrange a negotiated surrender that would, among other conditions, assure that its divine emperor, Hirohito, would not be punished or humiliated in any way. Unbeknownst to Japan, the United States had broken its secret code. The United States intercepted a secret message from Japan to its Russian diplomat, requesting Russia to negotiate Japan's surrender with the United States. The United States was well aware of Russia's plans to expand its Communist empire both eastward into China and Korea (and possibly also into Japan) and westward into Europe. When President Truman learned of Japan's intent to involve Russia in the surrender negotiations, he quickly made the decision to use the atomic bomb. Truman would settle for nothing less than Japan's unconditional surrender. He also wanted to prevent Russia from gaining any toehold in the governance of postwar Japan. By using the bomb in an industrial area, the United States could demonstrate its awesome destructive power and, hopefully, deter Russia from any thoughts of further expansion of its empire. Furthermore, if mediation was allowed in a protracted Japanese surrender, who knows how many more American and Allied lives would be lost in the interim? For better or for worse, history records that the United States dropped two atomic bombs on two major Japanese industrial cities (Hiroshima on August 6, 1945, and Nagasaki three days later), indiscriminately killing not only military personnel, but also many civilians, including the elderly, women, and children. And America has had to live with the results of that moral decision ever since.

It is obvious that the moral consciences of individuals as well as nations are likely to undergo radical revision during times of war. When it is either "them or us," the hammer is likely to come down on "them." This is the concept of "situation ethics." According to this philosophy, there is no universal set of standards against which moral decisions can be compared; it depends on the situation at the time (e.g., war versus peace, feast versus famine, normal health versus pandemics, etc.). Apparently there were times of collective morality when it was okay to burn witches at the stake, because of the evil spells they could place on people if allowed to run unchecked. It was okay to hack off the hand of a thief as a warning to all who might be tempted to break the law. Was it morally wrong for some members of the Donner party to cannibalize their dead comrades so they would not starve during the winter of 1846? Did most Germans think it was okay to gas millions of Jews to death to purify the Aryan race under Hitler's Third Reich? Was it a necessary evil to use the atomic bomb in 1945? In 1994, was it morally acceptable for parents to conceive and bear a child in hopes that it would be the same tissue type as its leukemic sister who was likely to die without a compatible blood transfusion? Whose sense of morality is correct? Who should be the ultimate judge over such matters?

Even a rudimentary acquaintance with history should make us sensitive to the facts that different societies often have different ethical and moral standards, that what we as individuals believe may not necessarily be the best course of action, and that the opinions of others deserve to be heard and considered. Those who have not learned the lessons of history in this regard may become zealots—so enslaved to their own concept of morality (or what they have been taught) that no other ideas can be tolerated.

Nowhere are ethical issues at greater stake than in medical science. One of the most interesting books on the subject of human experimentation has this to say:

> The vast majority of the medical profession are [*sic*] either genuinely ignorant of the immensity and the complexity of the problem or wish purposely to ignore the whole matter by sweeping it under the carpet. Even fewer lay people have any conception of the issues involved. But the medical pro-

fession must no longer be allowed to ignore the problems or to assert, as they so often do, that this is a matter to be solved by doctors themselves.[2]

The British physician and author of this book, H. Maurice Pappworth, claims that he had frequently been attacked in private discussions with doctors who believed that he was doing a great disservice to the medical profession, and that he was undermining the faith and trust that lay people have in their doctors. They tried to dissuade him from publishing his book *Human Guinea Pigs*, which would paint a less than flattering picture of medical scientists in terms that lay people could understand.

Pappworth makes a distinction between the physician-friend and the medical scientist. Although some doctors try to wear two hats, they are likely to become enmeshed in a moral quagmire. According to Pappworth,

> Some apologists for the kind of experiment I shall describe have quibbled about the meaning of the word "experiment." Their argument has been that every administration of medicine to a patient and every routine radiological or biochemical investigation is an experiment; that, therefore, experiment is inseparable from medicine; and that therefore any kind of experiment by any doctor is intrinsically justified. On this basis the empirical giving of an antibiotic to see if it will abate a high fever, or the performance of a barium X-ray to determine whether a patient's dispepsia is due to a peptic ulcer are experiments. So, in one sense, they are. But the experiments with which I am concerned cannot, by any stretch of the imagination, be deemed analogous. The first (such as administering an antibiotic in the case of a high fever) are directed solely to the treatment and cure of that one patient to whom they are administered; the second are directed, solely in some cases and chiefly in all cases, to the discovery of what may help other patients.[3]

Thus, Pappworth defines experimental medicine as anything done to a patient that is not directly related to helping that patient get well as its only justification. Whether any knowledge gained by this procedure leads to helping others should be immaterial as far as the primary mission of the physician-friend is concerned. Pappworth believes that "the ends of research and those of therapeutic medicine as generally understood are not only different, but actually opposed."[4] He then quotes the famous Hungarian research physician Dr. Albert Szent-Györgyi as having stated: "The desire to alleviate suffering is of small value in research—such a person should be advised to work for charity. Research wants egotists, damned egotists, who seek their own pleasure and satisfaction, but find it in solving the puzzles of nature."[5] There are two kinds of physicians: those who *accept* patients and are concerned only with their welfare, or those who *select* their patients because they have the kind of affliction about which they want to learn more and perhaps find a cure.

If I interpret Pappworth correctly, it would be unethical to subject a patient, even with his approval, to an unproven medical treatment in the hope that it might be of great benefit to the patient, when a medical treatment that has provided some relief in certain patients (but not in others) is already available. This is where things really get ethically murky. At that point, what it comes down to is "what is in the mind of the physician?" Is he more interested in curing his patient or in proving the effectiveness of the new treatment? Maybe the physician doesn't even know in her own mind which of these two motivations is the one which swayed her decision to try the new or stick with the old treatment. And what board of ethicists can read what is in the mind of any physician when she treats a patient? Is the fact that the doctor has many publications to her credit sufficient evidence to judge her guilty of putting science ahead of concern for her patient?

Any time a physician administers treatment, the patient should be informed about the possible negative consequences of that treatment, the chances that the treatment will be effective, and the possibility of alternative treatments. The physician should be as open and honest with his patient as possible and allow that patient to express her desires. The physician has to realize that most of his patients either will be unable (by physical health, emotional state, or educational background) to make an intelligent choice on their own. Most often, the patient will be swayed by whatever the physician tells her is the best option. Thus, the physician must know in his own mind that the course of treatment he is suggesting is one motivated by concern for his patient's well-being and nothing else. Only in this way can the intent of "informed consent" be given any credence.

Special moral problems arise when the patient is a child, a pregnant woman, one who is mentally sick or mentally defective, a prisoner, a ward of the state, or a person on welfare. Parents must usually make decisions for their children regarding the kind of treatment they should have. But families today are often so fragmented that sometimes the wishes of a child's mother and father are not in agreement. Then who decides what is best for the child? Even when both parents are in agreement, sometimes their decision has been overruled by the courts. For example, suppose that a team of medical experts agree that a child will surely die very soon (say, within a week) if not given a blood transfusion. The parents are both in agreement that their religion teaches that blood transfusions are morally unacceptable for their child. Is the court's decision to overrule the parent's wishes more ethical than those of the parents? When a decision must be made to save a pregnant mother or her baby, who is best qualified to decide whose interests should prevail? Obviously, the baby has no voice in the decision. If the baby is known to be genetically defective or to have some presently incurable defect, who should decide if the baby should be aborted or carried to term? Will such a baby be loved and nurtured if it is allowed to survive? Where will the money come from to provide the care (perhaps for the rest of his life) that the baby will require? Who is to speak for patients with mental illness or those who are mentally defective? Their wishes are often ignored by their doctors because what these patients want often conflicts with what the medical profession "knows" to be contrary to their best interests. This class of patient has been, and probably still is, subject to many experimental procedures that are not wholly (or even in part) for their sole benefit.

Prisoners have historically been subject to many abuses in the name of medical science. Until recently, prisoners essentially had no rights, and so scientists felt free to perform experiments on them without gaining their permission, and sometimes even against their wishes. In one sense, the use of prisoner subjects might be an improvement in experimental design over noninstitutionalized subjects. Since the environment and lifestyle of prisoners are highly regulated (they eat the same food; they go to bed and get up at the same time; they have less access to unauthorized drugs, alcohol, etc.), these sources of variation can be reduced to a minimum and thereby increase the power of the statistical tests (to find significant differences between experimental and control groups) employed to analyze the experimental data. Pappworth points out that offering money to prisoners or reduction in time served as inducements to participate in medical experiments is unethical. The only way to get around this problem is if the prisoner volunteers to participate without any form of compensation. There must be a guarantee that a prisoner's parole request will not be affected by his decision to participate or not participate in such experiments. Any other action could be considered illegal and/or immoral because the prisoner's decision was made under coercion or duress. They too would, of course, have to be given full disclosure of the known or possible risks they would be taking by consenting to participate in such experiments. Patients whose medical bills must be paid by the government should be afforded the same kind of ethical considerations as given to those who pay their own way. "A Persian prince at the time of Avicenna was giving advice to a young man who was going to join the medical profession. He said, 'Once you embark on a career as a physician if you wish to gain experience and a reputation, you must experiment freely, but you had better not choose people of high rank or political importance for your subjects.' "[6] I'm afraid that even today equal treatment is not usually extended to wards of the state or those stricken with poverty. There seems to be one kind or morality for those who can pay and another kind for the have-nots.

Probably the best-known example of medical research abuse of a socially vulnerable population in the United States is the Tuskegee syphilis study. Between 1932 and 1972 in rural Alabama, the United States Public Health Service studied the long-term effects of untreated syphilis in about four hundred poor African American men. These subjects were neither informed that they had been diagnosed as having syphilis, nor given information about its treatment or prevention. Even after penicillin became available in the 1940s, researchers and physicians involved in the Tuskegee study failed to use it to treat these men in order to further the studies' goals.

The experimental atrocities performed in the name of medical science against consent on prisoners of war is, of course, an aberration. The most highly documented examples came from the Nazi concentration camps during World War II. These political and military prisoners were subject to the most ghastly forms of inhumane treatment imaginable. Some of them were placed in tanks from which the air was partially removed to simulate an altitude of 5,000 to 8,000 meters. The pressure was then allowed to suddenly return to that of normal sea level. These studies were supposed to help the German air force in some way. No one was seen to leave these chambers of torture alive. Other prisoners of war were placed in tanks of ice water for three hours to observe the effects of freezing. Still others were purposely infected with streptococcus, gas gangrene, and tetanus microorganisms to test the effectiveness of sulfanilamide drugs. Poison was placed in the food of some prisoners so that the effects of various toxins and various dosages could be evaluated. Autopsies were performed immediately upon the death of some of these prisoners. Others were shot with bullets coated with poison to test the effects that such weapons might have on an enemy. Presumably, even a superficial wound with such a bullet might become fatal. Women were treated with X rays to induce sterilization. The racial hatred of the Nazi regime led to the starvation and/or gassing of millions of Jews. They were considered expendable and not worthy of even the most basic of human needs or rights. Even animals were treated better than these wretched prisoners. One of the first policies Hitler put into effect in Nazi Germany was an edict rendering animal experimentation illegal. This effectively opened the door to human experimentation as the only way to scientific investigation of the animal world in Nazi-occupied Europe. Today, many animal-rights advocates stress the point that animals are not humans. Anything that is found in a rat or other experimental animal does not necessarily pertain to humans. Thus, we should not be performing animal experiments at all. Something akin to this kind of mentality must have pervaded Nazi Germany because a large segment of the medical profession at the time in that country became caught up in the racial fervor and somehow lost their concern for every human's basic rights. They were up to their eyeballs in human experiments on captives, and they were free to do whatever they wished without any ethical restraints. Dr. Frankenstein had come home with a vengeance!

As a consequence of the excesses of that time, a set of guidelines was established for human experimentation after the Nuremberg trials. These guidelines made it necessary to obtain the voluntary consent of each human subject. Such experiments should yield fruitful results for the good of society, unprocurable by other means of study. The experiments should be so designed and based on results of animal experimentation and knowledge of the natural history of the diseases or other problems under study that the anticipated results will justify the performance of the experiment. These guidelines made such common sense that they became adopted into regulations by almost all nations of the world. And yet, violations of these rules continue at such a rate as to be unnoteworthy to the popular press. Pappworth believes that these guidelines are too lenient. He suggests that human experimentation should be concerned with at least the following principles: equality, valid consent, prohibited subjects, previous animal experiments, and experimenter's competence.[7] According to his thinking, no experiment should be done if the experimenter was in circumstances identical to those of the intended subjects, and the experimenter would even hesitate to submit himself or members of his own family to the proposed experiment. This is what Pappworth calls his "principle of equity." It is akin to the Golden Rule which says to "do unto others what you would have them do unto you." The principle of valid consent implies full disclosure of all experimental details (drugs, surgical manipulations, etc.); known and possible side effects; estimates of risk for pain, trauma, recovery time, etc.; purpose of the experiment (who will benefit?); and whether the procedure is experimental or proven. All this information should be given to the subject in terms that she can fully understand and under conditions in which she is fully able to make a rational, uncoerced, informed decision. Under no circumstances should experiments be performed on mentally sick patients or on the aged or the dying. These subjects, according to Pappworth, cannot make a rational, uncoerced, informed decision because of their abnormal psychological states.

In Britain, animal experiments require that a special license be obtained and complete records of all experimentation be kept. Animals must be anesthetized if they are to experience any pain, but no such rule protects human subjects. Thus, experimentation is much simpler in humans than in animals because record-keeping is unnecessary, a patient can be awake and responsive throughout the study, and the effects of the anesthetic can be eliminated from the results. England appears to be more concerned for the welfare of animals than for humans. Her Majesty's Stationery Office publishes an annual report on vivisection experiments which is open to the public. No similar list exists for each experiment done on hospital patients. Wherever possible, no experimental procedure should be tried on humans before success has been obtained in an animal model. Granted, rats are not humans, but all mammals share many of their biochemical and physiological processes in common, so that what is learned in animals usually serves as a good predictor of the same procedure in humans. At the least, it is better than nothing. Pappworth believes that experiments on humans should never be undertaken by the lone investigator. "If an experiment is conceived, carried out and answered for by a single physician, there may well be too little protection of the interests of those on whom it is performed."[8] If any members of the research team are not qualified in medicine, their roles should be limited to what they can safely do unsupervised.

The ethical aspects of medical science have probably been an impediment to progress in that field in the past, and are likely to be even more important in the future, not necessarily because scientists are becoming more ethical, but probably out of fear of lawsuits. In today's litigious society, patients are prone to sue at the drop of a hat (or scalpel). A patient who feels he has been medically mistreated (either in routine care or as part of an experiment) may wish to sue the physician or anyone remotely connected with the case, on up through the hospital administrators to the owners of the hospital. If an experiment was performed with the aid of federal funds, then the government could also be sued. The "deep-pocket principle" is regularly exploited. Lawyers love to go after the hierarchical level with the most money; e.g., the government or the corporation that owns the hospital. If responsibility cannot be established at these higher levels then attempts are made to tap the highest level at which culpability can be established. Thus today, even an experimenter with loose morals would be pressured to stay strictly within the established guidelines out of fear of being sued. One of the largest costs of medical care in the United States today can be laid squarely on the doorstep of the high cost of malpractice insurance. These higher operating costs of the physician are naturally passed along to their customers—the patients. Thus, on the one hand, the patient may benefit from better quality care, but the price of that care is more costly. This shouldn't surprise anyone—you only get what you (or somebody) pays for.

Pappworth concludes his book with an additional ethical concern. "The teachers in our medical schools are often more interested in research and research experiments than in the clinical care of patients. As a result those whom they teach, both under- and post-graduates, are easily infected with the doctrine that what matters first is science and what comes second is the patient."[9] He quotes from a speech given in 1954 by the doyen of British cardiologists Sir John Parkenson:

> Must every ambitious graduate be forced by custom or authority to prosecute research in order to obtain a University post or to succeed as a practicing cardiologist? Research ability used to be regarded as a rare gift, something of a phenomenon. . . . This is not the attitude today. . . . My question concerns the universality of the capacity for research; and I almost believe that the true investigator, great or small, is born not made. . . . In my view, we encourage good men, inept for research, to sacrifice their time and energy upon it, when they should be perfecting themselves as bedside physicians.[10]

GIVE ME SYPHILIS OR GIVE ME DEATH

The ethics of human experimentation has always been a hotly debated subject. And yet there is one aspect of human experimentation that remains essentially unassailable; i.e., self-experimentation. Who can attack the morality of an experimenter whose subject is himself? This and other relevant questions have been addressed at length in a very popular book by Jon Franklin and physician-scientist John Sutherland titled *Guinea Pig Doctors: The Drama of Medical Research Through Self-Experimentation*.[11] They drive their points home by the time-honored method of storytelling. The tales they weave are so interesting and provocative that, for me, it was hard to put the book aside to do necessary chores. Their premier case history of self-experimentation, or at least the main subject in the lead-off chapter, is that of John Hunter. He was born in 1728 in the Scottish farming village of Kilbride, near the city of Glasgow, at a time when "England's witchcraft statutes were still in effect, alchemy retained its influence, astrologers were more honored than physicians, and Charles Darwin would not be born for another century."[12] He was the youngest of ten children. His father died when he was thirteen years old. His mother was overindulgent, so John grew up to be a rough-and-tumble, red-haired kid, lacking in the social graces. John did not do well in school because he had a decided distaste for books. The townsfolk thought he was a dullard and certain to grow up worthless. He had no apparent pride in himself and became a drunkard at an early age. But what he lacked in book learning he made up for by his passionate interest in nature and the endless, fascinating questions it posed. "Books? What were books? Books were dead plants, with the shape pounded out of them, glued together with the hooves of worn-out horses, stained with cryptic symbols. There was more to be seen in a robin's egg than in books."[13]

His brother William was ten years older and quite different. William was a serious student who studied in London and eventually set up a school to teach anatomy to those who wished to become surgeons. Anatomy school was a prerequisite at that time to be an apprentice in surgery. John felt compelled to make something of himself regardless of what his family or the villagers thought. So in 1748 he went to London where his brother offered him a job. A steady flow of fresh human cadavers came to William's house. The corpses were usually bodies of those who had been hanged for some crime, or the poor and destitute. There of course was no artificial refrigeration in those days, so most of the dissections were done during the colder parts of the year. Even so, the rotting bodies were good only for a few days; the students would tear them apart in their dissections, and they had to be constantly replaced. Although he preferred healthy bodies, cadavers were always in short supply, so William sometimes had to purchase diseased bodies. It was John's job to embalm the bodies with a mixture of vermilion and the oils of terpentine, lavender, and rosemary. Regardless of what time of night the body would arrive, John would have to answer the door, inspect the body to be sure it wasn't already putrid, remove its clothing, cut into the leg to find a large vein, and splice an embalming hose onto it. He also had to open a vein elsewhere to allow the blood to flow out as it was being displaced by the embalming fluid. As the fluid was siphoned down from an earthenware bowl overhead, John would massage the body vigorously to work the embalming fluid into all of the tissues. He also had to clean up the place after each dissection class, scrub the floors, and dispose of all the bits and pieces of flesh, organs, bones, and the like. The job was demanding of energy but it didn't psychologically bother John at all. The human bodies that he worked with were no different from a dead pig or cat. The human spirit was no longer there; what happened to the body after death was of no concern to him. During William's classes, John would stand with the other students and listen to his brother's lectures and watch him do dissections. William knew of John's intense curiosity, and fed it by giving him special instructions as time allowed. John was always alert to find any unusual body part with some remarkable and instructive feature. He would save such items for William to decide whether they should be pickled in a jar or discarded. An anatomy school was always in danger of becoming a museum and being crowded out of space by a plethora of specimen jars.

Dissection of a corpse was only a means to an end for other students, but for John it presented many puzzles that were crying out for an explanation. For example, why were internal organs of the body, where no one could see them, variously colored? In attempting to satisfy his curiosity, John reduced many bodies to buckets of bits. Soon his expertise far exceeded that of the other students. After a year of work and study, John became a teacher in his second year. His interest was naturally aroused by pathological abnormalities. A common illness of the day was the "pox" (now known as syphilis). It took many forms, and horror stories abounded. For example, if a man got the pox his penis could rot away. Not usually, but it could happen. John wondered why the penis sometimes rots and not at other times. Why did the symptoms vary so much from one patient to another? Many doctors thought that the pox was more than one disease. The popular concept of disease in Hunter's day was of ancient origin, but it had been most profoundly expressed a century earlier by the great English physician Thomas Sydenham. According to Sydenham, disease came from "morbid and unwholesome particles" that sprung from putrefaction, either within or outside the body. These particles could be carried by the air, in what he called "miasmas." There was evidence, however, that the pox was transmitted by direct contact of the sexual organs. To Hunter's mind, this meant that the pox was not transmitted by aerial miasmas. It must be some kind of a liquid miasma. This still did not explain the variation in symptoms. Women were sometimes asymptomatic. In some men, however, the disease might begin in the soft moist urethra (the tube conveying urine from the bladder to the outside), making urination very painful for two or three weeks. The testicles might also become very sore. If this form of the disease affected females it could cause scarring in the Fallopian tubes. Eggs from the ovary could not then travel down the tube to the uterus, resulting in sterility. In other men it would start as a transitory chancre or open sore on the penis, followed by a variety of symptoms after the chancre disappeared in about two or three weeks. Then you could expect a sore throat, a skin rash, bone and joint pain, and pox on the skin. Sometimes a fever would develop and, if it didn't kill you, the pox would disappear and you would be cured. The treatment of the time was to rub mercuric compounds into the thighs. Today, mercury is known to be a highly toxic chemical, to be avoided at all costs. But this was the eighteenth century, and patients were being prescribed all kinds of substances (now known to be harmful) in the name of medicine. Sometimes the mercury treatment did not help, and the penis would begin to slough away. A pox-rotted penis was considered rare enough for William to have it preserved in a jar. John was convinced that the pox was a single disease with these two manifestations, depending on which tissues were touched by the disease-bearing particles. If the disease started on the dryer tissues on the outside of the penis, a chancre would develop. But if the disease began in the moist tissues of the urethra, a liquid discharge of pus would be produced. It seemed so logical that anyone ought to be able to understand it.

At an earlier time, the one who gave you a haircut and shave might also have been the same one who removed skin tumors and performed minor surgeries (barber-surgeons). By Hunter's time, surgery in England was beginning to separate from barbering and to become a medical profession in its own right. Surgery on internal organs and amputations of limbs were almost always fatal because the cause of infections was unknown, antiseptics and antibiotics were thus essentially nonexistent, and patients often died of shock because analgesics (pain relievers) and highly effective anesthetics (aside from opium and alcohol) were yet to be discovered.

Although John was advancing professionally, his personality wasn't ever going to change. Whenever he had free time, he could be found in the local pub, banging his beer mug on the table and associating with any kind of low life that cared to listen to his opinions. By 1754 he had obtained a position as surgeon's assistant at St. George's Hospital. The following year, on William's advice, John enrolled at Oxford, but he couldn't stand it and dropped out after two months. They wanted him to learn Greek and Latin. Books again! Now he was out of a job, so brother William took him back at the anatomy school until he could secure another residency. He continued to study comparative anatomy and became an expert on the lymphatic system that drains away excess tissue fluid and puts it back into the circulatory system. His wide-ranging interests led him to study the descent of testes in the male fetus, the nasal and olfactory nerve tracts, the formation of pus, and placental circulation.

In October 1760 John obtained a commission in the British army as staff surgeon. He participated in the Hodgson and Kappel expedition to Belleisle and the campaign in Portugal from 1761 to 1763, acquiring extensive knowledge of and expertise in treatment of gunshot wounds. Gangrene very commonly infected these wounds, and the only way to save such a limb was to amputate it above the wound. Speed was all important in these amputations and John could lop off a leg in as little as thirty seconds. The screams of his patients moved John so deeply that he began to make notes on his own experiences. He hated books, but apparently thought that his manuscript might prove useful someday to those who turned to books for answers. There were no anesthetics or pain relievers. Sometimes, if the patient did not die of shock due to the amputation, he might survive. But without antibiotics, many amputees died of infection anyway.

Following the end of the Seven Years' War in 1763, John set up his private practice in London. At first, business was slow. But gradually the remarkable success that he had with his patients became widely known and his business prospered. Now he had money enough to indulge his passion for collecting exotica and specimens for his comparative anatomy studies. When he heard there was an ostrich in London, he had to have it, and would pay almost any price for it. Next it was a zebra, and so on. In 1764, he purchased two acres of land about two miles from London where he could keep his growing menagerie of opossums, hedgehogs, jackals, water buffalo, and other animals. He also collected animal skeletons from India, Africa, and other faraway lands. Teeth especially interested him. One could tell from the teeth whether the animal was a vegetarian (herbivore) or a meateater (carnivore).

As Hunter's collections grew, so did the curiosity of men of learning such as those of the Royal Academy. They would come to his house (now more a museum than just living quarters) and marvel at the rare specimens. There were even some human skeletons in the collection, but what Hunter really wanted was the skeleton of a giant. The average human height in England at that time was a little over five feet. An Irishman by the name of Charles Byrne (1761-1783), who was living in London at the time, was seven feet seven inches tall. Not one to beat around the bush, John went directly to the man and offered him money if he would agree, after his death, to allow his body to be reduced to a skeleton as an addition to the Hunterian collection. Byrne was horrified. This little ghoul was after his bones! Hunter didn't understand Byrne's objection. After all, what difference does it make what happens to your body after you die? Although it was not in his nature to be patient, Hunter had no choice. Byrne was in poor health, and it was just a matter of time before he would die. He knew that Hunter was waiting to dismember his body. So Byrne went to the undertaker and paid him all the money he could spare. He left instructions that after he died, his body was to be under constant guard until a suitably large lead coffin could be obtained. Then his body was to be sealed in it, taken out to sea, and dropped into the depths where it would forever be out of the reach of Hunter or any other ghoul. Byrne died in 1783. Within an hour after his death, Hunter had been so informed. The fact that Byrne's body was being guarded presented no problem. Hunter simply found out where the guards went to drink. He then went to the tavern and offered them a sum of money they could not refuse—more than a year's pay. That night, as the story goes, the guards were overpowered and the giant's body disappeared. About two years later, the Hunterian museum had a new seven-foot, seven-inch skeleton on display under the title "O'Brien, The Irish Giant."

Despite Hunter's insistence that the pox was a single disease with two manifestations, many of his medical colleagues remained convinced that they were really two diseases, each with its own manifestation. Exasperated, Hunter decided to prove his theory once and for all. In 1767, he took a drop of liquid miasma from the penis of one of his patients—a typical "wet" case. The drop was transferred to the outside of his own "dry" penis. Using a lancet, he then punctured himself twice through the droplet and squeezed the small cuts open to allow the liquid miasma to enter the wounds. Two days later, he felt a tingling in his penis that was, according to his notebook, rather pleasant. There was also redness (inflammation) where the lancet had pierced the skin. After two more days, two small pimplelike chancres appeared. The chancres were firm, like the tissues from which they arose. Thus a "runny disease had produced a dry disease, proving, once and for all, that while it had many forms, the pox was but one disease."[14] He then placed a drop of acid on one of the chancres and it soon dropped off, only to be replaced by one just like it. He then rubbed some compounds of mercury into his thighs and the chancres soon disappeared, leaving two small dimples that retained a bluish cast for a long time. Nine months later, John developed a sore throat. He could see in his mirror that there was a small ulcer on one of his tonsils. Was this secondary pox? He rubbed some more mercury into his thighs and the ulcer healed. After another three months, the skin rash returned. He found that, by controlling the amount of mercury used, he could make the rash and the sore throat come and go at will. Finally tiring of the experiment, he rubbed a massive amount of mercury into his thighs and the pox went away for good.

In 1767, the same year in which his famous self-experiment began, Hunter was elected Fellow of the Royal Society. The next year he was elected one of the surgeons at St. George's Hospital and was able to instruct pupils in surgery. Many of his 449 pupils became famous in their own right. Perhaps the most famous of them all was Edward Jenner who, after Hunter died, showed how inoculation with cowpox could protect a person from getting smallpox. We now know that these two pox diseases are caused by closely related viruses, whereas the pox that Hunter studied in himself was caused by two unrelated bacteria. As luck would have it, the pus with which Hunter inoculated himself came from a patient who had both syphilis and gonorrhea. The symptoms that Hunter developed were characteristic of both diseases. In 1786, he published the results of his self-experimentation in a report titled *Treatise on the Venereal Diseases*. This convinced his critics that the pox was a single disease. Unfortunately, this erroneous conclusion stifled further research on syphilis for more than half a century after his death. Hunter was fond of saying, "Experiments should not be often repeated which tend merely to establish a principle already known and admitted. The next step should be the application of the principle to useful purposes."[15] So great was Hunter's influence that his self-experiment was accepted as definitive proof and, following his precept, no one else repeated the experiment on themselves. But then, who but Hunter would do such a painful and dangerous self-experiment?

In 1771 Hunter had published his first book, *Treatise on the Natural History of the Human Teeth*, and used the proceeds to finance his wedding to Anne Home. They had four children, of whom only one boy and one girl survived. Everard Home, Anne's brother, also became one of Hunter's pupils, and eventually his executor and his biographer. Hunter accumulated some fifty volumes of notes during his lifetime. His brother-in-law, Everard, published two of these volumes after Hunter's death, claiming them to be his own. Later, Everard became fearful that his plagiarism would become known and destroyed all the remainder of Hunter's notebooks. These works included eighty-six of Hunter's surgical lectures and the best series of drawings on the embryology of the chick before the nineteenth century. Hunter made contributions to many disciplines in addition to anatomy, surgery, and physiology, including dentistry, veterinary science, geology, botany, natural history, and psychology. In January of 1780 Hunter presented a paper to the Royal Society on the circulatory structure of the human placenta. A controversy arose when his brother, William, claimed to have previously reported the same information. John had made many discoveries during his employment under his brother, with William receiving the credit. The dispute over the placenta paper, however, continued to estrange the brothers until William died in 1783. Ten years earlier, John experienced the first of a series of chest pains (angina pectoris) that signaled heart trouble. It slowed him down a bit, but didn't stop him. In 1776 Hunter became Surgeon Extraordinary to the King. In 1785 he pioneered in tying off the main artery above a popliteal aneurysm (a bulging artery behind the knee), allowing the development of a collateral circulatory system to supply blood to the lower leg. He conceived of this idea by prior experimentation on animals.

Despite his renown as a surgeon, scientist, and naturalist, Hunter's dealings with his colleagues were marred by bitter animosities, acrimonious disagreements, and jealousies. At a meeting of the board of directors of St. George's Hospital on October 16, 1793, Hunter was, as usual, engaged in heated debate when he suffered his last and fatal attack of angina. The direct cause of his death was heart failure, but indirectly it was in part caused by his experimentation with syphilis. Hunter was aware of only two stages for syphilis, the chancre stage and the sore throat stage. There is a third stage of syphilis that is systemic; i.e., it affects many parts of the body including the heart, brain, spinal cord, and aorta. Thus, of the two diseases, syphilis is potentially more dangerous than gonorrhea with its manifestation as painful urination. For a long time before he died, John's heart had been causing him so much pain that he thought surely it must be a most unusual organ. He requested that when he died, his heart should be removed and placed in a bottle. A colleague removed John's heart, as requested, and passed it around to the others for their inspection. None of them saw anything unusual about his heart, so it wound up in the slop bucket. Thus, his hope of contributing something to science from beyond the grave came to naught. But aside from his many scientific discoveries and his famous collection, he left a more important legacy known as the Hunterian tradition. He instilled in his pupils a reliance on scientific observation, the experimental method, and the courage to try new ideas. His remains lie today in the North Aisle of Westminster Abbey under a brass plaque that reads, "The Founder of Scientific Surgery."

John Hunter was never well off financially because, even though he made lots of money during the latter part of his life, he spent most of it on expanding and maintaining his collections. Thus, his family was left with little except his real estate and his collections. It wasn't until 1799 that Parliament, plagued with the expenses of war, could afford to purchase the collection for fifteen thousand pounds and turn it over to the Corporation of Surgeons. All during these hard times, Hunter's student and secretary, William Clift, maintained the collection and what little was left of Hunter's writings. John had instituted something unique in his collections. He had organized them systematically. For example, he started with the organs of simple creatures and progressed up the ladder of complexity through insects and fishes, to birds and mammals. Each specimen was labeled. This set a pattern for the way that later museums would be organized. Hunter's collection was nearly destroyed by a direct bomb hit during World War II in the 1940 blitz of London. Today, the Royal College of Surgeons oversees some eleven hundred of Hunter's specimens (including the skeleton of the Irish Giant) that survived the blast.

John Hunter was not the last to engage in self-experimentation. It has continued to be practiced up to the present day. For example, Berry J. Marshall of the Royal Perth Hospital in Australia announced in 1985 that he had intentionally ingested bacteria (*Helicobacter pylori*) thought to be responsible for ulcers and cancer of the stomach. He developed a severe case of gastritis, but the painful inflammation disappeared without treatment. As reported in the February 1996 issue of *Scientific American*,[16] a male volunteer who tried the same thing in 1987 was not so lucky. After about two months, he started treatment with bismuth subsalicylate (Pepto-Bismol®). A biopsy taken five weeks later indicated that the medication had been effective. However, another biopsy at nine months after the first revealed that the infection and the gastritis had returned. Only after two different antibiotics plus bismuth subcitrate had been administered for three years could he be declared cured.

It might be thought that the morality of self-experimentation is unassailable. After all, a person should be free to do with his own body whatever he wants. Any harm that might come from self-experimentation would be restricted to the experimenter. But that conclusion is seldom, if ever, warranted because almost everyone has friends and/or loved ones. When a person is harmed or dies as a consequence of self-experimentation, the pain is shared to some extent by his family and friends. Their pain or loss must be considered as a heavy price to pay if a self-experiment produces unanticipated or undesirable results. Pappworth[17] has a section of his book devoted to self-experimentation, but no mention is made of the suffering that this may cause to others who know and love the experimenter. The sickness and death of every person is mourned by others, perhaps a multitude of others. There is no way to evaluate the total emotional distress of all their grief, but if you think about your own suffering at the passage of a close friend or loved one, and multiply that by the number of others who also grieve, the sum would appear staggering. How can something this important be forgotten even by the authors of *The Guinea Pig Doctors*[18] whose entire book is devoted to the consequences of self-experimentation? If they lost sight of this fact, then it is probably also true of the general public. In John Hunter's case, do you think that he gave consideration to the consequences, in this broader context, of his self-experimentation with venereal disease? And what about his wife? Certainly he knew that he could not have sexual relationships with his wife during the period of his self-experiment with the "pox." And what if he had been unable to cure himself of the pox at the end of the experiment? Had she been informed or given a say in the decision? I dare say not. Women were not held in very high esteem in those days. Even so, how could anyone with even a rudimentary moral character embark upon such a hazardous experiment without consulting (let alone requesting permission from) his wife who would be directly involved in the outcome of his experiment? The fact that these concerns are not addressed in the references I consulted to write this chapter indicates that science history has sadly neglected some very important moral issues that should concern everyone in every age. Great men of science tend to have a Hunterian type of personality. They let their curiosity get the best of them, and may even indulge in self-experimentation. If they die, or even if their life expectancy is shortened (as was the case with Hunter), as a consequence of their self-experimentation, the world of science loses out as well. Because it is from such scientists that great ideas and discoveries often spring, their premature demise thus diminishes the rate at which science could progress. This is another concern that science history has chosen to neglect. So, in my opinion, a person is not free to do whatever he will with his own body. There is much more at stake than is realized.

WILLIAM BEAUMONT AND THE FISTULA

A shotgun accidentally discharged not more than three feet away. The blast of powder and buckshot ripped a hole as large as a fist in a man's side just below his left breast. One of the bystanders yelled "Get the doctor" knowing full well that there wasn't much hope that anyone could survive that kind of wound. The date was June 6, 1822, long before the advent of the germ theory of disease. There were no effective anesthetics or pain relievers and no antibiotics to stop infection. The wounded man was Alexis St. Martin and the doctor was a U.S. Army surgeon's mate named William Beaumont.

Alexis St. Martin was a strong Canadian fellow, about nineteen years old at that time. He was born into a poor peasant family in a small village not far from Montreal. He was illiterate all his life. But his strength and stamina for work were instrumental in his gaining a job as a voyageur with the American Fur Company. He and seven other voyageurs made up the crew that paddled a heavy canoe, called a *bateau*, into the northern territories during the fall months in search of furs. A clerk who traveled with them determined the route from one Indian village to another where they would trade goods for furs. The voyageurs had to paddle tirelessly for hours upstream and carry the bateau overland with its load of furs and other equipment from one body of water to another. Often they would have little to eat and would sleep in the shelter of their bateau in the dead of winter. It was not a job for wimps. St. Martin and his crew had put ashore for a summer of fun and relaxation on Machilimackinac Island in Lake Huron, Michigan Territory. On this island was a U.S. Army fort and a village built up around the American Fur Company's headquarters and its general store. Every June, the local Indians (mainly Hurons, Chippewas, and Ottawas) would bring their furs to the American Fur Company about the same time as the voyageurs returned from further north and west. Indians and voyageurs would sing and dance, wrestle, box, get drunk, and generally blow off steam. St. Martin was loafing the day away in the company store when the accident occurred. He lay on the floor with tonguelets of fire licking around the torn edges of his jacket until blood spurting from his wound drowned the flames.

William Beaumont, born on November 21, 1785, was the son of a farmer in Lebanon, Connecticut. He didn't want to be a farmer and left home in the winter of 1806–7, at the age of twenty-two. He become a schoolteacher for three years at Champlain, New York. Eventually he became interested in medicine and decided to apprentice himself to Dr. Benjamin Chandler in St. Albans, Vermont. After two years in training, he had obtained his license to practice medicine and was about to set up his own business when the War of 1812 broke out. Being anxious for some excitement, he applied to the U.S. Army and received an appointment as an assistant surgeon (surgeon's mate) in the Sixth Infantry stationed at Plattsburgh, New York. On the battlefields of that war he soon had more excitement than he had anticipated. In his diary, he wrote that he was wading in blood, cutting off arms and legs, sometimes for forty-eight hours without food or sleep. He was awarded a citation for bravery under fire. Beaumont resigned from the army after the war, convinced that younger, less experienced men had been promoted over himself. But then he found himself trapped for four years as a small-town doctor in Plattsburgh. Technically speaking, he was not really a doctor, but had been licensed to practice medicine because of his experience as an apprentice to a doctor. Then one day in 1818 he received a letter from Dr. Joseph Lovell, one of his wartime buddies who had become Surgeon General of the Army. Lovell inquired if Beaumont wanted to serve another term as an army surgeon. Beaumont wanted some real action, so he agreed providing that he be sent to a post in the territories where Indian uprisings were common and men could become heroes overnight. He was sent to Fort Mackinac on the island of Michilimackinac, but the Indians there were too civilized to generate much excitement. There wasn't enough work to keep him busy. Aside from Robert Stuart, the Fur Company manager, and his charming wife, he found the rest of the island's personnel boring. About a year later he obtained a leave of absence, returned to Plattsburgh, and married Deborah Green. The following year, 1822, he returned to Fort Mackinac with his bride without expecting any great changes in his professional life. Everything changed when a messenger brought him the news that Alexis St. Martin had been shot.

Beaumont hurried down the hill from the fort. St. Martin had been placed on a cot in the rear of the store. Beaumont had never seen a wound like this. The gunshot had torn a hole the size of his fist through St. Martin's side. A piece of charred left lung hung out of the hole, and below it trickled out bits of his last meal. Beaumont couldn't believe that the hole also penetrated the man's stomach and that he was still alive, still breathing. Beaumont was certain that St. Martin would die, but his medical ethics demanded that he attempt to treat his patient anyway. He tried to return the lung to the chest cavity but it was hung up on the pointed end of a broken rib. By further manipulations he pushed the punctured lung back in place. Then he wiped the oozing food from the protruding stomach and pressed it back into the abdominal cavity. He bandaged the opening and stated that he would return in an hour. His wife was expecting their first child and he didn't want to leave her alone for any length of time. He fully expected that his patient would be dead by the time he returned. But when he returned, St. Martin was still breathing. So Beaumont began the process of deep cleaning of the wound. Fishing around inside the hole he carefully removed shreds of gun wadding, buckshot, slivers of bone, and pieces of cloth and flesh. Finally he had cleaned the wound as thoroughly as he could. If St. Martin survived long enough, Beaumont was sure the wound would become "fetid" (i.e., it would develop pus and he would become feverish). Today we would say that the wound had become infected by bacteria or germs, but germs had not been discovered as disease-causing agents at that time. It was popularly believed that pus and fever were necessary for the healing process. So instead of covering the wound with a bandage, he made a poultice by mixing yeast, flour, charcoal, and water, and laid it over the wound. This was done in the hope that the fermenting yeast would quickly induce the fetid condition.

Two days later, Deborah gave birth to a daughter named Sarah. Dr. Beaumont asked for and was granted permission to move his patient to the little hospital inside the fort. As expected, St. Martin ran a high fever for the next several days. Fresh poultices were made daily for the wound. Pneumonia and inflammation of the lungs were treated by opening a vein and extracting about a pint of blood, a common practice at the time. Although not fully conscious, St. Martin continued to survive, to the surprise of everyone. His fever abated and he gradually regained full consciousness. The food he ate dribbled out through the hole in his stomach if a dressing was not kept inside. Beaumont had made a plug of lint and pushed it into the hole like a cork into a bottle. By the end of 1822 the injured part of the lung had sloughed off, leaving a healthy remainder. Scar tissue had formed around the edges of the hole and was growing toward its center. Part of the stomach that had protruded through the hole had sloughed off and part of the flesh near the hole had become attached to the muscles between the ribs. The hole into the stomach was becoming a permanent anatomical peculiarity known as a *fistula*. It looked like a mouth pursed for whistling with a loose flap of tissue that hung down like a curtain that could be easily pushed aside for a look inside at the living stomach. Beaumont became fascinated with the opportunity to see what no other human had ever seen before. He decided to write an account of St. Martin's injury, the medical care he had been provided, and his remarkable recovery, with the idea of submitting it for publication to a medical journal. All of a sudden, Beaumont's life didn't seem so dull. He began keeping careful notes on his patient.

In the spring of 1823 the fort commander deemed that since St. Martin was not a soldier he was not entitled to use the fort's hospital. The American Fur Company had no more use for him either and refused to pay for his support. St. Martin became a charity case. The townspeople donated ten dollars per month for his room and board, but this didn't cover his medical supplies. After paying out about $150 the townsfolk said "enough!" They were going to send him back to his home, some fifteen hundred miles distant. Beaumont was furious when he learned that St. Martin was to be sent back to Montreal. He could not survive a trip that long in an open boat. Deborah suggested that they keep St. Martin in the Beaumont home. She knew that her husband would not be able to write his article until St. Martin was completely cured. In addition, she suggested that St. Martin could be of great help around their house as soon as he was cured. So in April of 1823 St. Martin was taken into the Beaumont home. But at that time, St. Martin could only move slowly and his wound still required daily attention. He did what he could—sweeping the floor and brushing off the doctor's uniforms. It made him glad that he could be of some use and it gave him something to do. By the second year after his injury, St. Martin had completely recovered. He could run errands and chop wood. He was planning on rejoining the voyageurs to make his own way in the world. When the voyageurs returned that spring, he stole away from Beaumont's quarters and headed on down to join his old comrades. After a few drinks with the boys he felt just like his old self again. But his former associates had heard of the famous hole in his stomach and lifted his shirt to expose it and him to ridicule. He was not the same man they had known before the accident. He would never again be anything but a freak to them. His only option was to try to get a job with another fur trader such as the Hudson Bay Company of Canada. They would not know about the hole in his stomach and he could hold his head high again among working men. But he had no money to make the journey to reach those agents. He was completely dependent upon the generosity of Dr. Beaumont for his daily sustenance. During all this time, Beaumont had, at his own expense, fed him, changed the dressing on his wounds at least once a day, and provided him with all the necessities of life.

Beaumont sent a report of his experiences with St. Martin to Surgeon-General Lovell in 1824. Lovell sent it on for publication and it appeared in the *Medical Recorder* under the title "A Case of Wounded Stomach by Joseph Lovell, Surgeon-General, U.S.A." The error regarding the author of the report was later corrected. Beaumont knew he had a unique opportunity to study gastric physiology in his patient, but it wasn't until May of 1825 that he decided to begin doing any experiments. It is not known why it took him so long to initiate these kinds of studies. Very little was known about the process of digestion in those days. It was known that food left the stomach and entered the small intestine as a gruel-like substance called *chyme*. There were several theories about what took place in the stomach. One theory suggested that the stomach grew warm enough to "cook" the food until it softened into chyme. Another theory proposed that the muscular movements of the stomach walls acted like a grinding mill. Most medical men of the time, including Beaumont, thought that food was dissolved into chyme by the action of a liquid called *gastric juice*, but they didn't know what it contained or how it worked. Beaumont requested Lovell to transfer him to a post where he could have contact with other doctors who could vouch for the authenticity of the fistula and the experiments that he was planning to perform. Lovell arranged for Beaumont, his family, and St. Martin to be transferred to Fort Niagara in New York State in the spring of 1825.

Beaumont began his experiments in August 1825. One of the earliest experiments involved tying a silk thread around bits of meat, fat pork, cabbage, and bread, and lowering them into St. Martin's stomach for various periods of time. The ends of the strings were left outside the hole and the lint plug was replaced. Thus, Beaumont could retrieve his specimens anytime he wished by pulling on the appropriate string. St. Martin had no say in the matter. And he didn't appreciate having things on strings placed in his stomach and then fished out, often at hourly intervals. Beaumont's notes reveal that at 1 P.M. he found the cabbage and bread was about half digested, but that the pieces of meat were unchanged. He returned the samples to the stomach. At 2 P.M. he withdrew them again and found that the cabbage, pork, and boiled beef had been thoroughly digested; the other pieces of meat (including raw salted beef) had been very little affected; he returned them to the stomach again. It was believed by some, including Dr. Lovell, that the stomach digested one kind of food first, then another kind, and so forth. Beaumont's observations destroyed that hypothesis. The stomach digested several kinds of foods simultaneously, although at different rates. Cooked beef digested faster than raw meat. St. Martin complained that Beaumont's experiments were causing his stomach to hurt. He said his head ached and that he felt very weak. Beaumont looked into the fistula and decided that the condition of the stomach was not right for further experiments. It took about a week for St. Martin to recover.[19] Beaumont was anxious for St. Martin to get well enough so that he could resume his experiments. On August 7, 1825, Beaumont put a thermometer into St. Martin's stomach and found that it registered one hundred degrees Fahrenheit. Normal body temperature is considered to be 98.6. He concluded that the stomach is not hot enough to "cook" food as some doctors believed.

Beaumont wondered if gastric juice could cause digestion outside the stomach (in what we would today call an *in vitro* system) or if digestion could only occur in the special environment of the stomach. Using a rubber tube, he withdrew an ounce of digestive juice and put it in a vial that was kept at body temperature in a bath of warm water. He placed a chunk of boiled salted beef into the vial of gastric juice and simultaneously put an equivalent chunk into St. Martin's stomach. After two hours he found that the meat in the stomach had been completely digested. The meat in the vial was being digested but at a slower rate. He shook the vial to simulate the motion of the stomach, and by eight hours later all the meat had been digested. Thus he was convinced that the process of gastric digestion was caused almost entirely by gastric juice.

Beaumont was eager to show his prize patient to other doctors, and had planned to do so during a furlough that was due him soon. St. Martin knew this would mean that he would have to expose his fistula to a never-ending series of strangers who would want to peer in and poke around his stomach. He had always been ashamed of his deformity and tried to keep it hidden. Beaumont expected him to be obedient when asked to disrobe for such inspections. The first stop on his tour to display St. Martin before the medical world was in Plattsburgh, where Deborah and her children had been visiting her family. That night, when the household was asleep, St. Martin slipped quietly away and headed north by the stars toward Montreal.

The next morning, Beaumont was distraught by the disappearance of his ticket to fame and perhaps fortune. He referred to St. Martin as an "ungrateful wretch." After all that Beaumont had done for him, was this the way he showed his gratitude? Beaumont wanted him back and made numerous attempts to locate him, but to no avail. Four years later, in 1829, Beaumont finally learned of his whereabouts. In the meantime, St. Martin had married, had two children, and had worked for a time as a voyageur for the Hudson Bay Company.[20] But then the company had to let him go because he was not able to paddle the bateau tirelessly for days on end and could not carry his share of the load when the boat had to be transported overland. He had to chop wood and do other odd jobs to support his family. Through correspondence with friends in the company, Beaumont learned that St. Martin would be willing to return as a guinea pig for a year or two for a reasonable wage and providing that his wife would also be in his employ. The letter also indicated that St. Martin had another job offer from a doctor in Montreal. In anger, Beaumont roared, "So! My patient is willing to help a strange doctor become famous by performing experiments on him! Am I not the one who saved his life, who supported him for years?"[21] Room and board was no longer enough for St. Martin. He wanted cash for his services and he wanted his wife to be paid as well. Beaumont was not happy with the deal, but he couldn't let another doctor get the credit that he himself was rightly due. With the help of his company friends, St. Martin was sent to Fort Crawford on the upper Mississippi River where Beaumont was then stationed. St. Martin arrived in August 1829 with his wife and two children.

In order to insure that St. Martin would be at his complete disposal, Beaumont had him agree that for a period of one year he would

> serve, abide, and continue with the said Dr. Beaumont . . . that he, the said Alexis, will at all times during the said term when thereto directed or required by said William, submit to assist and promote by all means in his power such philosophical, or medical experiments as the said William shall direct or cause to be made on or in the stomach of him, the said Alexis, either through or by the means of the aperture or opening thereto in the side of him, the said Alexis, or otherwise, and will obey, suffer, and comply with all reasonable orders of the said William in relation thereto and in relation to the exhibiting and showing of his said stomach and the powers and properties thereto and of the appurtenances and the powers, properties, and situation and state of the contents thereof.[22]

Let us all regain our breath after that short sentence and marvel at how difficult legal jargon can make a simple situation.

Now that St. Martin was under his control, Beaumont began his experiments again. One of the questions that he wanted to answer was: Is gastric juice always in the stomach or is it secreted by the stomach lining only when food is present? To answer this question, Beaumont made St. Martin go without food for many hours. Then he would have him lie in a position where sunlight could shine directly into the fistula. With a magnifying lens, Beaumont searched the stomach for traces of digestive juice, but could find none. Nor could he draw any gastric juice from the stomach with a tube. Then he placed a few pieces of bread in the stomach and noticed that they soon became moist. Gastric juice could then be extracted with the tube. In a separate experiment, Beaumont poked and rubbed the wall of St. Martin's empty stomach with the end of the tube. After about fifteen minutes of this "irritation," he was able to extract more than an ounce of gastric juice. Thus he was able to conclude that gastric juice is not always in the stomach, but it can be produced in response to food or irritation of the stomach wall. He performed many experiments to determine the time required to digest various foods. He also discovered that food digests more rapidly in the stomach when it is broken into very small pieces (as is usually done by mastication in the mouth). One day Beaumont had criticized St. Martin's wife for interfering in his experiments. The next day, after St. Martin had eaten his breakfast of fat pork, bread, and potatoes, he was still angry over the way Beaumont had spoken to his wife. Four hours later, Beaumont found that very little digestion had occurred. Furthermore, there was bile in the stomach.

Bile is a bitter, brownish fluid secreted by the liver into the small intestine. It is not usually found in the stomach. Beaumont incorrectly concluded that anger produces bile and that bile prevents digestion. We know now that bile is important as an emulsifying agent to increase the surface area of fats (lipids) and thereby accelerates digestion of fats by lipase enzymes secreted by the pancreas and the small intestine. We also now know that impulses from our sympathetic nervous system tend to inhibit the digestive organs. This system becomes active when we are emotionally stressed. At rest or under nonstressed conditions, impulses from our parasympathetic nervous system stimulate vegetative functions such as digestion. Both the sympathetic and parasympathetic systems are part of our autonomic nervous system that functions without our conscious control. But these facts were not known in Beaumont's day.

What is this stuff called *gastric juice*? Beaumont had tried to find out by tasting it. It tasted acidic to him. The only way to find out more was to enlist the help of chemists that knew how to analyze complex organic substances. Such expertise existed only in Europe at that time. So Beaumont planned to ask for a year's furlough and take St. Martin with him to Europe to have his gastric juice chemically analyzed. He told St. Martin that he would be able to visit Paris on their European trip. St. Martin, like most French Canadians, had always heard stories about the beauty of Paris, and thought of it as the center of the world. Beaumont knew of his longing to go there sometime before he died, and used it as a lever to send St. Martin's wife and children back to Canada. He allowed St. Martin to take his family back to Canada with the provision that he alone return in six months. Beaumont felt confident that St. Martin would return because of his intense desire to see Paris. Meanwhile Beaumont and his family visited Plattsburgh. He would leave his wife and children there when he was on furlough. St. Martin returned six months later as agreed upon. Beaumont had him sign (mark his X) an agreement that he would serve for one year and receive food, lodging, and clothes, plus $150 (a large sum of money for those times). Then they were off to Washington, D.C., where Beaumont proudly displayed his patient to any doctor who was interested. There he was able to read medical books that were unavailable at his army posts. He learned how little was really known about gastric digestion, and that he was probably the world's authority on the subject. He met many important people who encouraged him to write a book about his experiments with St. Martin. When he learned that his furlough had been cut to six months, Beaumont decided that a trip of that length to Paris would not be worthwhile. He knew that St. Martin would be crushed to learn that they were not going to Paris after all. With Dr. Lovell's help, Beaumont had St. Martin enlist in the army as a sergeant with duties as directed by Dr. Beaumont. It was then explained to St. Martin that deserters from the army were always caught and thrown into prison. Thus Beaumont was assured that St. Martin would not run away again when he was told they were not going to Paris. Another benefit to Beaumont was the fact that St. Martin would be earning an army salary, so he would not have to pay for his keep anymore.

After his furlough in Washington was over, Beaumont was appointed recruiting officer in Plattsburgh. There he had time to begin writing his book, which he confessed was an immense job. It was published in December 1833, eleven years after the gunblast that started it all. The book was entitled *Experiments and Observations on the Gastric Juice and the Physiology of Digestion* and was an instant success. It was praised by the popular press and by medical journals. It was first translated into German and then into several other European languages. Americans were considered to be incapable of doing scientific research. No American before Beaumont had made any substantive contribution to the field of physiology. For this reason, some historians have called Beaumont "The Father of American Physiology." He put St. Martin on demonstration before students at the Columbia Medical College in Washington and before doctors at a medical convention in Connecticut. The American ambassador to France asked Beaumont to bring St. Martin to Paris. Beaumont allowed St. Martin to visit his family before they left for Europe. But this time St. Martin did not return. He had been fooled before by promises of a trip to Paris. He was not going to be fooled again. Perhaps, too, he had learned that a deserter from the U.S. Army would be safe if he stayed in Canada. By correspondence, Beaumont tried many times, without success, during the years 1836 to 1852 to get St. Martin to change his mind. He even offered to pay for his family's support if he would come by himself, but to no avail.

After his book was published, Beaumont was ordered to a post in St. Louis, Missouri, where he remained for the rest of his life. He resigned his commission in 1840 and entered private practice. In March 1853 he accidentally fractured his hip. The hip developed a carbuncle and he died from its complications on April 25, 1853, at the age of sixty-eight.

Meanwhile, St. Martin had also become a very popular figure. He traveled around the country, exhibiting himself to doctors and at medical conventions as "the man with a lid on his stomach." The fees he obtained from these events helped to support his family, which eventually grew to include four married children and a flock of grandchildren. But in all his travels he never visited Beaumont. He died in Canada in 1880 at the age of eighty-three. Sir William Osler, who was then teaching in Montreal, attempted to obtain an autopsy on him, but St. Martin's relatives would not allow it. The "man with the hole in his stomach" had outlived his physician by twenty-seven years.

In Beaumont's day, it might have been considered quite ethical to do medical experiments on a person without his consent, as long as he was penniless and dependent upon his experimenter. The legalities of a contract such as the one agreed to by Alexis St. Martin might also have been acceptable at that time. But in the United States today, neither the ethics nor the legalities of Beaumont's treatment of his patient would likely be accepted. A person should be grateful beyond measure to anyone who saves his life. But today that would not entitle Beaumont to do anything to his patient that was not directly related to affecting better health for that patient. The experiments that Beaumont conducted were of great scientific value, but they did nothing to improve the health of St. Martin. Indeed, some of those experiments were irritating and in at least one case made him very sick. Even though he accepted pay for his services, should it be considered ethical to perform experiments on such a person when he had no other way to support himself or his family? Beaumont knew how ashamed St. Martin felt about his deformity. And yet Beaumont had no qualms about asking him to remove his shirt whenever another doctor indicated an interest in the case. Even today in teaching hospitals, patients are asked to allow themselves to be examined by groups of medical students. I'm sure that many of those patients become emotionally upset by such "public" displays, and their return to health is thereby retarded. What about the part of the Hippocratic oath that binds doctors to do nothing that might endanger the health of their patients? How should society weigh the benefits that are likely to accrue from such patient exploitation against the detriment suffered by the individual who is the subject of the research?

At one end of the spectrum of ethical arguments are those who believe the benefits of scientific research that *might be* gained by many people should outweigh the rights or supposedly minor discomforts of the individuals that are the subjects of this research. At the other end of this spectrum are those who believe that patients should never be subjected to any treatment that is not specifically intended to help that person regain health. According to this second philosophy, patients should not be given an experimental drug that *might* help them, but is sure to cause some unwanted side effects. Suppose that patients are dying, and they are offered an unproven drug or treatment that might help them survive longer. Are patients in this quandary really competent to make a decision to take or not take the treatment? If they are deemed not competent (by whom and by whose standards?) then who should have the right/responsibility to make that decision for them? If the doctor is involved in research, does he not have a conflict of interest? If the drug works, he can publish the results and perhaps get more funds to carry on further research. If the drug doesn't work, then the patient may have been put through needless suffering at a time when everything should be done to help alleviate the pain, both physical and mental. Physicians today encounter moral dilemmas far more complex than those that faced Beaumont. It's easy to look back into history and feel so superior about our moral stance. But who is to say that the standards that we establish today will stand the test of time? Perhaps not too far in the future, medical practitioners will look back on our times and wonder how we could have functioned under such a primitive code of ethics.

FETAL-CELL RESEARCH

In 1973, one of the great human rights issues of modern times was settled by the U.S. Supreme Court's landmark decision affirming a woman's constitutional right to abortion, but the ethics of "abortion on demand" continue to be widely debated. This decision inadvertently facilitated fetal cell research for a while.

> Embryonic cells have a number of properties (including their exceptional ability to adapt to a new environment, their capacity to stimulate the growth of new blood vessels, and their unique tolerance to long-term storage) that make them ideal as tissue transplants. Perhaps most important, fetal cells are immunologically 'naive'; that is, upon being transplanted they generally fail to stimulate a rejection response from the recipient's immune system.[23]

Arthur L. Caplin, director of the Center for Biomedical Ethics at the University of Minnesota in Minneapolis, remarked, "The transplantation of parts from the dead to the living is one of the strangest things human beings have devised."[24] It comes about as close to the Frankenstein model as anything yet developed by science.

Some folks may have serious reservations about harvesting parts from deceased adults, even though they have saved many a life. I, like millions of others, have a document in my wallet, attached to my driver's license, that authorizes the use of any of my body parts upon my death according to the Uniform Anatomical Gift Act. And why not? I surely can't use them anymore, and perhaps it might save the life or make living more enjoyable for someone else. But when it comes to transplants of fetal tissues, many folks consider it to be highly immoral. Of course, the parents of an aborted fetus must consent to allow their dead baby to be used in such research. They cannot direct the tissue to a specific recipient. These parents obviously hope that someone's life might be saved or some knowledge might be gained by their gift, so that their child's death will not have been in vain. No money or other forms of compensation can be given to such parents. Their donation must be motivated only by humanitarian concerns. The potential benefits of fetal cell transplants might be very great. Transplants of fetal cells into monkey brains have been able to reverse the symptoms of Parkinson's disease (a debilitating neurological disorder affecting 1.5 million people in the United States alone). The *substantia nigra* region in brains of those afflicted with this disease fail to produce a neurotransmitter called *dopamine*. Fetal transplants of these kinds of cells can produce dopamine. Some success has been achieved by fetal transplants in human Parkinson's patients in other countries. Swedish researchers believe that it is unethical not to try fetal cell transplants with human patients. Transplants of fetal pancreatic cells into Type I diabetics (suffering from low output of the hormone insulin that regulates blood sugar level) have likewise shown some success; early results indicate no adverse reactions, good cell-survival rates, with insulin requirements reduced in some cases. About two million diabetics stand a chance of being helped by this technique. Another two million U.S. sufferers of Alzheimer's disease; three hundred thousand victims of spinal injuries; and over ten thousand cases of hemophilia, muscular dystrophy, and Huntington's disease also might be helped by fetal cell transplants. If this technique proves highly successful, there will be a great demand for fetal tissues. But not all aborted fetuses make suitable donors.

There are two kinds of abortions: spontaneous (natural) and induced (elective). Scientists engaged in fetal-cell research can usually make better use of fetal cells from induced abortions. "Because fetal tissue is fragile and dies quickly after the death of the fetus, it is almost impossible to get tissue from a spontaneous abortion in time to use it, which is why doctors must rely on planned abortions."[25] In March 1988, "pro-life" (antiabortion) activists succeeded in lobbying the U.S. government to outlaw all research under federal funding on fetuses derived from induced abortions. Their argument was that the decision by some vulnerable women to have abortions might be swayed by the potential beneficial use to which their fetuses might be put. This moratorium slowed the pace of fetal research considerably, but did not stop it. The ban does not apply to research on spontaneous abortions or to privately funded projects. Nor does it prohibit fetal tissues from being transplanted into animals. But it does prohibit transplantation of human fetal cells into human recipients if funded by the government. One of the key recommendations of an NIH (National Institutes of Health) panel to study this problem was "that the decision to terminate pregnancy be kept independent from the decision to use the fetal tissue for transplantation research."[26]

Attempts have been made to establish cell lines from human fetal brains as a possible alternative to using cells taken directly from cadavers. Growing cells in vitro this way is not easy. Experience with pig brains has shown that beyond two passages (subcultures) the cells begin to differentiate, making it difficult to remove them from culture without damaging them. Most cell cultures usually die after about fifty or so cell generations. The few cultures that become "immortalized" can grow indefinitely with proper care. Immortalized cells have undergone some genetic changes that allow them to grow without restriction. Obviously it could be dangerous to transplant these abnormal cultured cells into human brains because they might continue to multiply and form a brain tumor or become cancerous and spread (metastasize) to other parts of the body. Nonetheless, cultured cells can be used for many scientific investigations other than as a transplantation source.

The United States has been in the forefront of medical research through most of the twentieth century. But in a bizarre twist of fate, its supremacy in this regard was hampered in the 1980s and 1990s by ethical concerns that have been converted into restrictive federal and state laws. So if a scientist wants to continue making rapid progress in a field such as fetal research, he might feel compelled to take his privately funded project to another country (such as Sweden or Mexico) where the laws are much less restrictive. Then the question arises as to whether journals published in the United States should accept research carried out in countries whose laws allow researchers to circumvent the intent of U.S. laws. Aside from ethical concerns, I wonder how much of a "brain drain" has or will occur as a consequence of U.S. laws that restrict certain kinds of research. Can America afford to lose the scientific talents and potential rewards of their efforts if these restrictions cause these researchers to move to other countries? Denmark has taken the strongest stance in Europe by passing a law in 1986 that outlaws all research on human embryos. The Vatican issued a directive in March 1987 which rejects all forms of "assisted human procreation" and "nontherapeutic research on embryos." Most European countries at that time were proposing legislation that would permit fetal research under strict limits (e.g., no experiments on embryos more than fourteen days old). In 1988, laws were being drafted in West Germany that would make it a criminal offense to conduct any research on an embryo unless the research is directed toward the embryo's own well-being. These restrictions would effectively isolate German scientists in this field from their colleagues in most of the rest of Europe. "Deeply etched memories of Nazi atrocities are digging a gulf between West Germany and other European nations over whether human embryos should be used for research purposes."[27] Karl-Heinz Narjes, EEC commissioner responsible for industry and scientific research commented, "We cannot have a situation in which the same research might lead to a Nobel Prize in some member states of the European Economic Community, and to prison in others."[28]

In April 1990, a baby girl was born at the City of Hope National Medical Center in Duarte, a suburb of Los Angeles. This baby was conceived to serve as a possible bone marrow donor for her eighteen-year-old sister who was dying of leukemia. Fetal tests had shown that the baby's marrow was compatible with the intended recipient. The chance that the bone marrow transplant would be successful was estimated at between 70 to 80 percent. Medical ethicists balked at the concept of creating one human to save another, moralizing that each child should be conceived for its own sake, not for any utilitarian purpose. The parents were hurt by such criticism, and said they would love the baby even if its cells later proved to be incompatible with those of her sister. But the question remains, Would the child have been conceived at all if her sister's life was not at stake? And would the child have been carried to term if earlier tests had shown it was incompatible? Twenty years ago, these kinds of questions would not have been raised because the immunological knowledge and clinical statistics were not available to make these kinds of transplants possible. Thus, the more medical science advances, the more complicated become the ethical and legal issues.

The whole field of human reproductive biology is struggling with these ethical problems. It has been over ten years since the world's first "test-tube baby" was born in Britain. That event, more than any other, caused the general public to reconsider the morality of this kind of technology. The practice was condemned by many on the basis of its being "unnatural." These kinds of religious arguments have an ancient history. For example, what if Edward Jenner's vaccination technique had been outlawed by the same kind of religious argument? How many lives would have been lost to smallpox over the last two hundred years? Fortunately, cooler heads prevailed, and today the world is free of that dreaded disease. Almost everything humans do is "unnatural." The first time a human egg is fertilized in vitro, and the resulting "test-tube baby" develops abnormally, I predict there will be a howl of protest that might stop all further experiments of that kind. People tend to forget that birth defects occur naturally too. It has been estimated that about 30 percent of all spontaneous abortions have chromosomal abnormalities, and chromosomal defects are almost invariably associated with severe physical and/or mental health problems. Perhaps heroic medical efforts to save all babies from spontaneous abortion should not be encouraged. Maybe in these cases we should just let nature take its course.

There seems to be broad agreement in Europe on a number of technologies that should be outlawed, including human cloning and the creation of chimeras between human and animal embryos. But it could be argued that human clones exist naturally; we call them "identical twins." Could something that occurs naturally be all that bad to produce artificially? Perhaps the day will come when the world gets so crowded with people fighting for daily survival (as much of the world presently is doing) that infertile couples will not be allowed to have children by in vitro fertilization (IVF) or other artificial means. It does seem foolish to be spending money on this kind of technology when overpopulation remains the world's number one problem. It also seems unfair that this technology can be used only for those who can afford to pay its high cost. China has had some success in controlling its population growth by allowing only one child per family. This policy sounds excessively restrictive to Western ears, but our ideas about what is moral may change radically over the next fifty or one hundred years as we run out of food and the infrastructure of society can no longer cope with public health and individual medical problems.

RECOMBINANT DNA

The Cold Spring Harbor Laboratory on Long Island in New York has, for over fifty years, been the mecca for students of molecular genetics. Summer symposia and advanced courses bring together outstanding faculty and unusually promising students from throughout the world. During the summer of 1971, Robert Pollack was an instructor there in a course on mammalian cell culture. In one of his lectures he devoted some time to safety and ethics, warning about the known and unknown dangers of indiscriminately mixing cells of different species under artificial conditions. One of his students, a woman from Paul Berg's laboratory at Stanford University in California, told Pollack that Berg was planning to combine DNA from an animal virus (SV40) with the DNA of a bacterium (*Escherichia coli*), using the latest techniques for recombining DNA molecules. The first *restriction endonuclease* had been isolated in 1968 by H. O. Smith, K. W. Wilcox, and T. J. Kelly. These enzymes cut double-stranded DNA molecules at specific base sequences called *recognition sites*. Many of these enzymes make staggered cuts within the recognition site so that the resulting DNA fragments have at each end one strand longer than the other. These so-called sticky ends or cohesive ends like to pair up by the rules: A pairs with T, and G pairs with C. These letter symbols represent the basic building blocks of DNA called *nucleotides*. Thus, two fragments of DNA from any source that were cut by the same restriction endonuclease would be able to recombine in this way. The discovery of this class of enzymes opened the door to the development of the entire field of recombinant DNA research. It was a turning point as great as any other in the history of science.

The virus SV40 is a monkey virus that can cause human cells in culture to transform into cancerlike cells. The bacterium *E. coli* is a common inhabitant of the human gut. Until 1995, it was the bacterial cell about which most was known genetically. In addition to its own circular DNA "chromosome," some strains of *E. coli* contain small circles, called *plasmids*, that can replicate independently of the cell's chromosome. Plasmids may be extracted from the cell and made to recombine with a DNA fragment from some other source, and then reinserted into a host cell by a process called *transformation*. Alternatively, a foreign piece of DNA can be inserted into the chromosome of a bacterial virus (phage), and be reinserted into a cell in a process called *transduction*. Pollack was shocked to learn of Berg's proposed experiment. Might not the SV40 DNA cause cancer if it got into the human gut? Once such a self-reproducing system got loose in the environment, might it not become pandemic? It sounded like a very risky experiment to him. So he telephoned Berg to voice his concern. Berg said that the experiments were only in the planning stage, so he would think about it. "Thus the concept of potential 'biohazard' from recombinant DNA was born."[29] After careful consideration, Berg became convinced that too little was known about these new kinds of genetic recombinants, and the potential risks were so unpredictable that caution would be prudent. So about six months later Berg called Pollack and invited him to join in organizing a conference on the safety of recombinant DNA research. A conference was held in January 1973 that led to publication of a book entitled *Biohazards in Biological Research* with Pollack as one of its editors.[30] This book raised the consciousness of many molecular geneticists to the possible risks of their research.

> This kind of seemingly casual interaction and highly individualistic behavior is characteristic of the way the scientific community works. It illustrates the intense concentration of scientists on particular objectives, the wide net of their informal communication, and the effectiveness of what is called "peer review." The last is really collective judgment, and it is brought to bear both formally and informally. It underlies a good deal of the success of science.[31]

Stanley N. Cohen, a colleague of Berg's at Stanford, discovered in 1973 that plasmids could act as recombinant vectors. Cohen made his plasmid available to other workers in the field. But he too recognized the need for caution, and requested that recipients of his plasmid would not make any recombinants with tumor viruses or with any DNA that would make *E. coli* more resistant to antibiotics. Furthermore, he asked them to refrain from passing his plasmid on to others without his permission. Soon, however, new plasmids became available from other sources, and the self-regulation system broke down.

In June 1973, the annual Gordon Conference on Nucleic Acids provided additional awareness of the growing concern over recombinant DNA experiments. Maxine Singer of the National Institutes of Health (NIH), cochairperson of that conference, reminded her audience that, "We have a responsibility to concern ourselves with the safety of our coworkers and laboratory personnel, as well as with the safety of the public. We are asking this morning to consider this responsibility."[32] The next day, a letter was drafted to the presidents of the National Academy of Sciences and the Institute of Medicine, requesting that an ad hoc study group be formed. This letter was published in *Science*[33] (the journal of the American Association for the Advancement of Science, AAAS) and was widely read by the scientific community. Paul Berg was asked by the president of the National Academy to chair such a committee. That committee met in April 1974. Among the recommendations in its report was one calling for an international conference to be held the following year. This letter, too, was published in *Science*.[34] Nuclear physicists had, some forty years earlier, also seen the dawning of a new perspective. J. R. Oppenheimer, who was the leader of the Manhatten Project that developed the atomic bomb, after viewing the first atomic explosion, declared "physicists have known sin; and this is a knowledge which they cannot lose." Molecular geneticists were going to try to avoid knowing "sin." They were going to have to face the new double image of fundamental knowledge; the results of recombinant DNA research might be harmful or beneficial.

The site chosen for the international meeting was the Asilomar Conference Center at the tip of the Monterey Peninsula in California. It was once a religious retreat, but now is run by the state of California. It was here, in this tranquil setting, amid the smell of pine trees and breathtaking views of the Pacific Ocean, that one of the most important events in the history of science occurred. Scientists on the cutting edge of new knowledge were going to put restraints on their own freedom of inquiry for the first time in history. About 150 researchers from many parts of the world assembled there in February 1975 to grapple with the intangible—the risks of probing a powerful new research field. No one wanted to stop all recombinant DNA experiments. All feared that if the workers in this field did not adopt some kind of guidelines that would minimize risks, someone else (legislation?) would do it for them in a way that would put them at a disadvantage. So the pressure was on the conferees to come to some consensus that they all could live with. No one raised social or ethical questions. Their job was to assess technical risks, to determine what kinds of experiments (if any) should be postponed, and to outline a system of biological and physical containment strategies that would provide the requisite level of protection for each type of experiment. One of the main problems revolved around the continued use of *E. coli* strain K12 as the workhorse for recombinant DNA expression. The fact that it was the best understood cell had to be evaluated in the light of its unknown ecological relationships. If recombinants of this bacterial strain got loose in the environment, would they disrupt the ecological balance of nature? Would its DNA be able to exchange naturally with that of other bacterial species? Once such an organism is set free, there is little hope of containing its spread. The consequences were unknown and could possibly be disastrous. How could decisions be made with so little real knowledge in hand? That was the dilemma facing the conferees at Asilomar. A session with four invited lawyers shocked the conferees with their assessments of the legal ramifications surrounding their kind of work and made them more willing to accept some inconveniences of self-regulation rather than risk the repercussions of legislative and judicial restraint.

The night before the third and final session of the conference, the leaders spent most of the night hammering out a five-page preliminary consensus statement. The document did not completely satisfy anyone, but it was an ingenious compromise. It was "a calculated effort to place the bridle between the conference's jaws without causing it to rear."[35] Each segment of the document was subject to vote. To the surprise of the conference leaders, no section received more than a half dozen dissenting votes. So the document was approved by an overwhelming majority. A consensus had been achieved. The vote was solidly moderate. "The organizing committee had read the situation correctly and done its job well."[36] In almost all countries represented at the conference, steps were taken to institute regulations of the kind that were agreed to at Asilomar. In the United States, the Asilomar document was sent to the Program Advisory Committee on Recombinant DNA Molecules, appointed by the NIH. After the third draft and a lot of public input, that committee's recommendations were sent to NIH director Donald Frederickson. These NIH guidelines, published in June 1976, were judged to be at least as stringent as the Asilomar consensus and in some cases even more so, and they provided the regulatory framework for recombinant DNA research supported by the NIH.

One of the developments that helped push recombinant DNA research forward was the subsequent production of grossly enfeebled strains of *E. coli* that require substances not found in nature and that can grow only at temperatures other than those of the human body. Minimal-risk experiments require only the procedures recommended for clinical microbiological laboratories: no drinking, eating, or smoking in the laboratory; wearing laboratory coats; use of cotton-plugged pipettes or mechanical pipettes; and disinfection of contaminated materials. Minimal-risk experiments generate novel biotypes that, based on present information, are judged incapable of altering the ecological behavior of recipient species, are unable to increase their pathogenicity, or will not prevent effective treatment of any resulting infections. In addition to minimal-risk procedures, there must be no mouth pipetting; lab access must be limited to authorized personnel; and safety hoods must be used for procedures likely to produce aerosols (e.g., blending and sonication). Progressively more enfeebled vectors and host cells must be used as the risk level increases. In moderate-risk experiments, there is a chance of generating a pathogenic agent or one capable of ecological disruption. Gloves must be worn, vacuum lines must be protected by filters, and negative pressure should be maintained in limited-access laboratories. High-risk experiments are those deemed to present significant biohazard to laboratory personnel or the public. Laboratories for these kinds of experiments must be isolated from other areas by air locks, and a negative pressure environment must exist so that air can come in but not leave the lab. Clothing changes and showers are required for personnel entering or leaving. Treatment systems must be installed to inactivate or remove biological agents that may be contaminants in exhaust air and liquid and solid wastes. Exhaust air from safety cabinets must be incinerated or passed through special filters.

Harvard University has always been on the cutting edge of new biological techniques. Both Harvard and the Massachusetts Institute of Technology (MIT) are in the city of Cambridge. A proposal to modify an old biology building on the Harvard campus to provide a P-3 (moderate risk) containment facility was met with a great deal of protest, both from within and outside the university. I think I know how these protesters might have felt. I remember the protests that my town raised to the erection of a nuclear power plant nearby. In the end, the arguments against it didn't make any difference. We got it shoved down our throat anyway. So we live in daily apprehension as to when some mistake at the power plant will occur and result either in our death or the loss of our town, our jobs, and our homes. The lessons of the 1986 Chernobyl meltdown in the USSR are not easily forgotten. The mayor of Cambridge, Alfred E. Vellucci, said, "We want to be damned sure that the people of Cambridge won't be affected by anything that would crawl out of that laboratory. . . . They may come up with a disease that can't be cured—even a monster. Is this the answer to Dr. Frankenstein's dream?"[37] The first of a series of meetings of the city council to discuss the matter was held on June 23, 1976, the same day that the NIH guidelines were formally issued. The safety concern would probably not have come to the attention of local politics had it not been for the rancorous disagreements among the scientists in Cambridge. The full range of opinions and levels of expertise were represented at that meeting. The city council, after taking all of the testimony into consideration, met again on July 7 to consider a proposal by Mayor Velucci to ban recombinant DNA research in Cambridge for at least three months or until control of all health hazards could be assured. Cambridge had a great responsibility to act fairly because there was a general feeling that other cities around the nation might follow the Cambridge model. A group, named the Cambridge Laboratory Experimentation Review Board, consisting of lay persons and scientists, was appointed to examine the problem and make recommendations to the council. A three-month extension of the moratorium was required for the review board to complete its work. The board submitted its findings and recommendations to the city manager on January 5, 1977, with an affirmation of its belief that "a predominantly lay citizen group can face a technical scientific matter of general and deep public concern, educate itself appropriately to the task, and reach a fair decision."[38] The board concluded that recombinant DNA experiments requiring P-3 level containment could be safely conducted in Cambridge under appropriate conditions and safeguards. Among the additions to the NIH guidelines was one requiring the establishment of a permanent Cambridge Biohazards Committee to monitor all DNA research in the city. These recommendations were endorsed unanimously by the Cambridge City Council on February 7, 1977.

The Cambridge guidelines, like those developed at Asilomar, dealt only with potential hazards to human health. The larger social and ethical issues eventually had to be faced. State and federal governments were being aggressively lobbied to also consider the ethical as well as environmental issues. Demonstrations were held in Washington D.C. Even some Nobel laureates such as George Wald of Harvard and Cambridge Universities and Sir MacFarlane Burnet of Australia were among those who called for a total moratorium on recombinant DNA research. The most highly publicized figure in this field was (and still is) Jeremy Rifkin, then leader of the "People's Business Commission" (and subsequently of other ecologically concerned organizations). Rifkin reminded his audience that Huxley's *Brave New World* was upon us, and if we don't act now to stop all DNA work, eventually we will be tinkering with the human genome. Who could forget the tragic consequences of the eugenics program in Nazi Germany to "improve human heredity"? These kinds of metaphors underscored a more fundamental fear than that of just genetic engineering; it was a fear of knowledge itself. A precedent had already been set in 1974 (under the National Research Act) for legislative regulation of scientific research in the United States—the use of human subjects in experimentation. Several bills were introduced in Congress in 1977 concerning recombinant DNA research. Time slipped away while the public was allowed to express its opinions. Meanwhile, expert reassessment of risks evaluated them to be significantly lower than was previously thought. So the urgency to push regulatory legislation forward was considerably reduced. Haste could now be replaced by deliberate caution.

The subsequent refinements of NIH guidelines and regulatory legislation have had a relatively minor impact on recombinant DNA work. In the epilogue to his book *A Double Image of the Double Helix*, Clifford Grobstein focuses on the more fundamental questions raised by the recombinant DNA debates.

> Accumulating knowledge has emerged as a major and essential factor in guiding human directions. But what determines the direction of knowledge generation itself? How and by whom are the choices being made? Does research sometimes set in motion trends that neither researchers nor such social planners as we have intend? . . . Is the march of human knowledge and invention as inexorable and yet unpredictable in its consequences as flood, drought, and earthquake? Or is the knowledge process amenable to purposeful control? If it is, how can we institutionalize the necessary judgment to control it wisely? These are the deep questions underlying the double image of the double helix, a current version of the Faustian dilemma.[39]

ANIMAL RIGHTS

Scientific experimentation on animals has an ancient history. Galen of Pergamum (131-200 C.E.) served as physician to the Roman emperor Marcus

Aurelius. Due to the prejudices and superstitions at that time, dissections of human bodies were not permitted. So most of his own firsthand experience was obtained by dissecting animals such as cattle, sheep, dogs, bears, and apes. Galen assumed that monkeys were similar to humans, so numerous anatomical and physiological errors were thereby made in medical applications of this knowledge. His monumental book *On Anatomical Preparations* became the standard medical text for the next fourteen hundred years, until the time of Vesalius (1514-1564). The animistic or other religious beliefs of some people have long made them mindful of the intrinsic value of animals, apart from any use to which they might be put by humans (either for meat, bone, milk, wool, skins, or draft). As examples, in Indonesia the elephant is considered sacred; in India, cattle are worshiped. But aside from these spiritual connotations, when human life is "hard" there is usually little concern about the welfare of animals other than pets. And life has been hard throughout history in most countries. Only in relatively recent times in certain industrialized countries have people been so affluent that their range of ethical concerns could broaden to include animal welfare. Cultural attitudes, however, continue to blind certain ethnic groups to the mistreatment of animals. For example, dogs are considered "man's best friend" in most parts of the civilized world. But in Korea there are dog-breeding farms to provide restaurants with dog meat for human consumption. Spain continues to mutilate and kill fighting bulls for sport. Malaysians still bet money on cockfights. The British still ride to the hunt and allow their hunting dogs to kill the fox. And North Americans continue to torment animals by riding them or roping them in rodeos, just for fun.

The world's first humane society was founded in London in 1824. Since that time, the new morality concerning humane care of animals has continued to grow. In the United States, most towns of even moderate size have a humane society or department to care for stray dogs and cats, and to painlessly "put them to sleep" (euthanasia) if no one wants to give them a home or if they are in constant pain from injury, illness, or old age. Within the past ten years or so, many of these humane departments have broadened their services to include neutering (castrating) males or spaying females to help control the animal population. They usually also carry on educational programs that encourage owners to be more responsible and provide better care for their pets. All of these services are laudable. But "animal shelters" have, in the past, been the source of many animals used in teaching and scientific experiments. Some of these shelters were operated by folks who thought of it as a business. Universities and industrial or governmental laboratories would pay these shelters to obtain the animals they needed for teaching or research. Those who dealt in such matters probably thought that at least the animal had served some purpose by its use in such experiments. Because most dogs and cats have been somebody's pets, there is a sentimentality attached to these animals that is seldom shared with other animal species. It would seem unlikely that any researcher would not have more kindred feelings toward these pets used in experiments than to a rat or a frog.

The most likely reasons why dogs and cats have been so widely used for research in the past are threefold. First, they are large enough so that surgical manipulations can be rather easily performed. Next, they are tame enough to allow easy handling, and will usually submit to a variety of manipulations (e.g., dosing of food or drawing blood) by humans without too much resistance. And finally, they could be purchased at little cost, and the supply of stray and unwanted animals was virtually unlimited. Aside from ethical concerns, dogs and cats have several disadvantages for scientific research. Large animals require more cage space and cost more to maintain (feed and care) than, say, rodents or other small animals. If the animals are being used as models for application in humans, there is a large genetic gap between carnivores and humans. The closer the genetic relationship between two species, the more probable that extrapolations from one to the other will be valid. Thus, primates would be better models than nonprimates. Chimpanzees make the best models because we share about 98 percent of our DNA sequences in common with them. They are our closest nonhuman genetic relatives. Chimpanzees have become an endangered species and can no longer be imported into the United States. The number of captive-bred chimps is severely restricted, and scientists are likely to lose that supply if the chimps do not breed well in captivity. They are also very expensive to maintain.

Since World War II, organizations representing animal rights have sprung up all over America, Britain, and several other countries. Their membership ranges from those who realize the necessity for some animal experimentation to those who want all such experiments to cease. People for the Ethical Treatment of Animals (PETA) is the largest animal-rights group in the United States, with a quarter-million members, a $7 million annual budget, and a headquarters in Washington, D.C. with sixty full-time employees. Its mission is to stop all animal use by society, not just those used in biomedical research, but also those used for food, clothing, and sport. PETA and all of the other animal-rights groups constitute a formidable adversary to scientific progress. Scientists do not get grant money to educate the public to the necessity of their work or to defend it from the often empty-headed accusations of animal cruelty made by irresponsible activists. There wouldn't be so many extremists in this regard if such a large segment of our citizenry was not so scientifically illiterate. The only hope of stemming this tide of irrational opposition to animal research is through doing a better job of science education in the schools. Science literacy is not mere memorization of facts and theories. It is also having an understanding of (and appreciation for) the manifold factors that influence its progress. That, in case you have lost sight of it, is the main objective of this book.

Most scientists, however, are moderates in wanting to reduce the number of animals used in experiments if possible. This desire is generated at least in part from the tight research budgets under which they usually operate. It is wasteful of both the experimenters' time and money to use more animals than needed. Federal guidelines, established under the Animal Welfare Act of 1985, requires each institution engaged in animal research to have an Institutional Animal Care and Use Committee (IACUC) to screen protocols for animal research, to suggest any possible alternatives to whole animals, to ensure that excessive numbers of animals are not used, and to monitor those projects that are approved for compliance with the guidelines. The guidelines call for use of anesthetics whenever a treatment is expected to elicit pain, and the humane disposal of animals at the end of the project (e.g., by euthanasia). Federal inspectors can drop in anytime to inspect animal facilities, and have the power to shut down labs that continue to violate guidelines.

Rats and mice constitute the models most often used in research for two major reasons. First, they are much cheaper to acquire and maintain than larger animals. And second, there are several highly inbred lines now available. After about twenty consecutive generations of brother-sister matings, a line can be made genetically uniform. That is, each same-sex animal within such an inbred line has the same genetic constitution. This is a great advantage in doing research because it rules out a potential source of error in experimental design, namely genetic variation. Such highly inbred lines are not yet available for most larger animals. Again, aside from humanitarian concerns, researchers should have a vital interest in maintaining their animals in good physical and psychological health. Animals that are stressed by poor nutrition, disease, parasites, lack of space, opportunity for social interactions, and other problems are likely to have their hormone levels out of control. These chemical messengers can affect many parts of the body, and potentially could give rise to widely fluctuating biochemical, physiological, and/or behavioral responses. This increases the meaningless "noise" in experimental data that tends to hide the effects of the experimental treatments from statistical analysis. Everything possible must be done to treat both test and control groups the same. The standard against which most experiments are compared would be that of optimal health. Thus, any experimenter worth her salt would want to provide such an environment for her animals. Furthermore, if there is a better or less expensive system for obtaining the requisite data, why would it not be used?

Institutional Animal Care and Use Committees, otherwise known as animal-welfare committees, look for ways to reduce the number of animals needed in experiments. If alternatives to animal models exist that would do as good a job in obtaining the objectives of the investigation, then they should be used. Unfortunately, there are few alternatives at present. Cell or tissue cultures may have several advantages over the intact animal for certain kinds of research. Several kinds of human cells or tissues can be grown in culture, but not all cell types might respond in the same manner to a given treatment. Even human nerve tissue, which in the intact body does not normally regenerate, has been made to grow for the first time in culture during the 1990s. Cells within a cultured line are genetically uniform, just like the highly inbred lines of rodents. The cost of using cultured cells might be significantly cheaper than using intact animals. Unfortunately, cells usually stop multiplying after about thirty to fifty generations. A few of them may become cancerlike, and can grow continuously under proper care. The problem with these continuous or permanent cell lines is that the cells are abnormal in one or more ways that may render them useless for certain kinds of experiments. Whatever may be discovered by using these abnormal cells may not be extrapolated to normal cells without caution. Furthermore, many biological problems cannot be adequately answered from cell cultures or tissue cultures. The response to a given treatment by the whole animal, with all of its complex of interacting components active, may be quite different than that seen in cells growing in relatively simple culture media. All of these factors must be borne in mind when making decisions about alternatives to whole animal research. Chicken embryos might be a useful alternative to in vitro tissue cultures for some purposes.

Pyrogens are substances that induce fevers in animals. One of the most promising alternatives for pyrogen testing of drug products in rabbits is the Limulus amebocyte lysate (LAL) assay. "Limulus" is the genus name for the horseshoe crab. Blood can be drawn from the crab without killing it. This blood contains wandering cells, called *amebocytes*, that when ruptured (lysed) release a proclotting enzyme that can be activated by bacterial endotoxins (the substances responsible for fevers), causing the blood to clot. The test can detect extremely small amounts of endotoxin, can easily be quantified by photometric instruments, and is much more economical than rabbit tests.

At least one bacterial system has been developed as an alternative to animal testing. The Ames test uses a mutant bacterium that cannot synthesize the amino acid histidine, and consequently cannot grow on unsupplemented medium. When exposed to a mutagenic substance, however, some of these cells will mutate back to the wild-type form that can synthesize histidine and thus can grow on unsupplemented medium. The number of such reverse mutations is a measure of the mutagenic activity of the test substance. Carcinogens are substances capable of inducing cancer. It is thought that all carcinogens are also mutagens, capable of altering DNA in some way. Hence, the Ames test can be used as a substitute for animals in testing various products for their carcinogenic potential.

Computer modeling might also prove useful, especially for some kinds of teaching exercises. But when it comes to research at the frontiers of knowledge, computer models are only as good as the data on which they are based. You know what they say about computers: "garbage in, garbage out." You may know enough about the variables in a teaching experiment to write a computer program, but what can the computer tell you that it wasn't programmed to do? So the utility of computer modeling in pioneering experimental research may be quite limited. For example, when Iraq pulled out of Kuwait in 1990, they left behind 610 burning oil wells. A group of scientists led by the noted astronomer Carl Sagan predicted that it might take several years to put out the fires. Meanwhile, black soot would rise into the stratosphere, spread over and cool the Earth, playing havoc with the world's weather. None of these predictions came true. The fires were extinguished in less than a year and no long-term harm to the atmosphere seems to have occurred. There were several reasons why the computer models failed to accurately predict the future. For one thing, the oil had mixed with more gas than the experts had expected, and it came out of the ground under tremendous pressure. Consequently, it mixed quickly with oxygen and burned very efficiently, producing nearly fifty times less soot than predicted by the computer models. Another factor was the assumption that oil and water would not mix. But the sooty particles contained sulfates that had also not been taken into consideration in building the model. These sulfates allowed the oily particles to interact with water in clouds and thus helped to clean the air. Perhaps the most important factor, however, was the unexpected high salt content of the plumes, causing the smoke to be only half as black as expected. Dark particles absorb solar energy, heat up, and rise like a hot-air balloon. The lighter-colored, salt-laden soot absorbed much less solar energy than anticipated and hence did not rise much above three miles, far short of the stratosphere which occupies a region about six to thirty-one miles above the Earth. The computer model failed because of insufficient hard data concerning the chemical and physical properties of the oil in these fires. Garbage in, garbage out!

With regard to teaching, however, the number of animals used might be reduced by computers, videotapes, or films of exercises that are repeated class after class. The outcomes are well known to the instructor and often serve only to illustrate a biological principle to the students. However, there is more to doing an experiment than the outcome. Students, whether they are science majors or not, need to gain experience and develop techniques by hands-on work with living systems. They need to gather the data for themselves and notice the sources of variation and potential points at which that variation can be controlled. They need to "learn-by-doing," and profit from the mistakes they or their fellow students may make. How else are they to obtain an appreciation for the skill, patience, attention to detail, and perseverance required in scientific work? Science isn't done as easily as one might think from playing computer games, watching a video, or reading a research paper from which all of the labor, the mistakes, and the blind alleys have been omitted.

In California, a sixteen-year-old vegetarian student named Jenifer

Graham had an ethical revulsion to frog dissection in her high school biology class in 1989. Such dissection was a local school requirement at the time to pass the course. Her parents tried to get the high school principle, the county superintendent of schools, and the school board to allow her to be excused from this requirement. They refused. In 1988, her case was influential in passage of AB2507 by the California state legislature. This bill mandates that schools must provide some way for students to demonstrate their biological knowledge other than through hands-on dissection. A movie of the incident was made the next year entitled *Frog Girl*. It was shown on national television both at the "after school" and "prime time" slots. Now the nation should be aware of the different sensitivities to ethical issues that our school systems have to deal with. The issue struck close to home because for many years I have been up to my eyeballs in dissections of commercially prepared frogs and other vertebrates in various zoology classes I teach at the university. But in all those years it was never a requirement for any student to dissect a dead animal. I had always looked upon it as a privilege and a chance for students to gain some skills that have led to many important discoveries in biology. Many students, however, prefer to let others do the dissecting while they watch, either because they don't want to get their hands dirty or perhaps out of squeamishness. Recently my university purchased a computer program that can be used as a kind of tutorial substitute for learning the parts of the frog. But there is no substitute for the real thing, and we require on laboratory exams that students be able to identify structures on dissected frogs. In this way we have been able to accommodate most of the philosophical objections to doing dissections. There are, however, a few students who object to the raising of frogs and other animals for teaching and/or research. At this point, most faculty members in the biological sciences would be unwilling to compromise. They refuse to allow a minority advocacy group to undermine the quality of science education. Some of these animal-rights advocates object to allowing a fish to die out of its water tank in a laboratory, but do nothing to lobby against the same practice by game fishing from a river, lake, or ocean. Some of them object to the pinning of insects in collections. What's next? Stepping on ants? Come on, folks! It's one thing to be sensitive about animal rights. It's another thing to be fanatical about it. What sensible person would want to go out of his way to harm any living creature that is of no personal threat?

There have been excesses on both sides of the issue. One of the most widely publicized instances of animal abuse is the Draize test. This procedure was introduced in 1944 by Food and Drug Administration (FDA) toxicologist John H. Draize to test each new drug, cosmetic, and household product as required by federal law. The procedure involves placing the test substance on one eye of an albino rabbit. The other eye remains untreated as a control. The irritation of the treated eye is then checked for the next three days. The lack of pigmentation in the eyes of albino rabbits makes it easier to observe adverse reactions. Rabbits do not produce tears, so the test substance cannot be washed away. Some substances merely cause the eye to become red or weep. But others cause the eye to swell, hemorrhage, ulcerate, or grossly deform. The torment suffered by these latter cases can only be imagined. Many years after the Draize test was introduced, it became clear that the results could vary markedly from lab to lab. The fact that many scientists eventually concurred that the Draize test was unreliable fueled animal-rights groups to demand a humane alternative. In 1989, the Noxell Corporation, international marketer of Noxzema skin products, as well as Cover Girl and Clarion cosmetics, announced that their use of a new non-animal screening test should eliminate 80 to 90 percent of the rabbits subjected to the painful Draize test. This test, called the *agarose diffusion method*, uses tissue culture with an overlay of agarose (a seaweed derivative related to agar). Products to be tested for toxicity are placed on filter-paper discs on the agarose surface. Survival rates of cells are then measured. Noxell reports that the results from thirty-eight products tested have shown an average 90 percent correlation with the results of animal tests. The company has met with the FDA and plans to share its scientific data with the government, the scientific community, and the public. FDA regulations demand that any substance that is to be sold for use on humans must first be tested on animals. It is bad enough that we have to use animals to test drugs that may improve human health; it is, in my opinion, unethical to use them for testing cosmetics. We don't need cosmetics. They are a luxury. Let those who want to make money from selling cosmetics test new products on themselves, not animals. Cosmetic companies have been asked to donate 0.01 percent of their gross annual sales to finding alternatives to animal testing.

A vast number of animals have been used in the past in establishing LD50 values. This is the amount of a substance (the lethal dose, LD) that will kill 50 percent of a test group of animals in a specified period of time. These kinds of experiments have been disappearing over the last decade as alternative and less expensive tests have become available.

As a recent example of the clash between science and ethics, there is the case of neurosurgeon Michael Carey at Louisiana State University. According to Carey, his is the only work in the world currently using an animal model to study the kind of brain wounds soldiers get in combat. Some sixteen thousand people in the United States also die each year from gunshot wounds to the head. This research involves shooting anesthetized cats in the head, about 125 a year. The outcry by animal-rights groups over this kind of research caused the Congress, in November 1989, to suspend Carey's funding from the Department of Defense (DOD) while a review of the case was made by the General Accounting Office (GAO). The GAO investigation was inconclusive and it made no recommendation to the DOD. Carey's research has become so politically sensitive that his funding may never be restored. Carey has been subjected to hate mail, threatening phone calls, and demonstrations since 1988.

The vast majority of those who fight for animal rights, however, operate within the law. Among the tactics used to accomplish their goal of preventing the use of animals in research is the practice of holding demonstrations and marches before the press and TV cameras. For example, U.S. Capitol Police estimated that twenty-four thousand animal rightists rallied on the steps of the Capitol on June 11, 1990. Celebrities such as singer Grace Slick and *Superman* movie star Christopher Reeve lent their support. However, the number of illegal actions taken by animal-rights activists has recently increased. These tactics include the vandalizing of research facilities and setting animals free. They don't seem to know or care that most of these animals cannot survive outside the laboratory. They will either be run over by cars or killed by pet dogs and cats. During the weekend of August 22, 1987, a group calling themselves the Band of Mercy (affiliated with the animal-rights group PETA) broke through a chain-link fence and cut several padlocks to get into the Protozoan Diseases Laboratory at the Agricultural Research Center in Beltsville, Maryland. They "cat-napped" twenty-eight cats and eight miniature pigs. Eleven of those cats were infected with toxoplasmosis, an incurable parasitic disease that could cause birth defects if pregnant women become exposed to the cats' feces. Members of these radical groups seldom know much biology. Their ignorance and their unlawful activity is not going to help find a cure for toxoplasmosis, but it may contribute to some deformed babies. If the research is being conducted in a federal institution or with federal funds, the FBI can get on the case, and the perpetrators of these crimes can be sent to federal prison.

Extremists in the animal-rights movement, such as those of the Animal Liberation Front, have been involved in more than eighty bombings in the last five years.[40] Many of these bombings are aimed at injuring or killing people, not just blowing up and/or setting fire to animal-research facilities. Some researchers have had not only their own lives threatened, but also those of their families. A thirteen-month-old boy was injured by a bomb planted under the car belonging to Max Headley, a veterinary surgeon who works at the University of Bristol in England. They don't care who they hurt or kill. For them, the ends justify the means. Who would ever think that one's life might be in jeopardy because of the kind of scientific research they were doing? It didn't make any difference in Headley's case that his research was devoted to finding ways to relieve pain and suffering in animals as well as humans. To the extremists, anyone using animals in research is a potential target.

In all fairness, however, mainstream antivivisectionists condemn bomb attacks or other forms of harming those who work with lab animals. Richard Mountford, chairman of Animal Aid, claimed, "It is ridiculous to use violent means when aiming for a nonviolent world. It is morally wrong and it does not achieve anything for the animals anyway."[41] Nonetheless, even the mainstream activities have slowed the pace of research and made it more costly without noticeable improvement of animal care. For example, Stanford University in California had planned to construct the most up-to-date and humane animal facility in the country in 1987. The local humane society in Palo Alto began a campaign with the county supervisors to block the building permit. The society's original objection was that animals to be housed in that building would be inhumanely treated. Since the issuance of the building permit was essentially a zoning question, that argument was not germane. Next, it was argued that since pending federal animal-welfare regulations would eventually require changes in the building, the permit should be delayed. That argument also failed because it had no effect on zoning. Then it was argued that dangerous substances, such as radioactive tracers, toxins, and recombinant DNA molecules, might be released from the new laboratory. California law requires that if there is a "serious controversy" over environmental issues, an environmental impact report (EIR) must be filed. The study took almost a year. The cost of that EIR and interim escalating construction costs ate up $1.3 million that could have been, in my opinion, used to advance knowledge instead of being poured down the drain as a nuisance cost.

Most of the animal-rights activists reveal an appalling ignorance of scientific methods and history. They know little or nothing of physiology, biochemistry, or genetics. They have no historical appreciation of how the medical knowledge we have today was obtained. For example, they are unaware of the role that guinea pigs played in the development of a method to prevent diphtheria, or the importance of dogs in the discovery of insulin as a cure for diabetes, or the part that monkeys played in the production of a vaccine for polio. They live in a dream world of self-righteousness where medical advances can be obtained without the use of animals. There's no free lunch in this world; everything has a price. They don't want animals to be used for these noble purposes, but they are not the ones that volunteer to be the "guinea pigs" on whom new treatments must be tried before they can be approved for widespread use. In one of his "perspective" articles, John Kaplan (editor of *Science* at the time) remarked, "Those opposed to research with animals have seldom stood on principle and instructed their physicians not to use the results of biomedical research on animals when it would benefit their loved ones or themselves. Nor have they been willing to forswear for themselves the advantages of any future advances from animal research."[42]

POSTSCRIPT

To show how deeply ethical concern for the welfare of experimental animals in science has grown, consider the case that prompted the title of this book. During the fall of 1994, a postdoctoral researcher, June Li, anesthetized a laboratory rat in preparation for surgical implantation of sensors into its skull. The animal "was left unattended in the surgery suite" of the Ames Research Center in Mountain View, California "while under anesthesia and later died from complications associated with inappropriate administration of anesthesia."[43] Ames is the main facility for animal research, on everything from tadpoles to rhesus monkeys, conducted by the National Aeronautics and Space Administration (NASA). However, the personnel at Ames responsible for animal care and preparing them for space travel was contracted out to Team Support Services of Corvallis, Oregon. Sharon Vanderlip was chief of veterinary services at Ames. When Vanderlip learned of the death of that rat and that Li had no previous experience working with rats, Vanderlip ordered Li to cease her animal work pending further training.

Under federal guidelines for safeguarding animal welfare and NASA's own standards, Vanderlip had the responsibility to interrupt or terminate any experiments in violation of these rules. However, Vanderlip was employed by Team Support Services, not directly by NASA, and thus had a conflict of authority to enforce these rules. One week later, the acting chief of Ames's life sciences division, Emily Holton, met with other life sciences managers without Vanderlip and decided to suspend the restraining order and to remove Vanderlip's letter to Li from the files. Vanderlip objected that her authority had been undermined by her status as a contractor. She also claimed that Ames officials had subsequently intimidated and harassed her. Vanderlip resigned after a few more months. At least partly in protest against Ames's handling of the affair, two other people also resigned (one was a manager of Team Support Services and the other was a member of the animal-care advisory panel at Ames).

On March 20, 1995, Vanderlip wrote an angry letter to Daniel Golden, a NASA administrator, charging that Ames had tried to circumvent standard procedures for obtaining permission from the National Institutes of Health (NIH) to conduct animal experiments funded by NIH. Upon receipt of a copy of that letter, NASA headquarters ordered a ban on all animal research at Ames and instituted an unaffiliated panel to investigate the problem. The panel found that Holton had improperly held the meeting in Vanderlip's absence. It recommended that NASA make the veterinarian's position a civil service job in order to establish clear lines of authority to enforce governmental rules regarding animal research. The panel charged that Team Support Services had hired incompetent animal technicians and that it should not have been allowed to oversee itself.

On May 24, 1995, Ames submitted to NASA a plan to rectify the problems identified by Vanderlip and the external panel. The ban on animal research at Ames was lifted by NASA on June 5. Ken Baldwin was a physiologist at the University of California, Irvine, who also chaired NASA's life and biomedical sciences and application advisory subcommittee. According to Baldwin, "NASA's involvement with animals [now] probably borders on a state of paranoia in assuring they are properly cared for."[44]

The accidental death of one rat probably doesn't stir much emotion from many people. Accidents can happen even with the best of intentions. But what if it had been a rhesus monkey or a chimpanzee that had died because of incompetence? Primates tend to evoke much greater empathetic public concern, as well they should, because they are so much more closely genetically related to humans than rats are, and because they have so many humanlike qualities. In addition to ethical concerns, however, are economic ones. Primates, and especially chimpanzees, are so expensive and/or endangered species that the loss of even one animal might jeopardize an entire experiment. Loss of these animals due to incompetence simply cannot be tolerated. It was probably a good thing that Ames was forced to get its animal-welfare program in shape by the death of a rat before a similar mishap resulted in the death of a primate. It remains a shame that the death of any experimental animal must occur before those involved in animal care do what is necessary to ensure the welfare of all these creatures.

NOTES

1. Christine A. Ruggere, letter to the editor, *Science* 286, no. 901 (October 29, 1999).

2. Maurice H. Pappworth, *Human Guinea Pigs: Experimentation on Man* (London: Routledge & Kegan Paul, 1967), ix.

3. Ibid., 9.

4. Ibid., 10.

5. Ibid., 11

6. Henry K. Beecher, *Research and the Individual* (Boston: Little, Brown and Company, 1970), 5.

7. Pappworth, *Human Guinea Pigs*, 189.

8. Ibid., 198.

9. Ibid., 217.

10. Ibid., 218.

11. Jon Franklin and John Sutherland, *Guinea Pig Doctors: The Drama of Medical Research Through Self-Experimentation* (New York: William Morrow and Company, Inc., 1984).

12. Ibid., 28.

13. Ibid., 29.

14. Ibid., 51.

15. Eldon J. Gardner, *History of Biology*, 2d ed. (Logan, Utah: Burgess Publishing Company, 1965), 171.

16. Martin J. Blaser, "The Bacteria Behind Ulcers," *Scientific American* 274, no. 2 (February 1996): 104B.

17. Pappworth, *Human Guinea Pigs.*

18. Franklin and Sutherland, *Guinea Pig Doctors.*

19. Sam Epstein and Beryl Epstein, *Dr. Beaumont and the Man with the Hole in His Stomach* (New York: Coward, McCann & Geoghegan, Inc., 1978), 38.

20. Ibid., 41; I. Bernard Cohen, ed., *The Career of William Beaumont and the Reception of His Discovery* (New York: Arno Press, 1980), 21.

21. Epstein, *Dr. Beaumont and the Man with the Hole in His Stomach*, 42.

22. Cohen, *The Career of William Beaumont*, 22–23.

23. Rick Weiss, "Forbidding Fruits of Fetal-Cell Research," *Science News* 134 (November 5, 1988): 296.

24. Ibid.

25. Barbara J. Culliton, "White House Wants Fetal Research Ban," *Science* 242 (September 16, 1988): 1423.

26. Joseph Placa, "Fetal Tissue Transplants Remain Off Limits," *Science* 246 (November 10, 1989): 752.

27. David Dickerson, "Europe Split on Embryo Research," *Science* 242 (November 25, 1988): 1117.

28. Ibid., 1118.

29. Clifford Grobstein, *A Double Image of the Double Helix: The Recombinant DNA Debate* (San Francisco: W. H. Freeman and Company, 1979), 17.

30. A. Hellman, M. N. Oxman, and R. Pollack, eds., *Biohazards in Biological Research: Proceedings of a Conference Held January 22–24, 1973* (Cold Spring Harbor, N.Y.: Cold Spring Harbor Laboratory, 1973).

31. Grobstein, *A Double Image of the Double Helix*, 26.

32. Ibid., 18.

33. Maxine Singer and Dieter Soll, Letters to the Editor, *Science* 181 (September 23, 1973): 1114.

34. Paul Berg et al., Letters to the Editor, "Potential Biohazards of Recombinant DNA Molecules," *Science* 185 (July 26, 1974): 303.

35. Grobstein, *A Double Image of the Double Helix*, 26.

36. Ibid.

37. Ibid., 66.

38. Ibid., 69.

39. Ibid., 107–11.

40. Jeremy Cherfas, "Two Bomb Attacks on Scientists in the U.K.," *Science* 248 (June 22, 1990): 1485.

41. Ibid.

42. John Kaplan, "The Use of Animals in Research," *Science* 242 (November 11, 1988): 840.

43. Andrew Lawler, "Key NASA Lab Under Fire for Animal Care Practices," *Science* 268 (June 23, 1995): 1692.

44. Ibid.

CHAPTER 2

TEAMWORK

Today, when someone writes a book, he may hope that it gets banned by some important person or noteworthy group because that will generate publicity and virtually ensure good sales. People just naturally want to read it for themselves to see what the fuss is all about. When British author Salmon Rushdie wrote *The Satanic Verses* in 1989, it offended Muslims and was immediately banned by the Ayatollah Ruhollah Khomeini, the Iranian leader of the Islamic faith, who also demanded the death of Rushdie. The author had to go into hiding to avoid being assassinated by some Muslim fanatic who thought it was his religious duty to commit murder. Rushdie made a lot of money from his book. It would have had a far smaller audience without all that free publicity. But what good is wealth without the freedom to enjoy it?

As I write this text, one of the books that I enjoyed as a child is being banned by some school boards in the United States. They consider *The Adventures of Huckleberry Finn* by Mark Twain to be racist and unacceptable for young minds. In my opinion, we are not going to abolish racism by book banning/burning. We do our students no favor by hiding historical truths from them. How better can children learn how minorities have been mistreated in American society? Instead of proscription of such a book, it should be a prescription not only for the great literature that it is, but also as an enticing way to get students to confront racial issues. What a marvelous opportunity for instructors to teach the lessons of history to our youngsters so that they will not be preprogrammed to carry the burden of racism into yet another generation.

Two of our greatest freedoms are freedom of speech and freedom of the press. Granted, there will always be a few radicals who will overstep the bounds of good judgment and say or write things that can do nothing but mischief. But to muzzle these aberrations in the name of decency would set a precedent that knows no limits. Where do you draw the line on such controversial issues? Our society must suffer the evils of allowing these radicals to have their say, in order to preserve the greater good of freedom of expression for all people. Ultraconservatives who would try to thwart these freedoms by banning certain works of oratory, literature, drama, art, dance, or music are as undesirable as the radicals they attempt to suppress. Equally despicable are those who pervert the works of others in these areas of human expression. As a student of history, I am not interested in any personal view that has been cleaned up or sanitized in some way to make it more palatable to the average reader. Inevitably this results in a loss of important information regarding either the facts of the case and/or insights into the personalities, motives, desires, biases, etc. of those involved.

One might think that we have learned our lesson since the days of Galileo and the Inquisition regarding the necessity for freedom in scientific endeavors and the accounting thereof. But one would be wrong. Even in these modern times, the specter of censorship and coercion still haunts the halls of science and academe. Unquestionably, the discovery of the structure of DNA in the spring of 1953 by James Watson and Francis Crick will go down in history as one of the ten most important events in all of science. Their discovery initiated the era of molecular genetics. Many reviews and books have been written about this discovery, mostly by folks who were not part of the event. In the mid-1960s, Watson decided to write what he called his "personal account of the discovery of the structure of DNA."[1] Harvard University Press had agreed to publish it. Watson circulated a draft manuscript of his recollections, entitled *Honest Jim*, among many of those who were prominently mentioned in his story. He was severely criticized for his unflattering remarks about certain people. You can see it in the very first sentence of his book where he says "I have never seen Francis Crick in a modest mood."[2] Apparently, the original title of Watson's book indicated that he was going to tell the story from his own point of view without pulling any punches about his personal feelings. However, in response to his critics, Watson either removed or watered down some of the offending passages, and added an epilogue in which he attempted to paint a more flattering picture of Rosalind Franklin than that presented in his story. The title of the revised book was changed to *The Double Helix*, perhaps because Jim Watson was not really being as "honest" with his readers as he had intended to be in his original manuscript. Even with all these changes, Harvard University Press (HUP) reneged on its agreement to publish his book. The reason given by the HUP for its action was that it did not want to take sides in a scientific dispute; a commercial publishing house would be more appropriate.

A slightly abbreviated version appeared in two parts in the January and February 1968 issues of the *Atlantic*. I still recall that time vividly. A friend told me that I should hurry down to the store and pick up the latest copy of the *Atlantic* because Watson had just spilled his guts about the history of his discovery. I rushed right down and got a copy and began reading it that afternoon. The story was so fascinating that I read right through the evening meal. Suddenly the story ended with "To be concluded." Darn! I was partly mad that they had broken the story in half and partly mad at myself that I had not checked it out before I began to read it. Now I had to wait a month to get the rest of the story. I knew from the first installment that this was a book of great importance for the history of science. Here for once was a historical account of a major scientific discovery written, not by secondary or tertiary sources that were not personally involved, but by one of the discoverers himself. There is nothing quite like getting the story from a primary source. And it was obvious that Watson was indeed relating some very personal feelings and viewpoints. He might not have gained my confidence had he just been a severe critic of other people. But he seemed to be equally critical of himself. For example, he revealed that he hoped that he could learn the structure of DNA without having to learn too much chemistry. He broke the terms of his fellowship by moving (between laboratories within the Danish State Serum Institute at Copenhagen, from the one directed by Herman Kalcar to the one directed by Ole Maaløe) without informing the fellowship board. He and Crick had obtained crystallographic data from the group at King's College, London, in a manner that may have been of questionable ethics. So I was convinced that at last I was going to get the whole story, warts and all. I tried not to lose perspective, however. I realized that Watson's version of the discovery (or that by any other participant) would not be free of bias. We all have biases because that is part of human nature. But at least I was able for once to see the history of this discovery unfolding before my eyes much as it must have appeared to at least one of the scientists who was a party to the entire event.

The commercial publishing house of Atheneum was glad to publish *The Double Helix* on February 26, 1968. The following year, a paperback edition was published by Signet and Mentor Books. It was a best-seller and was translated into at least seventeen different languages. Gunther Stent, who met Watson in Kalcar's lab, suggested to him that he should republish the book together with a historical overview, the recollections of two other major characters in the story (Crick and Linus Pauling), reprints of the original papers in which the double helical structure of DNA was first described, and a sample of some of the most interesting reviews of the book. Atheneum Press consented to allow W. W. Norton & Company to reprint *The Double Helix* under the subtitles *A Norton Critical Edition—A Personal Account of the Discovery of the Structure of DNA—Text, Commen-*

tary, Reviews, Original Papers; edited by Gunther S. Stent. This is the edition that has been most useful to me. I highly recommend it to anyone who wants to get the story from more than just Watson's point of view.

Harvard University Press lost a great opportunity—yes, to make lots of money—but more importantly, to educate the public about how intensely personal and human the pursuit of scientific knowledge really is. I dare say that no other book about the history of science has had such a mass appeal and been read by so many, both within and outside of science. I see this as a great public service. The public needs to become knowledgeable of and maintain an interest in science. Some critics feared that the public would get a warped opinion of how science is conducted by reading Watson's bizarre account of his own misconduct and did not want that personal account to be published. So here we are again, with some holier-than-thou types trying to influence what the American people should or should not be allowed to read. My only regret is that Watson allowed himself to be diverted from his original manuscript. I am outraged that any person or group has the gall and the influence to cause an author to change the style or preferred content of his work. You don't find critics telling an artist how to paint a picture *before* her work is completed. Why should literature be influenced in this way? How dare they take it upon themselves to pressure an author to clean up or sanitize his work to suit their own agendas? I, for one, am tired of reading insipid history books that have been gutted of the nuts and bolts of human emotion just so that no one's toes get stepped on. Today we must be politically correct! School boards in some states have the power to choose textbooks. Their decisions should demonstrate "intestinal fortitude" in the face of political and economic pressures from special-interest groups. If they do not resist, then it should not be surprising that our children find history unexciting. An eviscerated and sterilized history book that passes the test of toe-stepping, in order to please everyone, isn't worth reading. Our children deserve better than that. As soon as *The Double Helix* became available in paperback, I made it required reading for my history of biology class. It never fails to stimulate heated debates over the nature of science. It opens up a lot of innocent eyes. It is a marvelous educational tool. Thanks, Dr. Watson.

THE DOUBLE HELIX

James Dewey Watson was born in Chicago on April 6, 1928. He obtained his B.S. degree in zoology from the University of Chicago in 1947 and his Ph.D. degree from Indiana University in 1950. His doctoral thesis was supervised by Salvador Luria (who was destined to share a 1969 Nobel Prize for his contributions to viral genetics) and was concerned with the lethal effect of X rays on bacterial viruses (called *bacteriophage* or simply *phage*). Watson had become obsessed with the "holy grail of biology," i.e., an understanding of the nature of the genetic material DNA. Phage were considered to be naked genes in some respects and the simplest genetic systems available. The first experimental evidence that DNA (deoxyribonucleic acid) was genetic material came from the bacteriological work of Oswald Avery and his colleagues at the Rockefeller Institute in New York in 1944. It took several years before the idea caught on that DNA might be the genetic material, not just of bacteria but of all living things. So while others continued to pursue the idea that proteins were the hereditary material, Watson decided to "go for the gold" and try to discover the structure of DNA. With Luria's help, he obtained a Merck Postdoctoral Fellowship from the National Research Council that allowed him to go to the University of Copenhagen and work in the Danish Serum Institute with the biochemist Herman Kalckar. There he was supposed to learn the chemistry of nucleic acids such as DNA. Watson, however, found no stimulation in Kalckar's lab. Ole Maaløe and Herman Karckar were good friends and collaborators, and Watson could understand Ole's English much better than Herman's. So Watson and Gunther Stent (who was also a phage worker in Kalcar's lab) began regular visits to Ole's lab only a few miles away. Watson confessed, "At first I felt ill at ease doing conventional phage work with Ole, since my fellowship was explicitly awarded to enable me to learn biochemistry with Herman; in a strictly literal sense I was violating its terms."[3] Watson managed to justify it to himself that it was better to be productive in something at Ole's lab than to remain with Kalckar where he felt stifled.

Watson had to apply for an extension of his fellowship grant and was asked to outline his plans for the next year. He had no plans. But Kalckar suggested that Watson go that spring to the Zoological Station at Naples, where he (Kalckar) was going to be during April and May. Watson liked the idea of spending a few months in the sunshine rather than staying in Copenhagen "where spring does not exist."[4] So he wrote to his fellowship board to request permission to accompany Kalckar to Naples. The board granted his request and also sent him a $200 check for traveling expenses. "It made me feel slightly dishonest as I set off for the sun."[5] At Naples, he first met Maurice Wilkins, with whom both he and Crick would later share a Nobel Prize (1962, Physiology and Medicine).

Wilkins was a physicist who had worked on weaponry during the Second World War and had turned in disgust, from what he considered to be the misapplication of physics, to the crystallographic study of DNA. He worked in London at a division of the University of London known as King's College (not to be confused with King's College, Cambridge). Wilkins was a bachelor with little experience in the X-ray diffraction techniques needed for the crystallographic study of DNA. This technique uses the diffraction patterns produced by X-ray scattering from crystals to determine the three-dimensional structure of atoms or molecules in the crystal. Rosalind Franklin, at the age of thirty-one, was brought into his team because she was an experienced X-ray diffractionist. Unfortunately he found that she was unwilling to accept herself as Wilkins's assistant. She had been given to believe that she was to have DNA as her own research project. Her head-strong manner clashed with Wilkins's personality so that they were not able to work together effectively. She was also led to believe that her research appointment would last for several years. Despite their personal differences, Wilkins could not bring himself to discharge her.

Wilkins's boss, J. T. Randall, was scheduled to give a paper at a meeting on biomacromolecules in Naples, but finding himself overcommitted, he asked Wilkins to go in his stead. At the meeting, Wilkins showed an X-ray diffraction picture of DNA. It contained a lot of information that had not been fully inter-preted at the time. Most of the discussion went over Watson's head because he was not a crystallographer and didn't understand the jargon. Before that meeting, Watson was not terribly interested in the chemical approach to solving the structure of DNA because that had been going on for over fifteen years without making much progress. But the fact that Wilkins's team had obtained an X-ray diffraction picture meant that DNA must be capable of forming a crystalline structure. Until that time, Watson had thought that the genetic material might be fantastically irregular. Now, however, he knew that DNA could crystallize and hence must have at least some regular component in its structure. That regularity offered an opportunity for its structure to be solved in a straightforward manner. Watson suddenly became excited about chemistry! Watson was unable to talk to Wilkins after the meeting, but managed to meet him the next day on an excursion to the Greek temples at Paestum. He was telling Wilkins of his interest in DNA, but was cut short when the tour group had to board the bus. Watson's sister Elizabeth had just arrived from the United States. Jim had noticed that Maurice had shown some interest in his sister, and had hoped that a romantic relationship might develop between the two of them with the result that he would be invited to work with Maurice. But such was not to be. Watson had to face the prospect of returning to the Copenhagen lab and its unproductive chemical approach to structural analysis.

Shortly thereafter, Watson received the news that the great structural chemist Linus Pauling had deciphered an important three-dimensional fea-ture of many proteins called the *alpha-helix*. Linus was working at the Cal-ifornia Institute of Technology in Pasadena (Cal Tech). Watson felt that "Linus was too great a man to waste his time teaching a mathematically defi-cient biologist"[6] such as himself. And his attempts to impress Wilkins had failed. This left Cambridge, England, as the only other place to do X-ray dif-fraction work. Watson knew that Max Perutz was there working on the structure of the red blood-cell protein hemoglobin. So he wrote to Luria, requesting his assistance in getting him into Max's lab. Fortunately, Luria had met a Cambridge coworker of Perutz named John Kendrew who was looking for someone to join him in the study of a related muscle protein called *myoglobin*. So with Luria's help, Watson was able to go to Cambridge.

In the Cavendish laboratory at Cambridge, Watson was met by Perutz because Kendrew was on a visit to the United States. Max arranged for a meeting with Sir Lawrence Bragg, the head of the Cavendish unit. Bragg was the founder of crystallography and a Nobel Prize laureate. He gave Watson permission to work with Kendrew. Jim returned to Copenhagen to move his belongings to Cambridge. Kalckar was "splendidly cooperative"[7] in Watson's decision to move, and wrote a letter to the fellowship office endorsing it. On his return to the Cavendish, Jim met Francis Crick, a doctoral candidate who had been working on protein structures under Perutz for two years. The two of them established a rapport immediately.

Crick was born in 1916; attended school at Mill Hill; read physics at University College, London; and had embarked on an advanced degree when the Second World War broke out. He became part of the British Admiralty's scientific establishment, and made significant contributions to the development of magnetic mines. By the end of the war he had lost his desire to stay in physics and decided to shift to biology, where many significant problems lay begging for solutions. Perhaps the greatest influence in making this shift was his 1946 reading of *What Is Life?* (published in 1944) by the famous theoretical physicist Erwin Schrödinger. In his book, Schrödinger proposed that the gene is the key to life itself. He predicted that the gene would prove to be capable of forming an aperiodic crystal (a crystal whose component parts are not periodically repeated in a regular pattern). Many physicists who eventually moved to biology did so as a consequence of reading *What Is Life?* in the hope of finding new laws of physics in the molecules of heredity. With the help of physiologist A. V. Hill, Crick obtained a small grant to go to Cambridge in the fall of 1947. He first worked at Strangeways Laboratory for two years before moving to the Cavendish where he joined Perutz and Kendrew. Here he became motivated to try for a Ph.D. degree. He enrolled as a research student at Caius College, with Max Perutz as his supervisor. Crick's boisterous personality irritated Bragg, but Jim didn't seem to mind it. They both agreed that the model-building approach used by Pauling to solve protein structure was likely to be the best way to go about solving the structure of DNA. Wilkins was also brought up in the model-building tradition, whereas Franklin was trained to use the Patterson-vector approach, using the magnitude and direction of refracted X rays from a crystal to infer its three-dimensional structure at the molecular level. These philosophical differences were also part of the reason that those two did not get along well together. Kendrew soon realized that Watson would be of little help to him in deciphering myoglobin structure, so he gave Jim a lot of free time for pursuing his own DNA research.

Crick was interested in DNA, and occasionally dabbled in attempts to interpret X-ray diffraction patterns, but he had his own research to do and so could not devote nearly the time to DNA structure that Watson could. The only published X-ray photograph of DNA was one taken five years earlier by the English crystallographer W. T. Astbury. No one had been able to make much sense out of it. Watson and Crick thought that it might be a helix of some sort, but they wished they had the much better picture that Jim saw at Naples. At Crick's invitation, Wilkins gladly came to Cambridge for the weekend. Maurice had already been informed by his colleague, the theoretician Alex Stokes, that the picture was compatible with a helix. The first problem now was how many chains were in the helix—one? two? three? Wilkins thought there were three chains. But most of their discussions with Wilkins centered on Rosalind Franklin (or Rosy as Jim referred to her behind her back). She was now insisting that Wilkins should not take any more X-ray photographs of DNA. In trying to appease her, Wilkins "made a very bad bargain. He had handed over to her all the good crystalline DNA used in his original work and had agreed to confine his studies to other DNA, which he afterward found did not crystallize. The point had been reached where Rosy would not even tell Maurice her latest results."[8] The key to Franklin's success in obtaining excellent DNA photographs would later be explained by Wilkins as due to two main facts. They had received a superior source of DNA from R. Signer in Switzerland, and they had a feeling about how to treat the molecule right. They were able to align the fibers in parallel. But Wilkins could not learn anything else from Franklin until November 1951, when she was scheduled to give a seminar on her past six months' work. Maurice invited Watson and Crick to attend.

At the time that Pauling proposed his alpha-helix structure for proteins, there was no general theory by which its fine details could be tested. Then Bragg received a letter from the crystallographer V. Vand proposing a theory for the diffraction of X rays by helical structures. Bragg passed the letter on to Crick who quickly found a serious flaw in Vand's theory. He explained his objections to a quiet Scot, Bill Cochran, who was then a lecturer in crystallography at the Cavendish. Watson considered him to be "the most astute sounding board for Frances' frequent ventures into theory."[9] Cochran had independently found faults with Vand's theory and began to wonder what the right answer might be. Crick immediately began to work on that problem and by the next morning he had a solution. He was in the midst of explaining his answer to Kendrew and Perutz when Cochran walked in and told Francis that he too had a solution. They both had independently arrived at the same answer. "They then checked the alpha-helix by visual inspection with Max's X-ray diagrams. The agreement was so good that both Linus's model and their theory had to be correct."[10]

Franklin was convinced that the only way to establish the structure of

DNA was by pure crystallographic (Patterson-vector) approaches. Despite Pauling's success with model building in explaining protein structure, she considered the idea of using tinker-toy-like models to solve biological problems as a measure of last resort. Watson attended the report on DNA by Franklin in November 1951. There were only about fifteen people in attendance, and very few questions were asked. No one mentioned models, perhaps because (as Watson suggests) they all knew her feelings on the matter and did not want to elicit a sharp retort from her. According to Watson, everyone had to kind of "walk on eggs" around Franklin for fear of stirring up a hornet's nest. It was obvious to Watson, as he left the meeting, that "in spite of much elaborate crystallographic analysis, little real progress had been made by Rosy since the day she arrived at King's."[11]

The next day, Watson and Crick took an hour-and-a-half train ride to Oxford. Crick wanted to tell Dorthy Hodgkin, "the best of the English crystallographers,"[12] about his and Cochran's success in working out the helical diffraction theory. While enroute, Crick pumped Watson for what he had learned from Franklin's talk. Unfortunately, Watson did not know enough crystallographic jargon to understand all of what she had said. He could not report the exact water content of DNA she had presented. It was possible that whatever he told Crick might be in error by an order of magnitude. Crick was visibly annoyed by Watson's habit of trusting to memory rather than writing things down on paper. Apparently the reason why Crick did not go to Franklin's talk himself was because of the British sense of fair play. Everyone knew that Wilkins and Franklin had taken on the job of analyzing the structure of DNA. "The sight of Francis mulling over the consequences of Rosy's information when it was hardly out of her mouth would have upset Maurice. In one sense, it would be grossly unfair for them to learn the facts at the same time."[13] Wilkins had to learn what Franklin was up to along with others at her talk. "Certainly Maurice should have the first chance to come to grips with the problem."[14] This proprietary attitude toward science is, of course, completely foreign to American researchers. Scientific problems don't belong to any individual or group. The prize is up for grabs. But not between Englanders. There was, however, rivalry between the British and the Americans. Pauling's success in discovering the alpha-helix was a blow to the Cambridge group. About a year earlier, Bragg, Kendrew, and Perutz had published a paper on the structure of the polypeptide chain that missed its mark. If Watson had success in finding a structure for DNA, glory would accrue to English labs, not to those in America. Watson never did explain how he felt about that. The conversation that Watson had with Wilkins the previous night had led him to believe that Wilkins was not planning to try to use model building to solve the problem.

Using the water content of DNA and X-ray data that Watson reported, Crick began making calculations and soon came to the conclusion that "only a small number of formal solutions were compatible both with the Cochran-Crick theory and with Rosy's experimental data."[15] The number of chains in the molecule might be two, three, or four, depending on the angle and radii at which the DNA strands twisted about the central axis. Crick thought that the answer would become clear after a week or so fiddling with the molecular models. Upon their return to Cambridge, Watson began trying various arrangements of the parts of DNA to see what would work. It was known that DNA was made up of units called *nucleotides*. (See fig. 2.1.) Each nucleotide consists of three main parts: a nitrogen-containing base, a sugar, and a phosphate group. The bases are of two major types: pyrimidines and purines. Pyrimidines are single ringlike structures; purines consist of a pyrimidinelike ring and a second smaller ring. Thymine (T) and cytosine (C) are pyrimidines that differ by the side groups that stick out from the ring. Adenine (A) and guanine (G) are purines that also differ in their side groups. The sugar is five-carbon deoxyribose. The carbon atoms in the sugar are given primed numbers; the atoms in the bases are given unprimed numbers. The base connects to the 1' carbon in the sugar by a glycosidic bond. The phosphate group connects to the 5' carbon in the sugar. But just because we know that a building is made of bricks does not mean that we can infer what the building looks like. Similarly, the fact that DNA is made up of nucleotides says nothing about its architecture.

Watson and Crick believed that the backbone of each DNA chain consisted of alternating sugar and phosphate groups. This regularity is what they thought allowed the molecule to crystallize. They also thought that the backbone was in the center of the molecule. The bases were attached to the sugars and faced outward. It was assumed that the irregular sequence of bases along the chain carried the genetic information. One of the problems with these early models was how to neutralize the negative charges on the phosphate groups. The best guess that Watson and Crick could come up with was that divalent cations (ions with two positive charges) such as magnesium ($Mg++$) formed salt bridges between two or more phosphate groups. The crystallographic data was mute on this point. It didn't take long tinkering with the models, however, for them to conclude that this idea wasn't working. One of their models did look promising though. It consisted of three chains twisted about each other in such a way as to repeat every 28 Ångstroms (symbolized Å, after Anders Jonas Ångstrom; one Å is one-hundred-millionth or 10^{-8} centimeter) along the helical axis. This was one solid bit of information that could be gleaned from Franklin's X-ray photographs. A few of the atomic contacts were closer than theory predicted, but the wire models they had at that time were not as accurate as the ones that they would have later on. They called Wilkins to invite him to look at the model and give his opinion. He arrived the next day with his collaborator, Willy Seeds, but Franklin came along too with her student R. G.

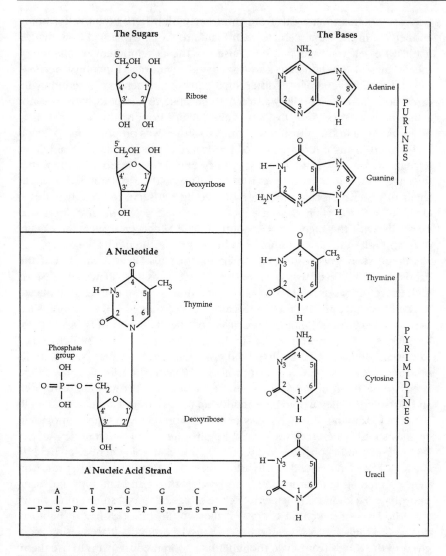

Fig. 2.1. Structural components of nucleic acids. The carbon atoms in the sugar are designated by primed numbers to distinguish them from the carbon atoms in the base. The #1 nitrogen atom of pyrimidines and the #9 nitrogen atom of purines are bonded to the #1' carbon of ribose or deoxyribose of nucleotides. From W. D. Stansfield, Jaime S. Colomé, and Raúl J. Cano, *Schaum's Outline of Molecular and Cell Biology*, fig. 2–13. © 1996 McGraw-Hill. Reproduced with permission of The McGraw-Hill Companies.

Gosling. Wilkins took a cab from the train station rather than suffer being cooped up in the same vehicle with Franklin. Crick explained how mathematical Bessel functions had been used in the development of his helical theory. Bessel functions are also known as "cylindrical functions" because they occur in several partial differential equations when expressed in cylindrical coordinates. Franklin was irritated because, to her mind, there was not a shred of evidence that DNA was helical. Furthermore, the idea that magnesium ions held the phosphate groups together was preposterous. Watson had goofed in his interpretation of the water content given by Franklin at her talk. It was a wet molecule, not a dry one. Franklin lectured them that the magnesium ions (if they existed at all) would be surrounded by tight shells of water molecules and therefore could not serve to hold their three-chain model together. The number of possible DNA models increased markedly with the higher water content. The King's group left the Cavendish lab with Franklin triumphant in her defeat of the model-builders.

Bragg soon learned of the humiliation that his lab had suffered at the hands of Crick. This was the time for Wilkins's boss, J. T. Randall, to discuss with Bragg the wisdom of having Crick and his American colleague duplicate King's heavy investment in DNA research. Bragg wanted Crick out of his lab as soon as possible anyway. But if Crick continued to dabble in DNA structure he might spend the next five years in the lab without collecting sufficient data to warrant an honest Ph.D. Bragg did not relish the thought of spending his remaining years as the Cavendish professor being responsible for "the outrageous antics of an unsuccessful genius."[16] Bragg told Perutz to inform Crick and Watson that they would have to give up their DNA research. "Crick could then buckle down to his thesis task of investigating the ways that hemoglobin crystals shrink when they are placed in salt solutions of different density. . . . With a Ph.D. in his pocket, Crick could then seek employment elsewhere."[17] Crick and Watson decided not to argue with Bragg and kept a low profile on their continuing interest in DNA. Any models they had built with the sugar-phosphate backbones in the center of the molecule forced some atoms closer than the laws of structural chemistry allowed. So this idea had to be abandoned. The jigs for construction of the models they had been using were given to the King's team, even though it had no immediate plans to use them. Otherwise it might appear that Watson and Crick were still in competition with Franklin and Wilkins. They would then not be invited to research colloquia to learn of Franklin's latest crystallographic data. Access to such data would be crucial for any further attempts at model building. Crick went back to hemoglobin. Kendrew gave Watson freedom to study theoretical chemistry and think about DNA even though he had no models to work with. The book that Watson used most frequently was Pauling's *The Nature of the Chemical Bond*.[18] The thought was always at the back of his mind that Pauling might any day announce that he had a structure for DNA.

Watson spent the Christmas of 1951 in Carradale, Scotland, at the home of his friend Avrion Mitchison. Late in January 1952, Watson received notice from his fellowship board that his fellowship was only valid for work in the designated institution. His violation of that provision caused the board to revoke his fellowship. In its place, however, Watson was happy to learn that he had been awarded a new fellowship, but only for eight months (until the middle of May 1952) instead of the customary twelve months. Thus, his punishment for not going to Stockholm was the loss of a thousand dollars. Watson then decided to work on tobacco mosaic virus (TMV) because it contained ribonucleic acid (RNA) rather than DNA. Wilkins was not working on RNA so there would be no conflict. Watson hoped that by studying RNA he might learn some vital clues about DNA. The fact that some viruses contained no DNA meant that DNA could not be the genetic material of all biological structures. This knowledge might have been partly responsible for perpetuation of the belief among most biologists that proteins were the genetic material rather than DNA. Watson did not indicate in *The Double Helix* how he resolved this dilemma. TMV had been studied with X rays by J. D. Bernal and I. Fankuchen in 1939. At that time, it was not known that proteins and RNA are constructed along radically different lines. Watson and Crick hypothesized that TMV consisted of a central core of genetic RNA that was surrounded by a large number of identical small protein subunits. Watson pilfered the Bernal/Fankuchen paper from the Philosophical Library so that he could show their X-ray picture to Crick in his office. Crick saw evidence that TMV's protein structures were arranged in a helix, making a complete turn every twenty-three Ångstroms along its axis. The way to reveal a helix was to tilt the oriented TMV sample at several angles to the X-ray beam. Fankuchen had not done this because no one took helices seriously at that time. Hugh Huxley taught Watson how to set up the X-ray camera for photographing TMV at several angles. Watson was able to obtain a sample of TMV from Roy Markham, who worked at the Molteno Institute in Cambridge. Watson labored for over a month without getting any pictures that were good enough to indicate a helix.

Word was out that Linus Pauling was coming to London in May 1952 for a meeting organized by the Royal Society on the structure of proteins. The State Department revoked his passport as he was about to leave the United States for that meeting. It didn't want him rolling around London like a loose cannon spouting off about peaceful coexistence with the Russians. The Gestapo-like treatment of Pauling came as no surprise to those who had recently been in Oxford for a Society of General Microbiology meeting on "The Nature of Viral Multiplication." Luria was scheduled to have been one of the main speakers. But two weeks prior to his scheduled departure he was notified that his request for a passport had been denied. The State Department gave no explanation for this action. In Luria's absence, the job of describing the recent experiments of the American phage workers fell upon Watson. Several days before the microbiology meeting, Watson received a letter from Alfred Hershey, then at Cold Spring Harbor. He told Watson how, with his coworker Martha Chase, his recent experiments with a DNA-containing phage provided powerful proof that DNA was the genetic material, and not protein. Watson read Hershey's letter at the microbiology meeting to an audience of over four hundred people. No one seemed to appreciate the Hershey report except Andre Lwoff, Seymour Benzer, and Gunther Stent. "They knew that Hershey's experiments were not trivial and from then on everyone was going to place more emphasis on DNA."[19] Watson learned from Wilkins that the models had not been touched since they arrived at the King's lab. The squabbling between Wilkins and Franklin was more intense then ever. Formerly she had maintained there was no evidence either for or against DNA helices. Now, however, she was insisting that her data indicated that DNA was not a helix. Any suggestions by Wilkins to her about trying to build helical models might result in her twisting the copper-wire models around his neck. The very presence of the jigs (that held pieces of the model together) at the King's lab was probably a thorn in Franklin's side. In any event, Wilkins asked Watson if he wanted the jigs returned. Watson said yes, knowing that Wilkins thought he was preoccupied with TMV and its RNA, not DNA.

A new, more powerful rotating anode X-ray tube had just been constructed at the Cavendish. It permitted Watson to take pictures twenty times faster than with the older equipment. Soon he had the X-ray patterns that proved the helical structure of TMV. Now there would be no problem in convincing Luria and Max Delbrück (the "father of phage genetics") that he should stay in Cambridge. However, it had become clear to Watson that the way to DNA was not through TMV. He needed to start thinking about the findings of the Austrian-born biochemist Erwin Chargaff. Chargaff had studied DNA samples from many different species and found some remarkable regularities. The number of purines always equaled the number of pyrimidines in DNA. Furthermore, the amount of adenine always was equivalent to the amount of thymine; likewise guanine always equaled cytosine. Chargaff did not know why that was the case. Watson reported "Chargaff's rules" to Crick, who didn't know what they meant either. Crick took the problem to the young theoretical chemist John Griffith. He interested Griffith by proposing that the perfect biological principle was the self-replication of the gene. Crick was thinking that gene copying proceeded directly by like pairing with like. But Griffith had the notion that gene copying was based on the affinity of complementary surfaces, analogous to a lock-and-key arrangement. This latter idea had been around for almost thirty years and was nothing new. Either way, Crick realized that it was important to pinpoint the attractive forces. Crick was convinced that hydrogen bonds were not the answer because his chemist friends had repeatedly told him that the hydrogen atoms in the purine and pyrimidine bases had no fixed locations. They could randomly move from one spot to another. He therefore had entertained the idea that DNA replication somehow involved specific attractive forces between the flat surfaces of the bases. After a few days, Griffith told Crick that according to his rough calculations it seemed possible that adenine and thymine could stick together by their flat surfaces; likewise guanine would be attracted to thymine. Crick was delighted, because that kind of base pairing would lead to Chargaff's rules. There was, however, no explanation why only one side of each base would pair in this manner.

News came early in July 1952 that Chargaff himself would soon visit Cambridge. When he arrived, Watson felt that Chargaff was not happy at the prospect that two dark horses might beat him to the structure of DNA. Chargaff immediately derided Watson about his long hair and Chicago accent. When Crick tried to explain Griffith's calculations, he forgot what the chemical structures of the bases were. Chargaff left the lab convinced that Watson and Crick knew neither where they were going nor how to get there.

Before Watson left Copenhagen for Cambridge, Max Delbrück had offered Watson a research position in the biology division of Cal Tech and arranged a Polio Foundation fellowship for him to start in September 1952. Watson wrote to Delbrück in March 1952 that he wanted another year at Cambridge. Delbrück approved and saw to it that Watson's fellowship was transferred to the Cavendish.

In July of 1952, Watson had his chance to meet Linus Pauling at a meeting of phage workers at the Abbaye at Royaumount, about 35 kilometers north of Paris. Delbrück mentioned that twelve months hence Watson would be coming to Cal Tech. When Watson brought up the X-ray work at King's, Pauling merely indicated that very accurate X-ray work would be vital to the eventual understanding of the nucleic acids. From Linus's wife, Ava Helen, Watson learned that her son Peter had been accepted by Bragg to work toward a Ph.D. with Kendrew.

Watson vacationed in the Italian Alps during August of 1952. After his return to Cambridge, Watson learned from Peter Pauling that Linus was preoccupied with supercoiling of alpha-helices in the hair-protein keratin. This was good news to Watson but bad news to Crick because Francis had been periodically working on coiled coils for over a year. Not wishing to be beaten again by Linus, Crick stopped his thesis work and bore down on his coiled-coil equations. With the help of George Kreisel, his logician friend at Cambridge, they were successful. A letter was quickly drafted and given to Bragg to send on to the journal *Nature* for publication. The Cavendish had beaten Linus to the punch this time. Crick's stature was in ascendancy. David Harker asked Crick to join him in Brooklyn for a year working on the structure of the enzyme ribonuclease, and offered him a six-thousand-dollar paycheck. In mid-October 1952, Crick wrote to Harker that he accepted his offer and would arrive in the fall of 1953. Watson, meanwhile, had learned from Wilkins that Franklin thought the sugar-phosphate backbone of DNA was on the outside of the molecule. But Watson had no way of checking it out because he was not privy to her experimental data.

Watson and Crick shared their office with Peter Pauling. In the middle of December 1952, Peter received a letter from his father containing the news that he had found a structure for DNA. There were no details. Watson wondered how Linus was able to solve the puzzle without access to the X-ray data from King's. Wilkins received the news blandly. He was much more excited about the fact that Franklin had told him that she wanted to transfer to J. B. Bernal's crystallography lab at Berkbeck College, London, and that she would not take the DNA problem with her. She would write up her work for publication in the next few months, and then be out of Wilkins's hair. Watson went to Zermatt, Switzerland for a Christmas skiing holiday. As soon as he returned to Cambridge in mid-January 1953 he asked Peter Pauling what latest news he had heard from his father. A manuscript on the DNA structure had been written and would be mailed to Peter soon. It arrived the first week in February. As soon as Watson saw Linus's model he knew something was wrong with it. It was a three-chain helix with the sugar-phosphate backbone in the center. But how could this be? Watson and Crick had tried this approach from all angles and could not make it fit the X-ray data. Closer inspection revealed that the phosphate groups on Linus's model were not ionized. Each phosphate group was chemically bonded to a hydrogen atom and thus had no charge. Without a negative charge, Pauling's nucleic acid wasn't really an acid at all. Furthermore, the hydrogen atoms formed hydrogen bonds to bind the three chains together. Without those hydrogens, the three chains would fall apart and the structure would vanish. The world's most astute chemist, Linus Pauling, had made a basic blunder. "If a student had made a similar mistake, he would be thought unfit to benefit from Cal Tech's chemistry faculty."[20] Watson took the manuscript to London to show to Wilkins and Franklin.

Watson found Wilkins busy, so he decided to drop in on Franklin in her lab. While she looked at Pauling's paper, Watson could not refrain from pointing out that the Pauling model resembled the three-chain models that Watson and Crick had abandoned the previous year. Instead of being amused, Franklin became increasingly annoyed by Watson's recurring references to helical structures. She reiterated that there was not one shred of evidence to support a helical structure for DNA. Wilkins had told Watson several months earlier that Franklin's evidence against a helical structure was a red herring. Watson then made the mistake of implying that Franklin was "incompetent in interpreting X-ray pictures. If only she would learn some theory, she would understand how her supposed antihelical features arose from the minor distortions needed to pack regular helices into a crystalline lattice."[21] Franklin came from behind the lab bench and began menacingly moving toward Watson. He quickly grabbed the manuscript and hastily beat a retreat out the door just as Wilkins was about to enter. Watson later told Wilkins that his presence might have saved him from her assault. Wilkins related that Franklin had made a similar lunge toward him several months earlier. "They had almost come to blows following an argument in his room."[22] Now Watson could understand some of the "emotional hell" that Wilkins had to suffer. Their mutual adversary cemented a bond between Wilkins and Watson. Wilkins then revealed that, in the middle of summer 1952, Franklin had found a new form of DNA when it was surrounded with a large amount of water. This new form, called the "B" structure, was much simpler than those previously obtained (dubbed the "A form"). Watson immediately recognized that it indicated a helical structure. Certainly Franklin must have known that the B form indicated a helix, but why she steadfastly insisted there was no evidence for a helix was not explained in Watson's book. Perhaps she did not want it known how close she was coming to solving the puzzle. Wilkins told Watson that Franklin wanted the bases in the center and the backbone on the outside of the molecule. Wilkins now agreed with her, but Watson remained skeptical. The King's group did not like two-chain models, for reasons that were not foolproof. The amount of water in the molecule was still not known with enough accuracy to make a clear decision on the number of chains. Watson decided to use this new information and see what kind of two-chain models he could build.

The next day, Watson told Bragg about the evidence that DNA was helical and it repeated every 34 Å along its axis. Watson said that he was going to ask the Cavendish machine shop to make very accurate metal models of the bases. Bragg gave Watson the go-ahead because he did not want his group to be scooped by Pauling again. Watson knew that the strong meridianal reflection at 3.4 Å was an indication that the flat bases were stacked on top of each other, like a pile of pennies, in a direction perpendicular to the helical axis. Evidence from both the electron microscope and X-ray pictures indicated that the diameter of the DNA molecule was about 20 Å. Crick was not convinced by Watson's argument that important biological objects come in pairs. He did not want Watson to rule out the possibility that DNA might consist of more than two chains. Watson, however, persisted for a short time in trying to construct a two-chain model with the backbone in the center because he suspected that a large number of models might be possible with the bases inside. It would then be very difficult to identify which one was correct. Furthermore, as long as the bases were on the outside he didn't have to worry about how they packed together. Nevertheless, he decided to at least try to build a two-chain model with the sugar-phosphate backbones on the outside and just ignore the bases. He found it was possible to build a stereochemically reasonable two-chain model with the backbones on the outside. It would be another week before the flat tin plates cut in the shape of the bases would be ready. Meanwhile, Wilkins came to Cambridge on a social visit. Crick seized the opportunity to ask Wilkins if he would object to Jim and himself starting to build DNA models again. When Wilkins replied no, Watson's pulse rate returned to normal, because even if he had objected they planned that their model building would have gone forward anyway.

The reason why Watson and Crick could be reasonably sure that the backbone was compatible with the experimental data was that they had gained access to the results from the King's group in a fortunate way. No one at King's College knew they had it. Perutz was a member of a committee appointed by the Medical Research Council to investigate the research activities of Randall's lab and to coordinate biophysics research within its laboratories. Wishing to make a good case for the productivity of his lab, Randall asked each of his research teams to write up a summary of their accomplishments. The comprehensive report was mimeographed and sent to each member of the committee. Since the report was not confidential, Perutz felt free to allow Watson and Crick to see it. Only a few minor modifications were needed to make the model backbone compatible with the data in the report. The mysteries of what held the two chains together and of how the bases fit in with the backbone, however, remained to be solved.

Watson had used J. N. Davidson's book *The Biochemistry of Nucleic Acids* as the source for the chemical formulas of the bases. Watson and Crick had been led by their chemistry colleagues to believe that one or more hydrogen atoms on each of the bases could move from one position to another (a *tautomeric* shift) and that all possible tautomeric forms of a base occurred with equal frequency. Thus for over a year Watson and Crick had dismissed the possibility that bases formed regular hydrogen bonds. This notion changed after Watson read a paper by J. M. Gulland and D. O. Jordan concerning their acid and base titrations of DNA. These researchers had concluded that a large fraction, if not all, of the bases formed hydrogen bonds to other bases. Unfortunately, Watson jumped to the conclusion that each adenine residue could form two hydrogen bonds with another adenine residue related to it by 180-degree rotation. The same was true for a pair of guanines, thymines, or cytosines. The pairing of identical bases, however, would disallow the formation of a regular backbone because a pair of pyrimidines would be wider than a pair of purines. The backbones on each chain would have to dip in and bulge out irregularly along its length and would not be conducive to crystallization of the molecule. Nevertheless, Watson became excited by the prospect that this kind of a structure would allow each strand to serve as a template for the formation of an identical chain during replication. He thought that the common tautomeric form of guanine could also hydrogen bond to adenine, and that other pairing mistakes could be made. So he hypothesized that there might be specific enzymes that only allowed the pairing of like with like.

The next day, Watson explained his latest model to the American crystallographer Jerry Donohue who, for the last six months, had been occupying a desk in the same office with Watson and Crick. In Donohue's opinion, the bases presented in Davidson's book were in the wrong enol tautomeric forms.[23] (See fig. 2.2.) Watson told him that other textbooks also pictured guanine and thymine in the enol form. Donohue explained that "for years organic chemists had been arbitrarily favoring particular tautomeric forms over their alternatives on only the flimsiest of grounds. In fact, organic chemistry textbooks were littered with pictures of highly improbable tautomeric forms." Watson thought that "[n]ext to Linus himself, Jerry knew more about hydrogen bonds than anyone else in the world."[24] Although he could give no foolproof reason for his belief, Donohue strongly preferred that guanine and thymine be in the keto tautomeric form. Watson knew that shifting the hydrogen atoms to their keto positions would exaggerate the size differences between purine pairs and pyrimidine pairs. When Crick was told about Watson's latest model, "[H]e immediately realized that a like-with-like structure would give a 34 Å crystallographic repeat only if each chain had a complete rotation of every 68 Å."[25] This would also require that the rotation angle between successive base pairs would be only eighteen degrees, a value that he thought was impossible. Furthermore, the structure offered no explanation for Chargaff's rules.

The metal models would not be ready for another two days. Watson couldn't stand still that long, so he spent the rest of the afternoon cutting out cardboard representations of the bases in the correct (according to Donohue) keto forms. The next day he was back in the lab fiddling with the new bases to see what kinds of hydrogen bonds between base pairs might be formed. "Suddenly I became aware that an adenine-thymine pair held together by two hydrogen bonds was identical in shape to a guanine-cytosine pair held together by at least two hydrogen bonds."[26] (See fig. 2.3.) He quickly asked Donohue if he saw any objections to the new base pairs. When Jerry saw no objection, Watson was elated because he

suspected that we now had the answer to the riddle of why the number of purine residues exactly equaled the number of pyrimidine residues. Two irregular sequences of bases could be regularly packed in the center of a helix if a purine always hydrogen-bonded to a pyrimidine. Furthermore, the hydrogen-bonding requirement meant that adenine would always pair with thymine, while guanine could pair only with cytosine. Chargaff's rules then suddenly stood out as a consequence of a double-helical structure for DNA.[27]

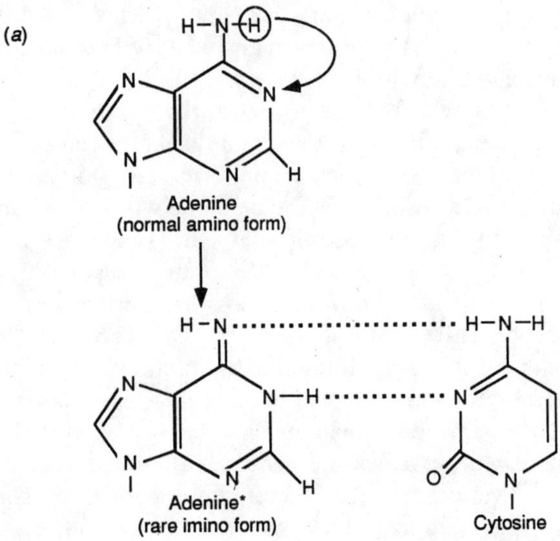

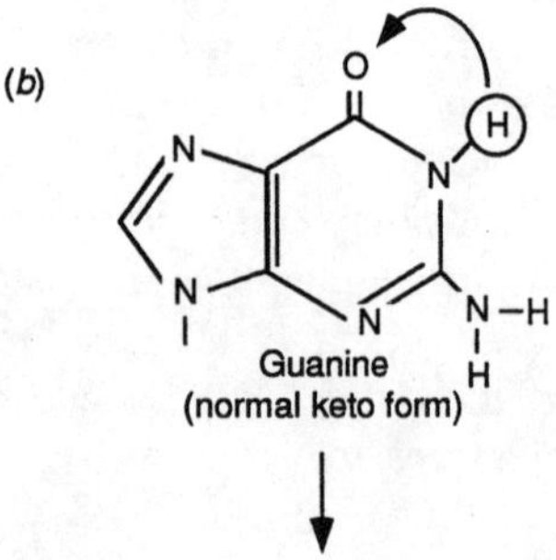

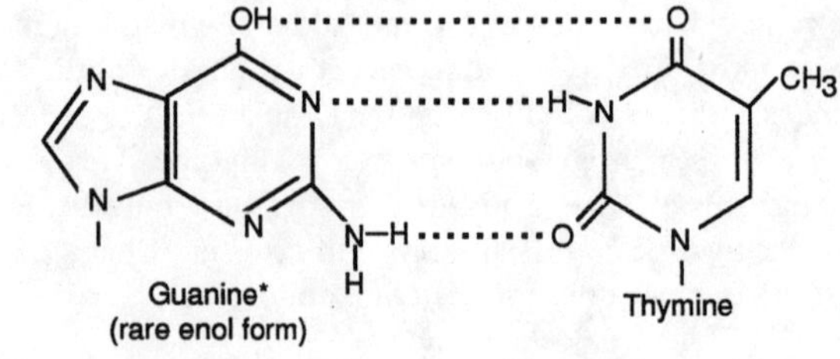

Fig. 2.2. Tautomeric shifts and abnormal base pairing. A tautomeric shift occurs when molecules redistribute their electrons and protons. Tautomeric shifts can cause the bases in nucleotides to become unstable and pair abnormally with each other. (*a*) The unstable (*) adenine forms two hydrogen bonds (dotted lines) with cytosine (A*=C). (*b*) The unstable guanine pairs by three hydrogen bonds to thymine (G*≡T). Conversely, unstable thymine bonds with guanine, and the unstable cytosine pairs with adenine. From W. D. Stansfield, Jaime S. Colomé, and Raúl J. Cano, *Schaum's Outline of Molecular and Cell Biology*, fig. 6-1. © 1996 McGraw-Hill. Reproduced with permission of The McGraw-Hill Companies.

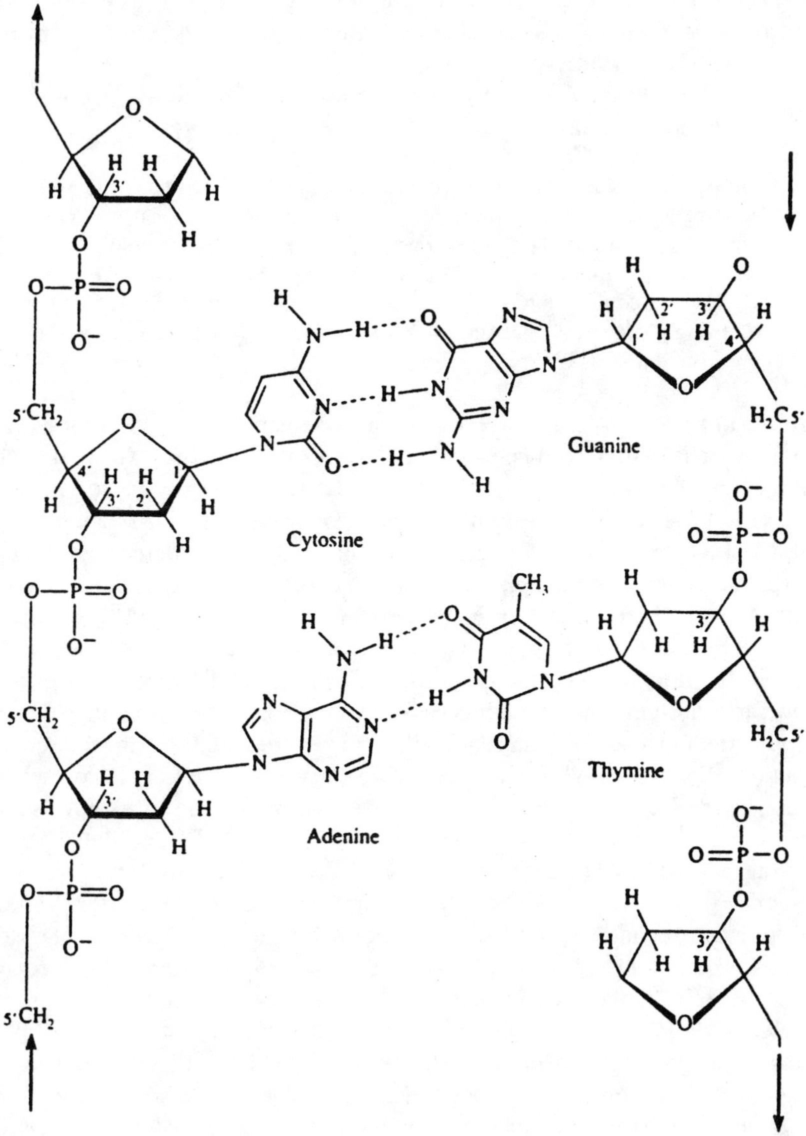

Fig. 2.3. Base pairing in DNA. From W. D. Stansfield, Jaime S. Colomé, and Raúl J. Cano, *Schaum's Outline of Molecular and Cell Biology*, fig. 2–15. © 1996 McGraw-Hill. Reproduced with permission of The McGraw-Hill Companies.

Not only that, but this type of double helix suggested that replication occurred by each strand serving as a template against which a *complementary* strand could be synthesized.

When Crick arrived he received the news with skepticism. Upon examining the model he noted that

> the two glycosidic bonds (joining base and sugar) of each base pair were systematically related by a dyad axis perpendicular to the helical axis. Thus, both pairs could be flip-flopped over and still have their glycosidic bonds facing in the same direction. This had the important consequence that a given chain could contain both purines and pyrimidines. At the same time it strongly suggested that the backbones of the two chains must run in opposite directions.[28]

It remained to be seen if the A-T and G-C base pairs would fit into the backbone structure that had been worked out over the previous two weeks. The bond angles had to be checked with the data from King's. Watson thought that the implications of their model were too important to risk a mistake at this point. He was therefore uneasy in the lunchroom when Crick blabbed to anyone within hearing distance that they had discovered the secret of life.

As soon as the metal bases became available, Watson and Crick assembled the model and began taking accurate measurements of bond angles for all of the atoms. It became obvious to them that all the previous concern over magnesium ions was unnecessary. It was highly probable that Wilkins and Franklin were right in insisting that they were working with the sodium salt of DNA. Now it was clear that it didn't matter which salt was present as long as the sugar-phosphate backbone was on the outside. Watson was confident that "a structure this pretty just had to exist."[29] (See fig. 2.4.) He wrote a letter to Luria and Delbrück to tell them of his DNA structure. Kendrew invited Wilkins to come inspect the model. After Wilkins inspected the model he returned to London and measured the critical reflections on the X-ray photographs. Two days later Wilkins phoned to report that he and Franklin found that their data strongly supported the double-helix model. They were going to write up their results and wanted to publish simultaneously in *Nature* with Watson and Crick. Franklin and Gosling would report their results independently of Wilkins and his collaborators. Strangely, the fierce annoyance that Franklin had with Watson and Crick disappeared. When Crick visited Wilkins in London, he found her willing to "exchange unconcealed hostility for conversation between equals."[30] Although she had been wrong about the helical nature of DNA, she was correct about the sugar-phosphate backbone being on the outside.

Delbrück was the first to give Linus Pauling the news about the Watson-Crick model. Pauling had made him promise to tell him as soon as he heard

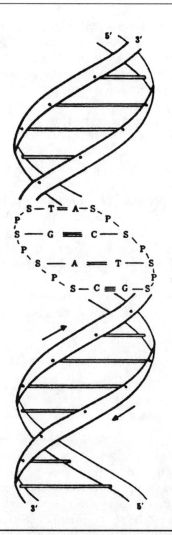

Fig. 2.4. Diagram of the Watson-Crick model of DNA structure. From W. D. Stansfield, Jaime S. Colomé, and Raúl J. Cano, *Schaum's Outline of Molecular and Cell Biology*, fig. 2–14. © 1996 McGraw-Hill. Reproduced with permission of The McGraw-Hill Companies.

any news from Watson. Besides, Delbrück hated any form of secrecy in science, even if a Nobel Prize was at stake. Pauling was genuinely thrilled, but wanted to see the evidence from King's College before granting full acceptance of the model. He would be able to do this three weeks hence while en route to Brussels for a meeting on proteins during the second week of April 1953. Watson returned from a trip to Paris on March 18. During his absence, Crick had devised a structure that was compatible with the data from the A form of DNA. Wilkins had shown that the relatively dry crystalline A form DNA fibers take up water as they go over into the B form. Crick hypothesized that the base pairs were tilted in the more compact A form, thereby decreasing the distance between the stacked bases from 3.4 to 2.6 Å. The following week the first drafts of the Watson-Crick manuscript were ready. Two copies were sent to Wilkins and Franklin for comment. Crick wanted to write at length about the biological implications of their model. But he was persuaded that this first paper should focus only on the structure itself. Crick left his audience in theatrical suspense when he added the following sentence: "It has not escaped our notice that the specific pairing we have postulated immediately suggests a possible copying mechanism for the genetic material."[31] The final draft was sent off to *Nature* under a strong covering letter from Bragg requesting rapid publication for this most important discovery. Bragg was thrilled that the X-ray method that he had developed forty years earlier was responsible for a profound insight into the nature of life itself. The final version of the paper was typed on the last weekend in March 1953 by Watson's sister Elizabeth because the Cavendish typist was not available. She was conned into doing the job by Watson telling her that "she was participating in perhaps the most famous event in biology since Darwin's book."[32] On Wednesday, April 2, the typed manuscript was sent to *Nature*. Linus Pauling arrived at Cambridge on April 4 to see his son Peter and to look at the model. After inspecting both the model and the photograph of the B form, Pauling acknowledged that Watson and Crick had an acceptable structure. The papers were published on April 25, 1953, and the world of biology was forever changed. A new era of molecular genetics had begun.

Analysis

This synopsis of some of the major events in *The Double Helix* has omitted most of the thoughts and activities that were peripheral to the scientific enterprise. One must read *The Double Helix* in its entirety to savor the more human, and in some ways more important, side of the story. This is a problem common to most history books that are written by historians rather than by those who were participants in the adventure. *The Norton Critical Edition* contains thirteen reviews of *The Double Helix* plus "A Review of the Reviews" by Gunther Stent.[33] These reviews should be read to get a more balanced perspective of what actually happened. Stent admits that he urged Watson not to publish his book because it would probably lack interest to a general audience. He must know now how wrong he was. Richard Lewontin, one of the reviewers, thought that Watson's book was too technical and required too much scientific literacy to be understood. While it may be true that many readers will not be able to understand all of the technical details, they cannot help but appreciate the deeper truths about science and scientists that lie between the covers of Watson's book. Some reviewers thought that Watson was a caricature of what motivates "normal" scientists and how normal science is conducted. They feared that Watson was a poor model for aspiring young scientists to follow and that he should not have written such a book. Here we are again, with our self-appointed guardians of good versus evil. While it may be true that Watson's scientific successes were achieved by a combination of good fortune, perseverance, and perhaps an unorthodox style, the suggestion that his tale should not have been told is unacceptable to me. I may not appreciate a Picasso painting, with an eye here and a nose there, but I certainly would not advocate that he should not have done those kinds of paintings for fear of the effects they might have on future generations of artists. Furthermore, how a person behaves outside of his job should have no bearing on the evaluation of his work. Several critics deplored Watson's conduct both within and outside of science. They somehow want to paint the scientist as a person who never has any conflicts with his fellow workers, never gets angry, never says or does anything that might rock the boat, never does anything wrong. Is this a human being we're talking about? Maybe Watson's lifestyle and problem-solving approach are atypical of science in general. But does that mean they should not be made known? In an age where the general public has become suspicious of scientists and somewhat antiscience, should a book such as this have been printed? Will it not add fuel to the fire of their suspicions and make it more difficult for scientific projects to get public support? Yes, that may happen. But I am more concerned about the public's right to know the truth. To suppress such literary works for the noble purpose of keeping the image of science spotless in the eyes of the general public is, to my mind, not only wrong, but criminal. It does a disservice both to the public and to science. Science is not something that people should worship. It is a structured way by which people (just like you and me) can go about finding answers to certain questions. Science is not done by saints who never make mistakes. Scientists are genuine people with all of the potentialities, foibles, and frailties that exist in the rest of society. The more people who understand these facts, the better electorate we will have. Scientific literacy is woefully lacking in the United States. Could it be that part of the reason is that the public has a warped sense of what science is all about? Have they been educated in schools, by the popular press, and by the boob tube to believe that science is something that it is not? Has science and science history been so sanitized by the "fearful Freddies" that it has crippled the educational process? We simply do not see these kinds of stories in most history textbooks. Why have we allowed this to happen? Isn't it about time we got our head out of the sand and see what the real world is all about?

One of the lessons that we can learn from *The Double Helix* is that the scientific method does not always (perhaps seldom does) lead straight to a discovery. For many questions, science has been unable to find an answer, despite years of research by some of the top brains in the business. Often, as illustrated in the search for the structure of DNA, the scientific approach leads down one blind alley after another. One has to give credit to scientists who can continue to pursue a project that yields nothing but failures. Think of the many attempts to produce a suitable three-chain model of DNA that were all failures.

Good fortune played an important role in the DNA story, too. Watson might not have obtained a fellowship to work at the Cavendish had not Luria and Delbrück known Perutz. Wasn't it lucky that Donohue shared an office with Watson and could give him the important clue about the common tautomeric forms of the bases? What if Perutz had not been assigned to a committee that investigated Randall's lab? Watson and Crick might not have had access to their report by which to confirm the details of their model. I have devoted a whole chapter (chapter 6) to serendipity in this book because it plays a much larger role in science than most folks are aware of, or than even scientists care to admit.

Many scientific projects have become so complicated that no one person has the expertise to solve even one of them on his own. Research teams are formed that contain specialists from different areas. For example, Watson was a geneticist and Crick was a physicist. They pooled their individual expertise and their own worldviews in a way that gave them a better chance of making the discovery than if they had worked in isolation. Watson and Crick confessed that part of their success was in being unafraid to criticize each other's ideas and being frank, even to the point of rudeness, in expressing their true feelings. If Watson thought the backbone was in the middle, Crick would play the devil's advocate and say to put it on the outside. One might think that Watson did the biology and Crick did the physics, but that was not true. Both of them switched roles and profited by the interaction.

One can become so engrossed in Watson's story that it is easy to miss appreciating how many events had to occur at the right time and in the right sequence to make this discovery possible. Watson had to meet the right people or read a particular paper at the right time. Any attempts at solving this problem would be doomed to failure before the chemistry of nucleic acids had been worked out and X-ray crystallography had been developed. The rivalry between the British and American workers was a constant source of incentive to push the research forward as quickly as possible. History usually remembers only its winners. Those who ran a good race but finished second do not get the recognition they probably deserve. After the second reading of *The Double Helix*, I began to appreciate the complexity of factors that were involved in the discovery of the double helix. I made a list of those factors and how they related to the discovery. Then I tried to make a flow sheet (see fig. 2.5) to show how these factors operated. Some of the factors acted in a positive fashion, some in a negative way. Some influences were one way affairs, whereas others were interactive. Some influences were dominant, others were of lesser importance. I sent my flow sheet to Watson for his comments. He was kind enough to respond in a letter to me dated February 24, 1982. He made only one suggestion which I incorporated into the flow sheet. I still have that letter signed by him. I show the flow sheet to my history of biology class after they have read *The Double Helix* and we have discussed its implications. It never fails to impress the students with the complexity of a scientific discovery. I suspect that most scientific advances have a comparable complexity of which the general public and most scientists outside the respective research team are unaware. How are we to know about this complexity unless we have lots more scientists tell it like it is, *a la* Watson. I can only hope that more such books important to the history of science will be forthcoming, for as Stent remarked, "And to what higher achievement can one aspire after having been anointed a Great Scientist? To make it as a Great Writer."[34]

NOTES

1. James D. Watson, *The Double Helix: A Personal Account of the Discovery of the Structure of DNA* (New York: W. W. Norton & Company, 1980).
2. Ibid., 9.
3. Ibid., 20.
4. Ibid., 21.
5. Ibid.
6. Ibid., 27.
7. Ibid., 29.
8. Ibid., 37.
9. Ibid., 41.
10. Ibid., 43.
11. Ibid., 45–46.
12. Ibid., 48.
13. Ibid., 48–49.
14. Ibid., 49.
15. Ibid.
16. Ibid., 61.
17. Ibid.
18. Linus Pauling, *The Nature of the Chemical Bond and the Structure of Molecules and Crystals: An Introduction to Modern Structural Chemistry* (Ithaca, N.Y.: Cornell University Press, 1960).
19. Watson, *The Double Helix*, 72.
20. Ibid., 94.
21. Ibid., 96.
22. Ibid.
23. Ibid., 110.

24. Ibid., 112.
25. Ibid.
26. Ibid., 114.
27. Ibid.
28. Ibid., 115.
29. Ibid., 120.
30. Ibid., 124.
31. Ibid., 129.
32. Ibid.
33. Watson, *The Double Helix*.
34. Ibid., 163.

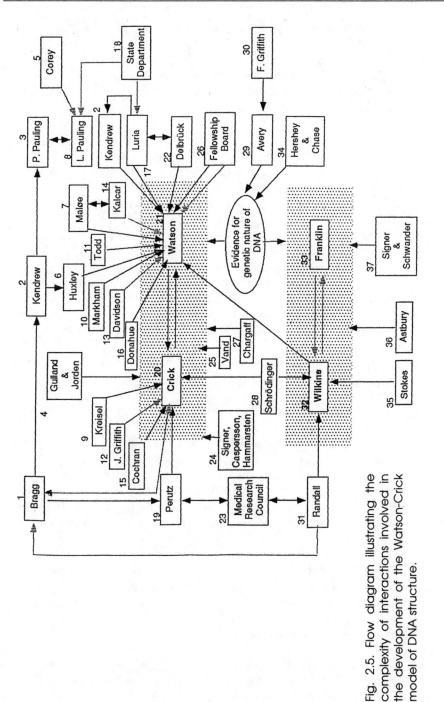

Fig. 2.5. Flow diagram illustrating the complexity of interactions involved in the development of the Watson-Crick model of DNA structure.

LEGEND TO THE FLOW CHART

On the flow chart, hatched arrowheads represent miscommunication, failure or unwillingness to communicate, or negative influences on solving the structure of DNA. Solid arrows represent positive interactions.

1. W. L. Bragg was director of the Cavendish laboratory of Cambridge University. He and his father W. H. Bragg invented X-ray crystallography. He was convinced by J. T. Randall to terminate the DNA work at the Cavendish. Bragg and Crick did not get along well together.

2. J. Kendrew was a coworker with M. Perutz on myoglobin; he supervised H. Huxley and P. Pauling. Kendrew allowed Watson to work with him on myoglobin structure, but also allowed him to work on DNA when he failed to be of help on the myoglobin project.

3. P. Pauling was a graduate student under J. Kendrew. He received letters from his father and kept Watson and Crick informed of his progress.

4. J. M. Gulland and D. O. Jordan's paper on acid-base titration of DNA gave Watson the idea that much base pairing by hydrogen bonds was involved. Gulland suggested that the polynucleotide chains might be linked by these hydrogen bonds to form multichain micelles.

5. R. B. Corey, a close colleague of L. Pauling, was not asked to check Pauling's interatomic distances on his DNA model, which later proved to be incorrect.

6. H. Huxley, a student of J. Kendrew, taught Watson how to set up the Cavendish's rotating anode X-ray camera.

7. Working with phage at Maaløe's lab was not in the terms of Watson's fellowship.

8. L. Pauling used the laws of structural chemistry to discover the alpha-helix structure of proteins. This stimulated Watson and Crick to use a similar model-building approach for solving the structure of DNA. Pauling developed a three-chain model for DNA structure. He wrote *The Nature of the Chemical Bond*.

9. G. Kreisel helped Crick find the correct equations for coiled coils.

10. R. Markham provided TMV for Watson's crystallographic studies.

11. A. Todd in 1951 envisioned each DNA chain to consist of nucleotides linked by 3'-5' phosphodiester bonds between the sugars.

12. John Griffith's calculations indicated that A and T should stick together by their flat surfaces; he favored the formation of complementary surfaces.

13. Watson used J. N. Davidson's book *The Biochemistry of Nucleic Acids* for his base formulas. The formulas for guanine and thymine were in the rare enol forms instead of the usual keto forms.

14. H. Kalckar's lab in Copenhagen was working on the metabolism of nucleic acids. Watson was first sent to work with Kalckar, but he found the work uninteresting. Watson then went to work with O. Maaløe on phage (a violation of his fellowship terms).

15. W. Cochran, a lecturer in crystallography at the Cavendish, was the most astute sounding board for Crick's frequent ventures into theory.

16. J. Donohue pointed out the keto vs. enol forms of bases and the errors in J. N. Davidson's book.

17. S. Luria was Watson's Ph.D. supervisor. He arranged for Watson to work with J. Kendrew on myoglobin structure. Luria and Delbrück published the first paper on bacterial genetics.

18. The State Department revoked the passports of S. Luria and L. Pauling.

19. M. Perutz was Crick's Ph.D. supervisor; he gave Watson and Crick the information on Franklin's DNA studies; it proved unnecessary for deriving its structure, but was necessary to check the accuracy of the model.

20. F. Crick was a British physicist working on a Ph.D. studying myoglobin structure under M. Perutz. His expertise in the interpretation of X-ray crystallographic data was important for Watson's model building.

21. J. D. Watson was an American molecular biologist trained in phage genetics. He came to the Cavendish in 1951 specifically to work on DNA structure, but was a novice in X-ray diffraction studies.

22. M. Delbrück, the "father of phage genetics," arranged for Watson to receive a Polio Foundation fellowship for an extra year at the Cavendish. He thought that Watson's paper on bacterial genetics was unsound. Luria and Delbrück performed the first genetic experiments on bacteria.

23. The Medical Research Council appointed a committee (of which M. Perutz was a member) to investigate J. T. Randall's lab; Randall gave Franklin's precise measurements from DNA crystallography to Perutz.

24. R. Signer, T. Caspersson, and E. Hammarsten showed that the bases in DNA lay with their planes roughly perpendicular to the length of the threadlike molecule.

25. V. Vand first proposed a theory for diffraction of X rays by helical molecules.

26. The Fellowship Board terminated Watson's grant because he did not stay in H. Kalckar's lab.

27. E. Chargaff established the base ratios in DNA (A=T and G=C); after visiting Watson and Crick he was convinced that they did not know where they were going.

28. F. Schrödinger wrote the book *What Is Life?* It became a catalyst for physicists (such as Crick and Wilkins) to shift their attention to solving the riddle of heredity.

29. O. T. Avery's in vitro transformation experiment proved that DNA was genetic material in bacteria.

30. F. Griffith's 1928 in vivo experiment was enigmatic at the time, but eventually led to the work of Avery.

31. J. T. Randall was Wilkins's boss. Randall asked Bragg if it made sense for Crick and Watson to duplicate the King's College heavy investment in DNA research. Bragg terminated the Cavendish work on DNA (or so he thought).

32 and 33. R. Franklin and M. Wilkins (at King's College, London) were supposed to collaborate, but had personal differences. Wilkins had been brought up in the model-building tradition whereas Franklin was schooled to use the Patterson-vector approach. Under the King's College administrative organization, Wilkins was Franklin's boss. Franklin wanted to be treated as an equal, not as a subordinate, and so she resisted collaboration with Wilkins. The high-water content of DNA used in Franklin's X-ray diffraction studies created ambiguity in deciphering its correct structure. She was initially against the helical theory as an explanation of both the A and B forms. In the spring of 1951, Watson became interested in the structure of DNA when shown the "A pattern" by Wilkins.

34. A. D. Hershey and M. Chase proved that DNA is genetic material in phage.

35. A. R. Stokes was a colleague of Wilkins and told him that the X-ray pattern of DNA was compatible with that of a helix.

36. W. T. Astbury published the first good photographs of DNA. He correctly interpreted the strong 3.4 Ångstrom reflections as being due to the planer bases stacked on top of one another.

37. R. Singer and H. Schwander supplied a superior source of DNA to the King's College group.

CHAPTER 3

DANCE FEVER

The popular view of science sees it as a logical progression of theories that explain more and more about the nature of the universe—what it is made of, how it is put together, and the regularities by which it operates. Objective evidence is supposed to be the criterion by which a theory stands or falls. Theories that conflict with empirical data deserve to be discarded. Scientists should be ready and willing to give up outdated theories when new ones come along that do a better job of explaining the data in hand and/or predicting the results of future experiments or observations. Many scientists, however, become too fond of their theories and fight like tigers to sustain them, sometimes even in the face of considerable evidence to the contrary. The philosopher of science Thomas Kuhn argues that scientists choose between competing paradigms in ways more like religious conversions. Explanations in terms other than objectivity, empiricism, and logical debate may sway the support given to certain theories. According to Kuhn, "we simply no longer have any useful notion of how science evolves" and the popular view of science as "the triumph of reason and method over ignorance and superstition . . . is not remotely possible today."[1] Bruno Latour, in his book *Science in Action*,[2] proposes that during a scientific debate, nature can never determine the outcome, since "no one knows what she is and says." Instead, scientists bring in her authority after a controversy has been settled, claiming that "it is Nature [*sic*] the final ally that has settled it and not any rhetorical tricks and tools. . . ." To what extent does evidence prevail in the choice of a scientific theory? This chapter may help us to answer this question.

The first time (and the only time thus far) that a Nobel Prize was awarded in the science of ethology (animal behavior) went to the Austrians Karl von Frisch, Niko Tinbergen, and Konrad Lorenz in 1973. Von Frisch is best known for his pioneering work in the 1940s on the dance of honeybees. His ingenious experiments demonstrated for the first time that symbolic communication existed in a nonhuman species.[3] His methodology helped to move ethology out of the amateur observation stage and into the mainstream of controlled scientific experimentation.

Von Frisch claimed to have noticed that when foraging honeybees return to their hive from a successful feeding trip, they perform a series of body movements (a "dance") that seems to convey to other bees in the hive the direction and approximate distance from the hive of the food source from which the foragers have just returned. Thereafter, most of the bees leaving the hive will go straight to the food source rather than search in a random pattern for it. Other bee species were known to dance, and many other insect species follow olfactory clues to find flowers with nectar. Von Frisch wondered if honeybees use odor, the dance, or a combination of both to transmit information about the location (and perhaps its richness) of a foraging area. Perhaps the dance was just a mechanism to excite other bees to leave the hive and go in search of the food that had just been found. So von Frisch set up a series of experiments to try to find an answer to these questions. He placed hives in positions that had physical barriers between the hive and the food source. He used odorless food so that the bees could not follow a scent trail up its concentration gradient. Any substance produced by one organism that affects the behavior of another organism in the same species is called a *pheromone*. Examples of such pheromones are sex-attractants, alarm substances, aggregation-promotion substances, territorial markers, and trail substances of insects. To prevent honeybees from marking the location of a food source with their own pheromone, von Frisch blocked the scent glands of foraging bees with shellac. To his great surprise, none of these manipulations prevented the foragers from communicating the location of food to their hive-mates. From his data he concluded that although honeybees normally use both odor and dance clues, the bee dance alone was sufficient to transmit the information needed to find a food source. For reasons that are still not clear, zoologists were quick to accept the bee-dance theory and it soon began to appear in biology textbooks as an accepted fact. The findings and methodologies of von Frisch had initiated a scientific revolution that stimulated a wave of research on animal communication. A new paradigm of bee language had become established. This was followed by a period of "normal science" (Kuhn's term) in which research was primarily directed at probing the ramifications of the paradigm (prevailing theory) rather than challenging it.

It was in this historical setting that the American zoologist Adrian M.

Wenner observed, in the early 1960s, that his honeybees were not behaving in ways predicted by the dance-language theory. This was about twenty years after publication of von Frisch's original work. The main factor that drives scientists to research is curiosity, and Wenner's curiosity had been aroused. He decided to repeat the original experiments of von Frisch, but with additional controls for the possible influence of odor. He published his first paper on the dance problem in 1962, at which time "he still endorsed the dance theory,"[4] but questioned whether von Frisch had adequate experimental controls to rule out odor clues. In subsequent experiments, he added the necessary odor controls and concluded from his observations that bees which had witnessed the dance could not locate a food source in a directed manner from just the dance. He proposed that the dance-language theory had, in his hands, been falsified. He then devoted himself to making this known to the scientific world. Through numerous publications, invitational lectures, attending meetings, and writing hundreds of letters, he was able to make his objections widely known to the public. Connie Veldink made a survey in 1974 (involving 182 respondents) to ascertain the success of Wenner's crusade.[5] Ninety-six percent of them reported being familiar with Wenner's research, and 70 percent of them had read his book *The Bee Language Controversy*.[6] All of the respondents were familiar with the work of von-Frisch, and 82 percent of them had read his major work, *The Dance Language and Orientation of Bees*.[7] Over 60 percent of the respondents said that the controversy had caused them to change their views on how bees forage. About one-third thought that they had changed in favor of Wenner's theory, whereas only 4 percent switched to the von Frisch theory. In 1974, Patrick H. Wells and Wenner published their strongest challenge in a major scientific journal.[8] They concluded that honeybees use only odor clues when foraging, not the dance. Of those who read the article, 17 percent believed that bees forage by odor or mainly odor, compared with only 2 percent of those who had not read the article. However, only 8 percent of those that had read the article believed that the dance contributed nothing to foraging information.

Not only were the respondents' views significantly related to having read or not read the Wells-Wenner paper, but they were also significantly related to the respondents' academic disciplines. Only 57 percent of psychologists believed that bees use both odor and the dance in foraging, whereas 82 percent of biologists did so. Of those who had read the article, 36 percent of the psychologists agreed that odor alone is the foraging clue, as compared to 2 percent of the biologists. Over 85 percent of all the respondents believed that bees get foraging information from the dance, either alone or in combination with other clues. Only eight out of 159 respondents, none of which were bee researchers, thought that the dance contained no such information; five of them were psychologists, two were biologists, and one was a linguist. Thus, Veldink concludes that bee scholars (not necessarily those active in bee research) were aware of Wenner's objections to the dance-language theory, but did not agree with him. The extent of belief in the dance-language theory at the time of Veldink's survey is remarkable because "*there was no adequate evidence* that bees use the dance clues" (italics hers).[9] Veldink also believes that the research of James L. Gould,[10] published in 1974, settled the controversy in favor of the dance-language theory. But my own personal communication with Wenner in 1989 revealed that he strongly disagrees; he feels that his data clearly refute that hypothesis, and the scientific community has stuck its head in the sand, refusing to acknowledge that its pet theory has feet of clay.

Gould thinks that the reason for the bee-language controversy was that the method used by Wenner to train bees relied on odor, whereas the method of von Frisch made them rely on dance-language clues. Wenner, however, disagrees because he had been able to reproduce von Frisch's results even when the bees were trained by the Wenner method. Gould admitted that honeybees probably depend on odor clues more than on the dance except when food is scarce. The dance normally plays only a minor role in recruitment of new foragers, but because of its extraordinary nature, it has been blown up out of proportion to its importance. Prior to publication of his own research, Gould granted that Wenner's locale-odor hypothesis could explain all of the results of von Frisch without recourse to the dance-language theory. Gould had done his doctoral work under Donald Griffith. It was Griffith who had arranged for von Frisch's first lecture tour to the United States. Griffith also thought that the dance-language theory rested on shaky empirical support before publication of Gould's experiments. Veldink leaves it up to the reader to evaluate the adequacy of Gould's evidence, preferring to concentrate on the question "How then was it possible that the bee researchers continued to support the language theory for twelve years (from Wenner's 1962 paper [*sic*] until Gould's 1974 article) despite a lack of clear evidence?"[11] Her analysis involved seven possible reasons.

1. The attractiveness and popularity of the dance-language theory

Wenner's olfaction theory is mundane, but the notion that bees can communicate information by means of a complicated dance can't help but capture one's admiration. Furthermore, according to Stuart Altmann,

> The twice-told tale for some curious reason is believed more than that same tale told once. The fact that the bee-dance story has been told over and over again by many people, popular writers, newspaper writers, people involved in the research, textbook authors, is, I think, a major factor con-

tributing to the inertia involved in challenging people's ideas. They believe it more because they have heard the story from so many sources.[12]

When a new theory stirs a lot of productive research, it is usually considered to be a "useful" device. It doesn't necessarily mean that the theory is correct or the best one. But in the continued use of the theory as a heuristic (helping to discover or learn) instrument, its credibility may increase even though no new corroborative evidence or real tests of falsifiability have been tried. The dance-language theory had been very productive in stimulating ethological research on bees and other species because it led to a reconsideration of animals' abilities. "Any theory which has such a record of service is going to have a group of loyal followers. As physicist Max Planck held, only death will separate some scientists from their beloved theories."[13]

2. The delightful and capricious honeybee

According to Wenner, one of the greatest barriers to obtaining rigorous results in animal behavior studies is the inclination of researchers to choose an animal that they like to work with, and then concentrate on the animal rather than on questions. Thus, animal behaviorists are more likely to succumb to the "experimenter effect" than are other biologists (such as molecular geneticists) whose experimental material is less compelling. By this effect, the researcher consciously or unconsciously "roots" for his animal, and thereby tends to attribute more capabilities to the animal than it really has. As interesting as the bee is, it often seems to behave capriciously when under scientific observation. Successful foragers do not always dance. A bee will only dance if the hive contains enough bees, if there is enough food available, if the temperature and other environmental factors are conducive, and so forth. The dance normally occurs in the darkness of the hive where it is difficult to observe by researchers. Relatively few bees can touch the dancer to receive the message. Furthermore, only a minority of the recruits are able to locate the feeding site indicated by the dance information. Hence the dance is a rather feeble means of recruitment. Von Frisch knew that some dancers regularly give poor directions, so he had to kill such deviants before running his experiments.[14] Some critics might view this as data tampering. Certainly much more subtle examples of data manipulations have been severely criticized in the history of science. But von Frisch seems to have escaped this kind of criticism.

3. Perceived personal qualities of von Frisch and Wenner

Veldink states that most scientists in the international bee-research community could not have known von Frisch or Wenner well, and thus perceptions about their characters had to be inferred from their writings, or based on others' opinions. No evidence is given to support this claim. I suspect that the number of bee researchers is so small that it might have been quite easy for most of them (especially the older ones) to have met both these men at various societal meetings that attract workers in the same field. Gould thought that Wenner might have been perceived negatively because of the tone of his arguments.

> The tone of Wenner's arguments became increasingly aggressive and defensive. In science, if you make yourself obnoxious, people are less likely to forgive your logical and technical oversights. There is no doubt that Wenner, by his almost paranoid writing—as though everyone who was reading it was going to be hostile—or, alternatively, by being very condescending in his explanations as he is in his 1971 book, has managed to alienate everyone I have ever talked to who has read it. That is too bad because as a result people have been wanting to dismiss him without considering his arguments.[15]

My own opinion of this criticism is to ask "What is a fellow supposed to do?" He's too harsh in some situations and too soft in others. It seems he's damned if he does, and damned if he doesn't. Wenner sounds like a real person that responds to different things in different ways. To me, that makes him a much more interesting character than the stereotype of the calm, rational, logical scientist of insipid history books. Gould is also quoted as saying,

> In this business, personal reputation is everything. If you ever blow it, let your reputation go down by being a little too careless, or a little too quick, then people react by taking you far less seriously than you deserve to be taken. Wenner has suffered from this. Being a little less cautious than he should have been has decreased the weight with which what he says is taken by many people I know.[16]

A graduate student in entomology who had just met Wenner for the first time told Veldink with surprise that he found him to be a very nice person, but in his department Wenner had been portrayed as an ogre. Others with whom Veldink had spoken who had not met Wenner sometimes regarded him as temperamental, unstable, egotistical, or paranoid. Now, how can anyone form a valid opinion of a person without the opportunity to get to know him? Wenner has given lectures on the bee controversy at least twice at the university where I work, and I can tell you from personal experience that he is not the ogre that he has been made out to be. He may twist your tail by making you think, and challenge you to propose experiments that have not been done that should be done. He might be a little exasperated (and rightly so) at having to put up such a long, lonely fight to have his work appreciated for its own merits without resorting to personal innuendoes.

As the battle wore on, Wenner found it increasingly more difficult to get his papers published. He estimated that it took up to four times longer to get his papers published after the controversy than prior to it. A friend wrote the following advice to him in 1971:

> You seem to have got yourself completely stymied with your opponents.
> . . . There is a strong social aspect in science, and we depend for our equi-
> librium on substantial congruence of our ideas with the ideas of our peers.
> If I develop an idea that is completely heterodox and if I am the only one
> who believes it, I think it is probably best to drop the matter after making
> a reasonable attempt to convince others. If I am right, sooner or later other
> evidence will turn up supporting me, and then I can safely return to the
> matter. But in the meanwhile, I had best do something else.[17]

Wenner had already begun to work in crustacean biology by the time he received that letter, but he still spent about ten hours per week in correspondence concerning the bee controversy.

4. Clinchers and authority

One of the major differences between religion and science is that religion depends entirely upon authority, while science continually challenges it. At one time, there were only three sets of experiments (besides those of Wenner) conducted in response to Wenner's objections against the dance theory. Unfortunately, repeated references to this small set of bee researchers by supporters of the dance theory gave them the appearance of authorities who had proved their case beyond a shadow of a doubt. Veldink called these works "clinchers" because they seemed to "clinch" (close) the case to further debate. Some scientists were concerned about such use of authority in winning scientific debates. For example, Stuart Altmann stated:

> The fact that major figures in the study of social insects, or in the study of
> animal navigation, people like E. O. Wilson and Donald Griffin . . . have
> stated so definitely what they believe the way to be the way [sic] the bee
> system works is unfortunate in the sense that a lot of young people, instead
> of being stimulated to pursue the question and ask what evidence would
> be sufficient, do not even think of doing that because after all, if Ed Wilson
> says it is one way and Don Griffin says it is one way, who are they to quarrel
> with it. This is the bio-politics.[18]

5. Use of the evidence

What seemed to matter most to the supporters of the language theory was the "weight" (volume) of the research done, none of which contradicted von Frisch. Wenner argued that you can run the same tests innumerable times and get the same results, but if they lack the proper controls they cannot be used to falsify the language theory. A theory can be confirmed in all kinds of ways, but all it takes to cause its rejection is one verified instance of falsification. Wenner called those who are swayed by the quantity rather than by the quality of research "cumulative scientists."

> If you are a cumulative scientist and know that von Frisch has done twenty years of research, and that there are a hundred other bee researchers throughout the world who have repeated his experiments, and you know that this is truth [sic], then someone like myself comes along and says, "Here is evidence that the hypothesis is not valid," you can see what it does to the whole psyche of all the people who are cumulative scientists. It is better to squelch me and keep the establishment intact than to admit that perhaps the cumulative approach is not the best approach.[19]

An eminent philosopher of science, Karl Popper, proposed falsification as a demarcation principle for distinguishing science from nonscience. Wenner claims that bee researchers have simply ignored the negative evidence provided by his experiments. "They noted with care the performance of bees that behaved according to the language theory, but ignored those, sometimes the majority, that behaved otherwise."[20]

Popper also pointed out how a theory can be protected from rejection by use of suitable ad hoc explanations. Wenner cited a case in which Lindauer had made up an ad hoc explanation of why his bees that attended two different dances (each indicating separate foraging sites) were found feeding at a third site. Lindauer proposed that his bees were able to average the dance information and thus be directed to a middling site. Wenner also argued that one of his critics, Richard Dawkins (of *The Selfish Gene* fame) used an ad hoc explanation when Dawkins claimed that the reason that the bees in Wenner's experiments could not correctly follow the dance information was that they had been "distracted" in some way.

6. Theory into fact

Soon after the dance-language hypothesis was proposed, it began to appear in biology textbooks as if it were a fact; e.g., "Honeybees communicate via a dance language." This kind of statement belies the fact that this subject is still far from being settled. It makes no mention of the contestants or the bones of contention. Such books pass up a marvelous opportunity to introduce students to some real science. Why not explain the experiments of von Frisch and those of Wenner and let students make their own evaluations? We do our students no favor by giving them the answers to scientific questions. Instead we should be offering them the opportunity to critically analyze data and either reach their own conclusions or (if not enough data exist) design further experiments to resolve the dilemma. Memorization of facts is not science. Analytical ability, problem solving, and critical thinking is what should be stressed in science classes.

7. Teleology

Teleology is the explanation of a phenomenon (such as the bee dance) by the purposes or goals it serves. Teleological explanations usually invoke supernatural powers and therefore are not scientific. Some biologists are strict adaptationists. They believe that every characteristic (anatomical, physiological, behavioral) helps the organism to adapt to its natural environment. Thus, if an animal expends energy in some type of activity, there must be a purpose or function to it that aids in survival and reproduction (adaptation). This metaphysical assumption has helped to maintain the dance-language theory. The reasoning goes like this: bees dance; dancing uses energy; the dance conveys information; therefore the purpose or function of the dance must be communication. Wenner agrees that the dance may contain information, but he argues that this is not sufficient to conclude that the function of the dance is communication. Since Wenner has failed to specify an alternative function for the dance, some critics see this as a fatal flaw in his argument. Wenner, however, has shown that some behaviors are nonadaptive; for example, the male fruit fly will drink the chemical methyl eugenol until it dies. He also points out that even if a behavior is adaptive we can never be sure that we have identified its true function. There are many examples of activities that are of no apparent value. For example, why do people scratch their heads when puzzled about something, or why do they drum their fingers on the table when nervously awaiting some event? These types of activities are referred to as "displacement behaviors." Animals have analogous displacement behaviors. A bird known as the spotted sandpiper runs along the ground, stops, flexes up and down, continues its run, and then repeats the pattern. What is the adaptive value of such behavior? Its bobbing action does contain information; e.g., it helps me to identify it as a spotted sandpiper, and maybe it is also an identification signal for others of its own species. But what is its value when no other birds are in its vicinity? Perhaps the dance of the honeybee is a displacement activity despite the fact that the dance is performed in the midst of other bees. Wenner does not care to speculate on why the dance is done. His argument is simply that although the dance contains information, that does not necessarily mean that the function of the dance is communication ("bee language"). Gould also rejected a teleological explanation, but for a different reason. He argued that if the dance of the honeybee communicated information about the location (distance and/or direction) and/or richness of a food source, then honeybees should be able to forage more efficiently than other species of bees that do not dance. The research of Landauer on stingless bees that do not dance indicated that they foraged more efficiently than do honeybees. Gould's critique of teleology appeared in 1976[21] along with his data that reconfirmed the dance-language theory. Veldink concludes that since Gould's paper of 1976 no new experiments have appeared in response to Wenner's challenge, and therefore the controversy was at an end. Because Gould's results fit so nicely into the already prevalent view (bees use both odor and language clues), his work was taken to be definitive.

For those who are interested in understanding how science operates, it is clear that evidence *alone* may not be sufficient reason to explain the "settlement of a controversy. . . . Evidence is silent: someone must speak for it."[22] Wenner's work was published for all to evaluate, but no major bee researcher publicly acknowledged its validity during the controversy for a variety of possible reasons, as outlined above. Some of you may wonder why anyone should make such a fuss over something so seemingly inconsequential. As a beekeeper wrote to Veldink, "I don't really care which researcher is right or which is wrong. Even if we never resolve this case, the bees won't mind. They will continue to make honey anyway."[23] That may be true, but it should matter to everyone what scientists do. If a theory can survive for at least a dozen years in a case of pure science, despite falsifying evidence, what are the implications for scientific controversies that might be very important for our own welfare?

Recognizing that the underlying epistomological issues of the dance-language controversy in some respects transcend in importance its scientific aspects, Wenner and his coauthor, Patrick H. Wells, were stimulated to write a book entitled *Anatomy of a Controversy*[24] that directly addressed those issues. In order to understand this controversy, Wenner and Wells launch into a discussion of philosophy. They explain that before Thomas Kuhn's book *The Structure of Scientific Revolutions*[25] was published in 1962, it was widely held that science progressed in a gradual fashion by incremental accumulation of ever more bits of knowledge. Kuhn proposed, however, that major scientific advances occur by a punctuated process in which a new way of looking at the same old data replaces the established view (paradigm). This shift to a new paradigm may not be rapid (perhaps seldom is); it may take several years or perhaps decades for the shift to occur. Because the old view is held so tenaciously by the old guard, the "paradigm hold" may only be broken as the members of that group die off. This is sort of analogous to the "punctuated equilibrium" concept of speciation in modern evolution theory. The fossil record seems to indicate that most species remain relatively unchanged over most of their phylogenetic (evolutionary) histories in what has been called "stasis." Then, suddenly in geological time, the old species seems to disappear and one or more new ones take its place. Thus, relatively long periods of stasis seem to be "punctuated" (interrupted) by rapid speciation events. In actuality it may require thousands of years for these speciation events to be completed, but on a background of millions of years of geological time, these events appear to be essentially instantaneous. So it is with Kuhn's "paradigm shifts." These revolutionary changes in science history may only appear to be instantaneous on the background of thousands of years of recorded history.

Wenner and Wells view Kuhn's paradigm shift (or Michael Polanyi's "crossing of a logical gap") as a conceptual change from the philosophies of realism to that of relativism (antirealism). Neither Polanyi (1958) nor Kuhn (1962) *advocated* any particular research approach.[26] The realism school believes that truth exists and can be discovered or proved. Rudolph Carnap is cited as a champion of the verification approach, whereas Karl Popper is noted for his falsification approach to providing "proof" of the correctness of a scientific theory. Realists believe that through scientific objectivity we can come ever closer to understanding "reality." In contrast, the relativists insist that knowledge of objective reality can never be fully achieved. In 1985 James Atkinson, a leader in this school, argued that, "What we accept as 'reality' at any time is based upon shared perceptions . . ." and that "current teaching about science 'fosters naive realism' in students and laypeople alike."[27] In 1890, Thomas Chrowder Chamberlin provided the relativism school with the method of "multiple working hypotheses."

> The effort is to bring up into view every rational explanation of new phenomena and to develop every tenable hypothesis respecting their cause and history. The investigator thus becomes the parent of a family of hypotheses: and, by his parental relation to all, he is forbidden to fasten his affections unduly upon any one.[28]

John Platt advocated his "strong inference" (a simple case of "multiple inference") approach of pitting one hypothesis against another in 1964. Both Chamberlin and Platt advocated repetitively using an inductive-deductive approach in which one hypothesis would be rejected by the results of a "crucial" experiment (providing "strong inference") while simultaneously retaining the other(s).

In order to give the reader some feeling for the kind of data that allowed such a strong inference to be made, let's look at what Wenner and Wells called "Our 'Crucial' Experiment."[29] They defined this approach as involving an *a priori* experimental design that permits the validation of only one of two or more predicted and mutually exclusive interpretations for any forthcoming set of results. That is, one set of results from an experiment would validate one of the hypotheses while simultaneously negating the other(s). The converse would also be true. The "crucial experiment" thus precludes ad hoc rationalizations at the conclusion of the experiment about any "unfavorable" results that might have been obtained.[30]

They designed a twenty-four-day experimental sequence of three hours each day. The majority of days would predict results similar to those reported by von Frisch because they had a single-control design as he used. On predetermined days, however, conditions would be provided that would exclusively predict results according to either the "dance-language" or odor-search hypotheses. On a relatively level open grassland with no trees nearby, they set up a hive and three feeding stations (fig. 3.1). Stations 1 and 3 were so far apart that the middle control site (2) was "well outside the area supposedly indicated by dances of foragers feeding at the two outside stations." Ten experienced bees (numbered foragers) were then allowed to visit each of the two outside stations for at least two weeks "in order to eliminate any unmarked strays left over from the training period." At no time did these experienced foragers visit the middle station. All recruits (unmarked visitors) were captured and killed. Thus, regular foragers were allowed to indicate the location of the food sources at the two outside stations by performing dances back at the hive (if bees really do communicate in this way).

> The mutually exclusive feature of the experimental design hinged upon the fact that we never provided the standard marking odor (clove oil scent) at all three sites on the same day. That is, clove oil-scented sugar solution was *either* placed at the two outside stations or at the central station (or a few times at neither). Also . . . experienced foragers visited only the two outside stations throughout the twenty-four day period.[31]

According to the dance language hypothesis, recruits should always arrive at the two outside stations regardless of odor location. However, if bees use odor in their search pattern, recruits should arrive at the central station when odor was present only at that location (thus ignoring any information that they may have obtained from witnessing a dance). If *some* recruits use odor and/or dance-language information *some* of the time, then recruits should arrive at all three stations in an unpredictable manner. The results of the experiment show that this did not happen. On the twenty-fourth day, all foragers were killed, so none of them could return and recruit others; a single recruit arrived at site 1; none arrived at the other two sites. The inference derived from this last control is that forager trips are necessary for recruitment. On day 14, no scent was provided at any of the stations. Very

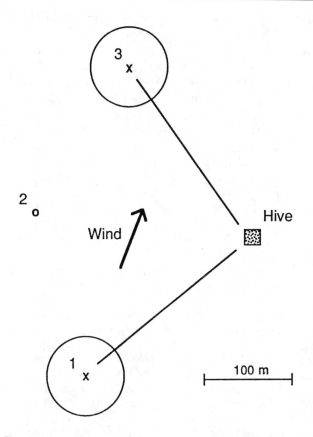

Fig. 3.1. Hive and feeding station arrangement for the "crucial experiment" (after Wenner, Wells, and Johnson 1969). Foragers were permitted to collect food at sites 1 and 3 but never at 2. They could thus presumably provide direction and distance information in their hive for the outer sites but not for the middle site. The circles denote the predicted area of language effectiveness if searching bees *used* information obtained from dancing bees in the hive. From Adrian M. Wenner and Patrick H. Wells, *Anatomy of a Controversy: The Question of a "Language" Among Bees*, fig. 10–2. © 1990 Columbia University Press. Reproduced with permission.

few recruits were successful (6, 2, and 5 bees arrived at stations 1, 2, and 3 respectively). "One or two days of scented sugar solution at the two outside stations preceded a day on which the same scented solution was furnished only at the central test station." On control days (1, 3, 5, 6, 8, 10, 11, 13, 15, 17, and 23, with scent at the two outside stations), when the dance-language hypothesis was *not* put to the test, recruitment occurred as expected according to both hypotheses. For example, on day 5, the number of bees arriving at stations 1, 2, and 3 were 87, 0, and 90, respectively.

"Results obtained on experimental days, on the other hand, contrasted sharply with predictions of the dance language hypothesis (when odor clues *of the day before* competed with presumed 'use' of dance language information). If searching recruits had used their 'dance language' information on those experimental days (2, 4, 7, 9, 12, and 16) they should have arrived only at the two outside stations. If they had used the odor *of the day before* on those days, they would have arrived primarily at the central station" because the previous days had scent at the two outside stations. "The arrival of recruits predominantly at the central station during experimental days . . . indicated to us that recruits, while conducting their search, had ignored any 'dance language' information they might have obtained before leaving the hive." For example, on day 12 the number of bees arriving at stations 1, 2, and 3 were 4, 91, and 5 respectively.

> Results obtained during a particular three-day sequence (days 15–17 . . .) were especially revealing. On the first and third of those three days, the marking scent (clove oil) was provided at the outside stations; recruitment was then heavy. On the intervening day, however, while the clove scent was provided at the central (test) station, peppermint scent was added to the food at the two outside stations routinely visited by foragers. Regular foragers now returned to their hive *with* an odor stimulus and could also presumably indicate by their dances the location of the two outside stations. However, no appreciable peppermint odor accumulation would have occurred in the hive during such a short period of time. That new condition did not alter the results. Recruits again arrived preferentially at the central station that was marked by the odor used in the food the day before during the intervening day (day 16), rather than at the two outside stations. That happened despite the fact that dances presumably indicated those two outside stations and despite the fact that both outside stations were now marked with a specific odor as well.[32]

Wenner and Wells contend that the results of this "crucial" experiment give strong support for an as yet ill-defined odor-search model of honey bee recruitment while simultaneously falsifying the dance-language hypothesis.

Atkinson proposed that much of what we consider to be "discovered truth" may actually be nothing more than ideas generated or created in our minds.[33] He used the term "exploration" for this process. If a research project is begun without any preconceived notions of what "truth" might be, we are in a discovery or exploratory mode. If one is thinking as a relativist, she should be proposing multiple hypotheses from which to choose by strong inference. But if the investigator already has accepted the ruling paradigm, then she is likely to be attempting either to confirm and/or falsify that one theory. A person does not have to be irreversibly locked into one of these ways of thinking in all of his scientific work. An individual may use one philosophical stance when working on one research project and another school of thought on a different project. Or he may use different philosophies on the same project at different times in its history. This somewhat schizophrenic approach probably characterizes the minds of most scientists. Controversies are most likely to arise, however, when scientists are unaware of what philosophical mind they are in at the time they do or discuss research.

Wenner had begun his bee experiments under the dance-language paradigm of the realist school. He went about doing "normal" science, trying to extend the implications of the ruling paradigm. Among his many investigations were the possible role of sounds (e.g., wing frequency of vibration) in the dance and the part that conditioning plays in recruitment by the dance. He had accepted an invitation to participate as a "discussant" at a seminar to be held at the Salk Institute in California in March 1966. During the interval between the acceptance of the invitation and the seminar itself, Wenner's own experiments had convinced him that if a dance language existed it would be useful primarily for the recruitment of naive bees rather than experienced ones. These ideas ran counter to the tenor of the main seminar speaker, for whom he was supposed to provide positive discussion. Wenner describes the moment of his philosophical conversion as follows:

> As I walked to the front of the (seminar) room, my dilemma became resolved. For the very first time, I realized that von Frisch's original experiments had lacked adequate controls against forager flight paths and extraneous odor clues. I had 'created the image' [of a new way to interpret all of the work on the question of bee dance behavior] at that instant, on the way to the front of the room, in the sense meant by Atkinson.[34] (Brackets mine)

In the summer of 1966, "We set out to repeat von Frisch's original experiments. Our eyes had been opened wider now (we suddenly had new 'spectacles'). We moved back to the exploration approach . . . and away from our former 'verification' approach of attempting to 'prove' that bees used sound signals as part of their 'dance language.' "[35]

Wenner and Wells assert that either a bee "language" hypothesis or an odor-search hypothesis can explain virtually all of the experimental evidence that has been gathered on honeybee recruitment to date. Both explanations have been with us (at least in a romantic sense) for centuries. During this controversy it was commonly asserted that more evidence was needed to resolve the matter. Wenner and Wells, however, feel that "when experimental results are evaluated without sufficient attention to basic assumptions or without regard to the suitability of the experimental approach used, issues cannot and will not be resolved. . . ."[36] Thus, it appears that the honeybee forager-recruitment controversy is not about the nature of evidence but rather about the nature of hypotheses. It is not what investigators observe (the data) but what they believe (infer) that is at the heart of the controversy.[37] Widely accepted hypotheses (such as that of bee language) may nonetheless be unjustified by the data.

> I cannot give any scientist of any age better advice than this: the intensity of the conviction that a hypothesis is true has no bearing on whether it is true or not. The importance of the strength of our conviction is only to provide a proportionately strong incentive to find out if the hypothesis will stand up to critical evaluation.
>
> Peter Medawar[38]

Atkinson recognized that researchers belonging to different philosophical schools are unlikely to be able to communicate effectively with one another. He uses the ancient legend of the blind men and the elephant to make his point. "Each of the men having hold of a portion of the elephant proclaims knowledge of the whole; each is partially right but all are wrong. So, too, the various historians of science, philosophers of science, sociologists of science, and psychologists of science seemed to have grasped some part of the intellectual pachyderm we call science yet none of the various schools or 'isms' seems entirely satisfactory."[39]

> How am I to persuade Sir Karl [Popper] . . . that what he calls a duck can be seen as a rabbit? How am I to show him what it would be like to wear my spectacles when he has already learned to look at everything I can point to through his own?
>
> Thomas Kuhn[40]

Kuhn will never have the opportunity to persuade Sir Karl of anything now, because Popper died in 1994.

NOTES

1. Karen J. Winkler, "Histories Fail to Explain Science to the Laymen, Scholar Says," *Chronicle of Higher Education* 30, no. 23 (August 7, 1985): 7.
2. Bruno Latour, *Science in Action: How to Follow Scientists and Engineers Through Society* (Cambridge: Harvard University Press, 1987), 97.
3. Connie Veldink, "The Honey Bee Language Controversy," *Interdisciplinary Science Reviews* 14, no. 2 (1989): 166.
4. Adrian M. Wenner, "Sound Production in the Waggle Dance of the Honey Bee," *Animal Behavior* 10, nos. 1 and 2 (January–April, 1962): 79–95.

5. Veldink, "The Honey Bee Language Controversy," 167.

6. Adrian M. Wenner, *The Bee Language Controversy: An Experience in Science* (Boulder, Colo.: Educational Programs Improvement Corporation, 1971).

7. Karl von Frisch, *The Dance Language and Orientation of Bees* (Cambridge: Harvard University Press, 1967).

8. Patrick H. Wells and Adrian M. Wenner, "Do Honeybees Have a Language?" *Nature* 241 (January 19, 1973): 171-75.

9. Veldink, "The Honey Bee Language Controversy," 168.

10. James L. Gould, "Honey Bee Communication," *Nature* 252 (November 22, 1974): 300-301.

11. Veldink, "The Honey Bee Language Controversy," 170.

12. Ibid.

13. Ibid.

14. Ibid., 171.

15. Ibid.

16. Ibid.

17. Ibid., 172.

18. Ibid.

19. Ibid., 173.

20. Ibid.

21. James L. Gould, "The Dance-Language Controversy," *Quarterly Review of Biology* 51, no. 2 (June 1976): 211-44.

22. Veldink, "The Honey Bee Language Controversy," 174.

23. Ibid, 175.

24. Adrian M. Wenner and Patrick H. Wells, *Anatomy of a Controversy: The Question of a "Language" Among Bees* (New York: Columbia University Press, 1990), 366.

25. Thomas S. Kuhn, *The Structure of Scientific Revolutions* (Chicago: University of Chicago Press, 1962).

26. Wenner and Wells, *Anatomy of a Controversy*, 40.

27. Ibid., 15.

28. Ibid., 38.

29. Ibid., 178-85.

30. Ibid., 175-76.

31. Ibid., 179.

32. Ibid., 183-85.

33. Ibid., 40.

34. Ibid., 357.

35. Ibid., 360.

36. Ibid., 10.

37. Ibid., 11.

38. Ibid., 208. Peter B. Medawar, *Advice to a Young Scientist* (New York: Harper and Row, 1981), 39. Medawar began research in Professor H. W. Florey's laboratory in Oxford in the early days of the development of penicillin. He subsequently investigated causes of rejection by human beings of organs and tissues transplanted from other human beings and sought possible remedies. For this research he was awarded the Nobel Prize in 1960.

39. Ibid., 48.

40. Thomas Kuhn, "Logic of Discovery or Psychology of Research," in I. Lakatos and Musgrave, eds., *Criticism and Growth of Knowledge* (Cambridge: Cambridge University Press, 1970), 3; Wenner and Wells, *Anatomy of a Controversy*, 151.

GRAND ILLUSIONS

"**S**cience, like any other area of human endeavor, has had its grand illusions: compelling concepts that have excited a substantial segment of the scientific community and yet have turned out to be wrong."[1] Perhaps one of the best-known instances of such an illusion is N rays. Even so, the embarrassment to science of such a fiasco makes it easier to bear if we don't talk about it and just try to erase it from our minds. As a historian of science, however, I think that science education is biased whenever only the successes of science are taught. There is much of value to be learned from studying the failures of science. How is each generation of scientists to avoid repeating past mistakes if they have not been told about them? The flurry of scientific research on "cold fusion" that began late in the 1980s might have been avoided had the story of N rays been known and appreciated.

N RAYS

The setting for our story is near the beginning of the twentieth century. X rays had been discovered by W. K. Röntgen (alternatively spelled Röentgen, 1845–1923) in 1895 in emissions from cathode-ray tubes. The following year, A. H. Becquerel (1852–1908) found that uranium was an unstable radioactive element that released gamma rays comparable to X rays. Alpha rays and beta rays were also known before 1900. Thus, physicists at that time had been psychologically prepared to find even other kinds of electromagnetic radiations. René Blondlot (1849–1930) was, at that time, a member of the French Academy of Sciences. Like Röntgen, he had studied electric discharges in various gases and had become very interested in X-ray research. It was not known then whether X rays were particles or electromagnetic waves. Of course, now it is known that any form of matter or energy can exhibit the characteristics of both a particle and a wave. But at that time, physicists thought they could distinguish rays from particles. For example, it was known that particles coming from an electrified source normally carry an electric charge. By causing such particles to pass in a straight line between two metal plates with opposite electrical charges, these particles would be deflected from their straight course. Electromagnetic waves, on the other hand, were not deflected by this procedure, but they could be polarized, i.e., they could be made to oscillate in a single two-dimensional plane. So Blondlot planned to investigate the true nature of X rays by this criterion. If they were waves, they might be polarized as they emerged from an electric-discharge tube. He placed a pair of sharply pointed wires in various orientations in the path of the X rays. A short electric spark was made to jump between the wires. He reasoned that if his detection system could be oriented such that the line along which the spark was jumping could be made to coincide with the plane of the polarized X rays, the electric component of the X-ray wave should reinforce the energy of the spark so that it would increase its brightness. To his delight, Blondlot found an orientation of his detection system that seemed to increase the spark's intensity. But his joy was short-lived. He soon found that the radiation could be bent by passage through a quartz prism. It had already been established that X rays could not be bent in this way. So Blondlot reasoned that since he had proved that his radiations were waves and they could be bent by a prism, they must not be X rays. Thus they must be some new type of electromagnetic radiation. He called them N rays, for the University of Nancy, France, where he was employed.

Blondlot's research developed more sensitive detection systems for N rays, the most successful of which depended upon substances (called *phosphors*) that emit light when more energetic radiation strikes them. In this way, he was able to further investigate the properties and sources of N rays. The first of his publications on N rays appeared early during 1903 in *Comptes rendus*, the annals of the French Academy of Sciences. An army of physicists, physiologists, and psychologists soon were researching N rays. As with X rays, most substances were transparent to N rays. However, unlike X rays, water and rock salt were opaque to N rays. These substances could then be used to screen out N rays, allowing the design of new kinds of experiments. Other sources of N rays besides electric discharge tubes were discovered. Heated pieces of sheet iron and silver produced them. So did the Welsbach mantle (a type of gas burner commonly used in households around the turn of the century); but oddly the Bunsen burner did not. N rays also were produced by the Nernst glower, a lamp in which a thin rod of rare-earth oxides was heated by an electric current. In nature, Blondlot found that the sun emitted N rays. Augustin Charpentier was a respected professor of medical physics at the University of Nancy. In 1903, he submitted a paper to the Academy of Sciences reporting that parts of the human body gave off N rays, particularly the nerves and muscles. He also proposed that N rays could be used to explore the body for clinical purposes, such as detecting the outline of the heart, through the emissions accompanying its motor activity. It was even claimed (by a Monsieur Lambert) that even enzymes isolated from the body produced N rays.

Whenever a major discovery is made in science, a bunch of "Johnny-come-latelys" emerge from the woodwork to take credit. In this case, Blondlot received a letter in 1903 from Gustave le Bon, claiming that seven years earlier he had discovered a type of radiation that could penetrate metals. Later that same year, P. Audollet petitioned the Academy of Sciences to recognize him as the discoverer that living organisms produced N rays before Charpentier. Within a month, a spiritualist named Carl Huter made a similar claim. The matter was settled in the spring of 1904 when the Academy of Sciences issued a decision that Charpentier had precedence in this matter. Also in 1904, the academy awarded to Blondlot the prestigious Prix Leconte and fifty thousand francs. The prize was for "the whole of his works." His "new ray" was mentioned only at the end of the three-page citation accompanying the award. With this kind of endorsement from the academy, research on N rays became a hot subject. Over fifty papers on N rays were published in *Comptes rendus* during 1904 alone.

R. W. Wood (1868–1955), an American professor of physics at Johns Hopkins University, had an international reputation as an expert in optics and spectroscopy (the branch of physics concerned with detection and analysis of electromagnetic radiations). Among his many other talents, Wood was a relentless debunker of frauds such as spiritualistic mediums. Upon learning of Blondlot's N rays, Wood set about trying to duplicate his experiments, but was unable to do so. He was not alone. Other physicists such as Otto Lummer and Heinrich Rubens in Germany, and Sir William Crooks and Lord Kelvin in Britain, also could not replicate Blondlot's findings. So Wood decided to go to the University of Nancy and observe Blondlot's experimental materials and procedures where, as he put it, "the apparently peculiar conditions necessary for the manifestation of this most elusive form of radiation [appeared] to exist."[2] Wood was cordially welcomed by Blondlot and his colleagues, and he witnessed a series of experiments intended to demonstrate the various properties of N rays.

In the first demonstration, N rays from a Nernst lamp were supposedly concentrated by an aluminum lens on the spark of a detector apparatus. A ground glass plate was placed behind the detector opposite the N-ray source.

This was supposed to diffuse the light of the spark and make it easier to observe changes in its brightness. Blondlot claimed that if a hand was inserted between the spark-gap detector and the glass plate, some of the N rays would be blocked and the brightness of the spark would be diminished. Wood published a paper in the September 29, 1904, issue of the journal *Nature* on his observations at Blondlot's laboratories. He did not mention Blondlot's name in that article, but Wood's biographer, William Seabrook, claims that all the events in Wood's article occurred during his visit at the University of Nancy. Wood reports the first experiment as follows:

> It was claimed that [the fluctuation in the brightness of the spark] was most distinctly noticeable, yet I was unable to detect the slightest change. This was explained as due to a lack of sensitiveness of my eyes, and to test the matter I suggested that the attempt be made to announce the exact moments at which I introduced my hand into the path of the rays, by observing the screen. In no case was a correct answer given, the screen being announced as bright and dark in alternation when my hand was held motionless in the path of the rays, while the fluctuations observed when I moved my hand bore no relation whatever to its movements.[3]

The second demonstration was designed to give photographic evidence of N rays. The apparatus employed consisted of a horizontal photographic plate with a screen of ground glass above it. A spark-gap detector was placed over the glass. The assembly was enclosed in a light-tight cardboard box. The N-ray source was placed over the box. N rays were supposed to be able to penetrate dry cardboard, but not wet cardboard. The apparatus was constructed so that two spots on the photographic plate would be alternately exposed, one with and one without wet cardboard between the source and the detector. Several developed plates from previous experiments were shown to Wood, and he witnessed one such experiment. All of these plates had a distinctly more intense image on the spot with the dry cardboard screen in place. Wood, however, failed to be convinced by this evidence. He estimated that the natural brightness of the spark could vary by as much as 25 percent, a value so high as to make "accurate work impossible." Furthermore, the images were made by alternately moving from the dry- to the wet-screen side several times. Wood claimed that if a person knew the design of the apparatus, it would be easy to introduce (perhaps unconsciously) a bias in the exposure time favoring the dry-screen side. The photographic evidence was actually consistent with Blondlot's thesis, but Wood was "unwilling . . . to believe that a change of intensity which the average eye cannot detect when the n-rays [*sic*] are flashed 'on' and 'off' will be brought out as distinctly in photographs as is the case on the plates exhibited."[4] Wood suggested a way to render the operator incapable of knowing what screen was in place, thereby removing any possible bias. Unfortunately, Wood only spent about three hours observing experiments in Blondlot's laboratory and a photographic experiment mod-ified according to his suggestions was not performed in his presence.

In the next experiment viewed by Wood, the N rays from a Nernst lamp were passed successively through screens of aluminum foil, black paper, and wood in order to remove all kinds of electromagnetic radiation except the N rays. A narrow vertical slit was cut in a wet piece of cardboard that was placed in front of the other screens. The N rays emerging from the slit were then passed through an aluminum prism. Blondlot claimed that N rays had different wavelengths that could be separated into a spectrum by the prism. A narrow strip of phosphorescent paint was applied down the middle of a small piece of dry cardboard. This detector system was mounted in a device that could control the movement of the strip very precisely. Blondlot asserted that there were at least four positions at which a detector movement of less than 0.1 millimeter would cause the phosphorescent strip to go from glowing to dark. Wood again could not accept that a beam of electromag-netic radiation coming from a slit three millimeters wide could be resolved into individual wavelength widths as small as 0.1 millimeter. Blondlot and his associates were aware of Wood's criticism, but for them this was just one of the inexplicable and astounding properties of N rays that made them so fas-cinating to explore. When Wood was allowed to move the detecting device up and down the N-ray spectrum, he was unable to see any change in the brightness of the phosphorescent strip. Since the room had to be in total darkness to run this experiment, Wood decided to remove the aluminum prism without the others' knowledge. "The removal of the prism . . . did not seem to interfere in any way with the location of the maxima and minima in the deviated (!) ray bundle."[5] Wood was convinced that those who claimed to have seen the effects of N rays were allowing their beliefs to govern what they were seeing; i.e., "believing is seeing," rather than "seeing is believing."

Other experiments were also witnessed by Wood, but when he left Blondlot's laboratory he did so "with a very firm conviction that the few experimenters who have obtained positive results have been in some way deluded."[6] The publication of Wood's 1904 report in *Nature* essentially brought research on N rays to a halt outside of France. In 1905, Blondlot responded to Wood's criticisms of his experimental procedures. He claimed to have instituted procedures for automatically controlling photographic exposure times. He even allowed more time for exposure of the photo-graphic plate in the absence of N rays. His new photographs still proved that N rays caused an increase in brightness of the spark. He also promul-gated a set of instructions on how best to observe the manifestations of N rays. For example, when making judgments on the brightness of the lumi-nous source, the observer was advised to avoid looking directly at the source, to avoid straining of vision, and to be completely passive in percep-tion. The ability to make observations in this way would require practice.

Some people might never be able to attain these skills. Smoke, particularly tobacco smoke, must be avoided because it could distort or mask the manifestations of N rays. All unnecessary noise must be kept to a minimum while making these observations. Apparently, then, failure to detect the effects of N rays was due to a defect in the observer rather than to a defect in the theory of the existence of N rays. So fiercely did some French scientists defend the theory of N rays that it was claimed only the Latin races had the sensory and intellectual sensitivities necessary to detect the manifestations of N rays. It was suggested that the perceptive powers of Anglo-Saxons had become eroded by continual exposure to fog and those of the Teutonic races had been shriveled by the continual excessive consumption of beer.

Of course France wanted N rays to exist because they had been discovered by a Frenchman. After publication of Wood's critique, some French scientists performed N-ray experiments instituting the checks and balances against fraud or self-deception that he had proposed. They soon discovered that they were unable to detect any sign of N rays. The French journal *Revue scientifique* became anxious to avoid any further embarrassment to French science by N rays and wanted to resolve the matter once and for all. The journal challenged Blondlot to perform an experiment. A piece of tempered steel (considered to be a source of N rays) and a piece of lead were each to be placed in small boxes of identical size and weight. They would be sealed and marked with an identification number known only to a disinterested party. Blondlot would then be allowed to use any of his detection devices to determine which of the boxes was producing N rays. Blondlot finally answered the challenge in 1906. "Permit me to decline totally your proposition to cooperate in this simplistic experiment; the phenomena are much too delicate for that. Let each one form his personal opinion about N rays, either from his own experiments or from those of others in whom he has confidence."[7] Blondlot obviously had not abandoned his belief in N rays. His unwillingness to do the kinds of experiments that might shatter his belief effectively removed him from the fellowship of other scientists. His credibility had been irreparably damaged. It is said that a scientific theory dies only when the last of its supporters dies. If this is true, the theory of N rays died in 1930 with the demise of Blondlot.

CONFUSION IN A JAR

The Englishman William Thomson (Lord Kelvin, 1824–1907) was, according to some historians, the most eminent physicist of the nineteenth century. He calculated the amount of energy that the sun produces and, based upon the conversion rates of conventional fuels here on earth, came to the conclusion that the solar system was not very old (at least by today's 4.5-billion-year estimate). He of course was unaware that the sun's energy is derived from atomic fusion reactions deep in its core. The surface of the sun is a mere 5,500 degrees Celsius, whereas its core is over 10 million degrees Celsius. According to conventional wisdom, fusion reactions require temperatures like that in the sun's core. In this process, four hydrogen nuclei are required to produce one helium nucleus. A small amount of mass is lost during this multistep conversion, but it results in a very large release of energy (recall that Einstein's formula for the interconversion of mass [m] and energy [E] is $E = mc^2$, where c is the speed of light). At these high temperatures, hydrogen is not a gas but rather a fourth state of matter called a *plasma*, in which negatively charged electrons have been stripped away from hydrogen atoms leaving only positively charged nuclei.

Current methods for attaining fusion temperatures in the laboratory may involve bombardment by high-speed beams of neutral particles, exposure to microwaves, or induction of electrical currents in the plasma. The hot plasma can be contained by a magnetic field. One such "magnetic bottle" maintains the plasma in a sausage shape. Charged nuclei race back and forth along the sausage, being reflected at each end by a "magnetic mirror." A second kind of "magnetic bottle," called a *tokamak*, maintains the plasma in a donut shape. Two sets of powerful magnetic windings are required to keep the plasma in the center of the container. The combined effects of the two magnetic fields cause the nuclei to spiral around the donut shape without touching the walls of the containment chamber. *Inertial confinement* is an alternative approach to the magnetic bottle. Powerful lasers, arranged around a spherical chamber, fire in unison at a falling frozen pellet of fuel when it reaches the center of the chamber. In less than a billionth of a second, the surface of the pellet is compressed by the laser beams to 100 million times that of atmospheric pressure. The outer layers of the fuel pellet may reach 100 million degrees Celsius. For a very brief time, fusion conditions like those of the sun's core are produced and fusion energy is released. A fraction of a second later another fuel pellet is dropped into the chamber and the process is repeated over and over again.

Although high enough temperatures have been produced in the laboratory, it has not yet been possible to maintain the plasma at high enough concentrations for long enough times to reach the break-even point. That is the point at which as much energy is being released by fusion reactions as that required to heat and maintain the plasma. To be commercially useful, sufficient energy must be released to sustain the plasma without need for further heating, so that once ignition has started, the fusion reaction maintains itself. The engineering break-even point will come when the total fusion power generated is greater than the power needed to run the plant. Fusion power will only have commercial viability if it competes favorably with other forms of energy. So there is a long road ahead before fusion reactors become practicable (if ever).

The only viable nuclear power source today is fission rather than fusion. The first self-sustained nuclear chain reaction occurred at the University of Chicago on December 2, 1942. Enrico Fermi (1901–1954) and his colleagues constructed the first "atomic pile" and created the first nuclear reactor. Uranium nuclei are split within the pile, releasing a large amount of energy. The heat from such a reactor is constantly removed by a closed-circuit cooling system. The heat in the coolant can then be transferred outside the reactor to a heat exchanger that produces steam to drive a turbine and generate electricity. Nuclear plants of this kind have several advantages over generators that use fossil fuels such as oil or coal. The amount of fuel required to run a nuclear plant is minuscule compared to fossil fuel plants. Burning of coal produces large quantities of ash that create a disposal problem. Furthermore, burning both oil and coal releases gases containing sulfur, nitrogen, and other substances that, when oxidized, can be harmful to the environment. These products, when dissolved in atmospheric water, form acid rain. Whole forests are dying because living plants cannot tolerate these acid conditions. Fish and other aquatic animals die with the acidification of the ponds and lakes in which they live. The carbon dioxide produced by these fossil-fired furnaces accumulates in the atmosphere, preventing some of the earth's heat from radiating away into space. A global warming trend can be produced by this "greenhouse effect." Some climatologists predict that we are in the early phases of just such a period, and the results may be disastrous. Climates will be radically altered worldwide. Usable cropland will diminish. Deserts will expand. The polar ice caps will melt, causing a rise in sea level that will inundate many cities near the coasts. But nuclear fission does not produce carbon dioxide or the gases that lead to acid rain. However, the cost of uranium fuel is high and the spent fuel will remain radioactive for thousands of years. What to do with all that dangerous waste is a big problem. Many people think it is morally unconscionable to dump these wastes in the lap of future generations to care for in perpetuity.

That is part of the reason why technologically advanced nations are pumping so much research money into nuclear *fusion* projects. The fuel for a fusion reactor can be obtained from either fresh or sea water. A small percentage of normal water is "heavy water," made of an oxide of hydrogen called *deuterium*. Whereas a normal hydrogen nucleus consists of just a proton, heavy hydrogen has a proton and a neutron. Of course some energy must be expended to extract heavy water from normal water, but the supply is essentially boundless and relatively cheap compared to fission fuels. Furthermore, fusion reactions do not generate any gases that would contribute to acid rain or the greenhouse effect. Fusion reactions cannot run out of control and cause meltdowns the way that atomic piles can. Nuclear fusion reactors produce neutrons and gamma rays, both of which are very penetrating. Concrete shielding several feet thick is necessary to protect workers against these fusion products. As soon as a tokamak is turned off, production of these hazardous products ceases. But the neutrons that have bombarded the tokamak and its shielding make these materials radioactive, so they continue to emit harmful radiation after the instrument is shut down. Nevertheless, the amount of radioactive waste produced by this conventional-type fusion would pose much less of a problem than that of fission fuel wastes. We are going to run out of fossil fuels, especially oil, very soon. Then what will we do? If we haven't developed fusion reactors by then, technologically advanced civilizations, as we know them, will collapse. Little wonder that so much effort is being devoted to fusion research. The tokamaks and other machines by which we hope to attain sustainable fusion reactions have already cost billions of dollars, and we haven't even reached the break-even point yet.

A ray of hope appeared over thirty years ago when the Nobel Prize–winning physicist Luis W. Alvarez thought that he had discovered a way to make cheap and abundant energy. This new process did not require extremely high temperatures and was to become known as "cold fusion." Alvarez was working at the Los Alamos Meson Physics Facility in 1956 on the fundamental forces that acted on elementary particles. One of these elementary particles is called a *muon*. A beam of muons was passed into a container of liquid hydrogen and liquid deuterium. Alvarez and his colleagues were looking at the tracks that the particles left on film when they saw something quite unusual. A muon track was interrupted by a gap followed by a new muon track of exactly 1.7 centimeters. There were other muon tracks with the same gap followed by the 1.7 cm track. The result was highly repeatable. With the help of Edward Teller (father of the U.S. hydrogen bomb), Alvarez finally figured out what was happening. Somehow, the muon was catalyzing the union of a light hydrogen nucleus and a deuterium nucleus. They hypothesized that the muons were helping the two hydrogen nuclei to get close enough for them to overcome their very strong mutual repulsion forces. This repulsive barrier can theoretically be breached under special conditions by a process known as *quantum tunneling*. The 1.7 cm track was the unmistakable sign that cold fusion was taking place. Alvarez later discovered that his observations had been predicted several years earlier by F. C. Frank and Andrei Sakharov (Sakharov was the father of the Russian H-bomb and went on to win a Nobel Peace Prize for his contributions to human rights). Disappointment came, however, when Alvarez did the input and output calculations. He found that it requires much more energy to generate the muons than can be produced by the resulting fusion reactions. This then was the historical background that existed in the spring of 1989.

A press conference was held at the University of Utah in Salt Lake City on March 23, 1989. It was called by two chemists. One of them was B. Stanley Pons, chairman of the University of Utah's chemistry department, and the other was Martin Fleischmann, emeritus professor of electrochemistry at the University of Southhampton, England. I remember that day vividly. I was driving to work and listening to the news on my car radio. These two guys claimed to have evidence for cold fusion taking place in a jar of heavy water by simply passing a current of small voltage (3 to 25 volts) through it. A palladium rod was the negative electrode; a platinum wire wrapped around the palladium served as the positive electrode. Their proof was indirect. They claimed that, in addition to a lot of heat, they had detected three of the by-products of nuclear fusion: neutron particles, tritium, and gamma rays. Tritium, or hydrogen-3, is a radioactive isotope (its nucleus contains one proton and two neutrons). I could hardly believe what I had heard. Neither could most of the scientific community, including leading physicists. But the world of science had been stood on its head many times in the past by completely unexpected discoveries. I had lived through some of them and learned my lesson about rejecting new findings out of hand. I needed to reserve judgment until the experiments had been replicated and the results either confirmed or not. Nevertheless, I was overjoyed at the prospect that it might be true. The implications for society were staggering. A cheap, inexhaustible supply of energy could make even the poorest nations better able to feed and care for their people. As soon as I got to my first class I asked my students if they had heard the news yet. Most of them had not. So, with unreserved enthusiasm, I proceeded to tell them about the possible technological breakthrough. One scientist had asserted that "it is as important as the discovery of fire." That night I stayed up until midnight to hear of any late-breaking reports about cold fusion on the TV. The newspapers were full of reports and speculations the next day. I knew that this was either going to be the beginning of a golden age for mankind or one of the biggest boo-boos in the history of science. Either way, I did not want to miss out on any of it. I began to collect every article I came across that dealt with the cold-fusion issue. Very soon I had scores of articles that formed a folder almost an inch thick. I have continued to add to that folder right up to the time of writing this section of the book, and still have not lost my interest in the project even though my initial hopes for cold fusion have been dashed. To a historian of science, this has been a golden opportunity to see so many facets of what makes science tick. Watching the process of science is at least as interesting as the results of that process. If a Hollywood writer had produced this script, it would have been rejected as being too unrealistic.

It was very unusual for the results of scientific research to be delivered in this way. Normally scientific work is published in a refereed journal. The paper would give complete details on how the experiment was conducted, the results that were observed, and an interpretation of those results. As it was, Pons and Fleischmann had chosen to tell the world through a press conference. No one can learn all of the details of an experiment from newspaper articles or TV newscasts. Yet that was all that was available to the scientific community for a long time after the original announcement. Scientists around the world, anxious to test the new hypothesis, began to construct electrolytic cells with what little information they could glean from the popular press. Naturally, many of these attempts to replicate the Pons-Fleischmann results were doomed to failure from the start because of incomplete information. This is a terrible waste of time, effort, and money. Any negative results obtained in this way would be meaningless because the experiments could not exactly duplicate the original Pons-Fleischmann setup. Nevertheless, many attempts to find cold fusion were made almost immediately despite the odds against their success. The heavy water used by Pons and Fleischmann had been obtained from the world's major supplier, the Canadian firm of Ontario Hydro. Its phone lines were tied up night and day by calls from commercial and academic laboratories. Even schoolchildren wanted to try to make heat by cold fusion. The price of the rare metal palladium soared as the law of supply and demand kicked in.

Before the end of March 1989, partial confirmation of cold fusion came from a group led by Steven E. Jones at Brigham Young University (BYU), only forty miles away in Provo, Utah. The BYU team had not observed large amounts of heat, but had detected neutrons. Their report was more guarded than that of the Salt Lake City duo. The two groups had been working along similar lines for several years, but only became aware of that fact near the end of 1988. They had agreed to submit papers to *Nature* simultaneously on March 24, 1989, and in this way the independence of their discoveries would be made known to the scientific world. Contrary to their agreement, Pons and Fleischmann held their news conference on March 23, a day earlier. So Jones submitted his paper to *Nature* that same day. Pons submitted his paper the next day, as scheduled. Pons explained that he had decided to announce his results to the press before submitting his paper to Nature to address the rumors surrounding his work and because the University of Utah had already applied for a patent on his fusion process. If cold fusion was really possible, the patent holder stood to become fantastically wealthy. The discoverer of such a process would certainly be in line for a Nobel Prize. The temptation was too great for Pons and Fleischmann not to cheat on their agreement with Jones.

Jones had been working since the early 1980s on muon-catalyzed fusion, but was finding that it took more energy to make the muons than could be recovered from fusion, just as Alvarez had predicted. He coauthored a popular article entitled "Cold Nuclear Fusion" in the July 1987 issue of *Scientific American*. The problem was to get the deuterium nuclei close enough for a sufficient time to allow quantum tunneling to take place. Jones decided to forget the muons and try to pack the deuterium nuclei together within the lattice-like structure of a palladium or titanium electrode, and hold them there by an electric current. It stretches credulity to believe that two groups, working in the same field and only forty miles apart, would not be aware of one another. However, they belonged to different disciplines (Jones was a physicist whereas Pons and Fleischmann were chemists) and they published in different journals. Perhaps the most famous example of independent discovery in science is that of the theory of natural selection developed independently by Charles Darwin and Alfred Russel Wallace. They, like the two fusion teams, decided to resolve the problem of priority by publishing their separate papers simultaneously in the same source. The BYU group built its first fusion cell on May 22, 1986, but had little initial success. They tried various combinations of metals and salt solutions. By the fall of 1988, their results had become more encouraging. Paul E. Palmer had been a member of the BYU cold-fusion team since 1986. Palmer and Jones reasoned that if deuterium nuclei could be fused by electrolysis, neutrons should be emitted, but the overall effect would be very small. They would need a very sensitive neutron detector. Fortunately just such an instrument was being developed in the physics department of BYU by Bart Czirr and Gary Jensen. It wasn't long before Jones and Palmer detected low levels of neutrons issuing from their electrolysis tubes.

Steven Jones was a grant reviewer for the U.S. Department of Energy (DOE). On September 20, 1988, he was sent an application from DOE that had been submitted by Pons and Fleischmann. It explained their proposal for scaling up their cold-fusion work. They had tried to keep their work under wraps since its inception. They even financed their research ($100,000 worth) out of their own pockets. However, with some evidence of success, they now had to ask for federal funds to carry the project on toward possible commercial scale. They, of course, had no idea that their request would wind up in the hands of a man who had been doing parallel work on the same project. Jones was one of five reviewers of their grant. Four of them were skeptical. Jones felt he needed more information to fairly evaluate the merits of the proposal. Pons exchanged several phone calls with Jones to expedite the review. In the process, Jones offered Pons the use of BYU's sensitive neutron detector. Pons and Fleischmann finally met with Jones and Palmer at BYU on February 23, 1989. Over lunch, Jones mentioned that he was about to publish the results of his experiments thus far and suggested that the two groups should publish a joint paper. Fleischmann did not like the idea of sharing his work with another group and felt that Jones had nothing to contribute beyond his neutron detector (which the Utah group never did use). Furthermore, the "Utah group" (i.e., the University of Utah) thought it was too early to publish. They wanted another six to eighteen months longer to be more certain of their findings. But the American Physical Society was having its spring meeting in Baltimore on May 1, 1989, and Jones had already submitted a paper on cold nuclear fusion to the society for publication in its bulletin. Copies of the *Bulletin* would soon be mailed to the society's membership. On May 1, the cat would be out of the bag anyway. So Jones suggested that if the Utah group didn't want to publish a joint paper with the BYU group, perhaps they would agree to submitting independent papers simultaneously to the same journal. Finally, on March 6, 1989, with the intervention of the administrations of the two universities, it was agreed that simultaneous publication was the only solution. The journal *Nature* was chosen because it can, when conditions warrant, publish important papers in minimal time. It was agreed that on March 24, 1989, both teams would go to the Federal Express office together and send off their papers to *Nature*. Until that time, neither team was to say anything about their work on cold fusion. Unbeknownst to Jones, the Utah group decided on March 11 to send a "preliminary note" to the *Journal of Electroanalytical Chemistry*, announcing its success with cold fusion. As late as March 21, the Utah group was still in communication with Jones, but never revealed to him that it had already sent off a scientific paper. Fleischmann said that the Utah group was under tremendous pressure to go public with its results because the patent lawyers needed assurance that the University of Utah had priority in the discovery of cold fusion. Pons and Fleischmann buckled under the pressure by the weekend of March 17. Without informing Jones, they set March 23 as the date for a press conference.

The day before the press conference, an article on cold fusion appeared in the *Financial Times of London*. It seems that Fleischmann contacted an old friend, Richard Cookson, for advice on how to get the best coverage of their discovery in Britain. His friend told him to contact his son Clive Cookson, who was a writer for the *Financial Times*. Clive Cookson told Fleischmann that the *Financial Times* would not be published on March 24, the day planned for his paper to be sent to *Nature*, because it was Good Friday, a bank holiday. If Cookson had to wait until the results of the March 23 press conference were known, his report would not be published until the following Monday. After consulting with Pons, they agreed to give Cookson the information for publication a day before anyone else. Reporters immediately began phoning the U.S. Department of Energy and the two universities. That same day, BYU learned "with shock and dismay" of the press conference scheduled on the morrow. The day of the conference (March 23), one day prior to the agreed upon date, Jones sent his paper to *Nature* without informing the Utah group. The next day Pons and Fleischmann went to the Federal Express office at the time agreed upon and stood around waiting for Jones to show up. Growing tired of waiting, they eventually called BYU to find out what had happened to him. They then learned that Jones had sent his paper the day before. From that point on, each group felt it had been deceived by the other and a gulf of enmity existed between them.

Some weeks later, Fleischmann apologized to Jones for calling the press conference. Both men submitted to shaking hands in front of Italian photographers, but the mutual distrust could not be so easily forgotten. On April 2, the Salt Lake City newspaper *Desert News* published an article intimating that Jones had stolen his ideas on cold fusion from Fleischmann and Pons by having reviewed their grant application. Pons confessed that, "In all my scientific life, I have never seen a situation where a proposal was sent to a certain person, who calls up and says, 'Tell me more,' and who then immediately reveals himself as the reviewer and suggests collaboration. I had no idea when he was going to go public."[8] It is well documented, however, that Jones had been researching cold fusion for many years before he reviewed the Utah proposal. The electrolytic cell that Jones was using was very different from that of the Utah group. For example, the composition of the electrolytic fluid contained chemicals that would have rendered cells of the Utah group nonfunctional. The Utah group ran its cell for about one hundred hours before detecting the excess heat of fusion. The BYU group detected evidence of fusion after just one hour, but the cell's activity fell off markedly after eight hours without detection of heat. The Utah group theorized that only 1 in 10 million fusions emitted a neutron; the remainder was producing heat rather than nuclear radiations. They were thus proposing a new kind of fusion unknown to science. Jones, on the other hand, assumed that each neutron was produced by each act of fusion. Thus Jones was detecting a fusion rate more than one million times smaller than that claimed by Pons and Fleischmann.

The "preliminary note" submitted by Pons on March 11 to the *Journal of Electroanalytical Chemistry* was accepted for publication on March 20, 1989, three days before their news conference. The article lacked many of the details that would have accompanied a full research report. The paper submitted by Jones to *Nature* was accepted for publication on April 12 and was to be published in the April 27 issue. The paper submitted to *Nature* by Pons and Fleischmann, however, was unacceptable in its original form. They withdrew it because (according to Pons) he was now too busy to make the changes requested by the editors. To my knowledge as of this writing, they have not resubmitted it for publication.

One of the first confirmations of cold fusion came from two Hungarian researchers on March 31, 1989. An early experiment at the Lawrence Livermore National Laboratory in California exploded, as forewarned by Fleischmann and Pons. By April 4 in Japan, Tokyo University of Agriculture and Technology reported observing large amounts of heat in their experiments. On April 10, Texas A&M University confirmed that 60 to 80 percent more energy was given off as heat from their cells than the electrical energy being pumped in. Georgia Tech reported detection of neutrons but no heat. Was each group seeing different aspects of the fusion reaction? Fleischmann left Britain to continue his research at a secret location, claiming, "I didn't want everyone sitting on my tail."[9] At a meeting of the American Chemical Society in Dallas on April 12, Pons was so mobbed by reporters and scientists that he had to change hotels twice and register under an alias. The Soviet news agency Tass announced on April 12 that the physics department at Moscow University had detected neutrons in twenty experiments and were able to "assert with confidence that the nuclear fusion reaction actually takes place."[10] On April 14, the Massachusetts Institute of Technology (MIT) announced that none of their twenty experiments had confirmed the Pons-Fleischmann hypothesis. Researchers at AT&T's Bell Labs in California were also unable to observe cold fusion. At the same time, however, scientists at MIT and the University of California at Berkeley were devising theories to explain cold fusion that made the positive results more palatable to scientists. On April 15, Georgia Tech scientists withdrew their claim that they had repeated a key part of Utah's nuclear-fusion experiment. They blamed a malfunction in their neutron-detection instrument. Pons dismissed the failed experiments as probably due to lack of proper preparation of the electrodes. Nathan Lewis of the California Institute of Technology (Cal Tech) claimed to have tried to get more specific information from Pons and Fleischmann without success. Lewis didn't even know the exact dimensions of the Utah cell. He had to guess the dimensions from the length of Pons's hand in a photograph. He later learned that the photograph was just for publicity and was not an actual fusion cell.

Pons and Fleischmann's funding application from the DOE was approved soon after the March 23 press conference. Now, however, the University of Utah refused the $332,000 grant. Norman Brown, director of technology transfer for the university, said, "As soon as we take one dollar of federal money, any discoveries we make become the property of the U.S. government." But by mid-April, Utah University president Chase N. Peterson was suggesting that a national fusion center be established in Utah. This would require about $100 million to put into operation, and the federal government should supply about $25 million as seed money. Ira Magaziner, president of Telesis, Inc., lobbied Congress for funding of cold-fusion research by warning them against the threat of another Japanese takeover of an American discovery: "America is prepared to fight to win this time. . . . I have come here to ask you, for the sake of my children and all of America's next generation, to have America do it right this time."[11]

A meeting of the American Physical Society was held in Baltimore on May 1, 1989. The main topic was cold fusion. Pons and Fleischmann had been expected to attend, but they failed to show up. Jones reported on his evidence for low-level fusion and predicted that commercial exploitation of it would be difficult if not impossible. Most of the papers given on cold fusion at that meeting were on experiments that had taken great care to rule out the effects of background radiation on sensitive detectors. Time and again the bottom line was no neutrons, no characteristic gamma rays, and no excess heat. The gamma rays that had been reported by some research teams could have been due to such natural sources of background radiation as radon gas. Neutrons could also come from natural background radiation and from cosmic rays from space that are continually bombarding the earth. The neutron flux from day to day is quite variable. While low levels of nuclear fusion could not be discounted by the neutron counts, it was deemed to be nothing like the levels reported by the Utah group. Pons and Fleischmann apparently had not taken enough care to stir the electrolytic soup and distribute its heat evenly. If the thermometer was placed too close to the palladium electrode, their calculations would indicate an excessively high net heat yield. The consensus of the conference was decidedly negative against cold fusion, served to dampen the enthusiasm for further research, and tended to dry up funding for it. A year and a half later, some scientists were still researching the unexplained aspects of earlier experiments such as the source of the low level of neutrons. They continue because of their curiosity, not because of dreams of wealth and glory. They simply must, in the interest of pure science, try to find an answer to these mysteries. And they are not afraid that their reputations may be damaged for following an idea whose time had come and gone. Science needs these kinds of workers even though they probably do not receive the recognition from their peers that they deserve.

A meeting of Utah's Fusion/Energy Advisory Council was scheduled for October 25, 1990. Either Pons or Fleischmann was expected to attend and present their latest data to convince the panel to release the final $1.3 million portion of the $4.5 million grant to the state-funded National Cold Fusion Institute. Fleischmann, however, was under medical treatment at his home in Southampton, England. Two days before the scheduled meeting, local newspapers reported that Pons had disappeared. His phone had been disconnected and his house was for sale. Neither his friends nor his employer knew where he was. A month earlier, Pons had written a letter to his chemistry department head, Peter Stang, stating that he had arranged for someone else to teach his classes and that he would be absent from the campus for the rest of the quarter. Stang received a fax from Pons's lawyer, C. Gary Triggs, on October 24 requesting a sabbatical leave beginning November 15, 1990. The university told Triggs that more details of Pons's plans would be needed before a decision on his sabbatical could be made. Since June, Pons had been virtually incommunicado, and the only way to contact him was through his lawyer.

It is obvious that Pons, Fleischmann, Jones, and the scores of others who claimed to have seen evidence of cold fusion were not purposefully being fraudulent. After all, their reputations (and very likely their livelihoods) were at stake. Rather, it seems that they had become swept up by their desire to believe that some such process might exist and the excitement of contributing to the beginning of something really important. It seems incredible that such collective mass self-deception could ever occur among modern scientists. But even a nodding acquaintance with science history should have prepared us to be ready for it because it has happened many times before and probably will do so in the future. In the story of N rays presented earlier in this chapter, we learned that over a hundred papers in France alone were published confirming the existence of this new type of radiation. "As it turned out, the N stood for nothing."[12]

Shortly after N rays were disproved, respected American scientific journals began publishing papers describing the Allison effect, a complicated method employing magnetic fields to detect isotopes. A half dozen new chemical elements were reportedly discovered using the Allison effect, all of which were eventually shown to be due to experimental errors. About the same time, "mitogenic rays" were reportedly produced by plant and animal cells. In one experiment, the mitogenic rays emitted by an onion seemed to affect the growth of another one nearby. These "rays" could be blocked by placing a glass between them. Over five hundred journal papers were devoted to the study of such rays before their existence was disproved. British researchers announced in 1958 that they had constructed a hot-fusion reactor called Zeta. Neutrons from the hot plasma were detected, making the noted scientist Sir John Cockcroft 90 percent certain that fusion had occurred. The neutrons were soon found to be produced by a different process. One of the most spectacular cases of collective scientific error was "polywater." It was first discovered in the Soviet Union in 1968. Polywater was more viscous than regular water, and it was very difficult to purify. Even the U.S. National Bureau of Standards confirmed its existence. There was concern that if polywater escaped from the lab it might spontaneously convert all of the earth's waters to the new form and destroy all life. Over the next few years, many scientific papers were published on polywater. In 1972, however, polywater was shown to be just ordinary water with some impurities. Only one year before the Pons-Fleischmann cold-fusion announcement, dozens of labs had been able to attain superconduction at the very low temperatures of liquid nitrogen. At the same time, several reports of superconduction at room temperatures appeared, mostly from obscure labs in distant countries. Eventually all these claims to "warm superconduction" were falsified.

The point to this brief romp through science history is to emphasize the fact that self-deception is nothing new in science. There are probably more cases of it in very recent times because of new communication technologies. High-speed electronic mail and fax machines make it possible for today's "discovery" to be in the hands of scientists on the other side of the world that same day. According to Rolf Sinclair, a National Sciences Foundation physicist, "Now people can't think about their results overnight and then retrieve the letter from the outbox."[13] In other words, scientists can now "shoot from the hip" without giving proper thought and reflection to what they have just observed. In the normal course of science, a researcher usually discusses one's latest findings and conclusions with others in the same lab or at an informal seminar. Then a paper is written and submitted to a journal. One or more expert referees are assigned by the journal editor to evaluate the merits of the paper for publication; this is called "peer review." When the paper is published, then the entire scientific community has a chance to make its own evaluation and begin the process of attempting to duplicate or falsify the experiments and observations. But with the new communication devices, a copy of the paper submitted for peer review can be in the hands of other researchers before it reaches the journal office. When the normal process of science is subverted, the consequences can be very undesirable. A University of Oregon psychologist, Ray Hyman, says: "The so-called rationality of science comes from the fact that it's usually all cleaned up before the public hears about it.[14]

NOTES

1. Irving M. Klotz, "The N-Ray Affair," *Scientific American* 242 (May 1980): 168.

2. Ibid., 170.

3. Ibid.

4. Ibid., 173.

5. Ibid., 174.

6. Ibid.

7. Ibid., 175.

8. F. David Peat, *Cold Fusion: The Making of a Scientific Controversy* (Chicago: Contemporary Books, 1989), 74.

9. Ibid., 92.

10. Ibid., 95.

11. Ibid., 102–103.

12. Michael Rogers, "The Follies of Science," *Newsweek* 113 (May 8, 1989): 56.

13. Ibid.

14. Ibid.

CHAPTER 5

STRANGE BEDFELLOWS

R eligion, politics, and science might seem like such diverse subjects that they would have little to do with one another. But science had its beginnings in societies where religion, superstition, and magic were major determiners of what constituted truth or conceptual reality. Fundamentalism and secularism have influenced, to various degrees, the advancement of science during all of its history. Probably the most famous example of religion's influence on science is that of the Inquisition involving the mathematician, physicist, and astronomer Galileo Galilei (1564-1642). Although just about everyone has heard of this event, it has been widely misunderstood and its importance in the history of science has been grossly underestimated. If you think that Galileo invented the telescope you're wrong. If you think that he originated the idea that Earth goes around the Sun and not vice versa you're also wrong. If you believe that Galileo was accused of heresy by the Inquisition you're wrong again. The real story of this remarkable scientist "still has much to say about the practice and the philosophy of science. What was at issue was both the truth of nature and the nature of truth."[1]

THE GALILEO AFFAIR

Galileo Galilei was born in 1564, the son of an unprosperous cloth merchant in Pisa, Italy. At the age of twenty-seven, he obtained a professorship in mathematics at the University of Padua. As a physicist, he believed that there are natural laws (mathematical regularities) governing falling objects and flying projectiles that can be applied equally well to those on Earth as well as to the heavenly bodies. To understand the cosmological theories on which his work was based, it is necessary to go back to the time of Aristotle (384–322 B.C.E.). Aristotle believed the heavenly bodies were perfect, incorruptible, and unchanging, whereas Earth was imperfect and subject to decay. For him, movements on Earth were in straight lines, but the natural state of things was to be at rest. Thus any motion of things on Earth were forced motions; the only exceptions were those of the four elements. Fire and air were naturally light and therefore ascended; earth and water were heavy and thus sank. All things that were composed of these four elements were subject to forced movements and therefore tended to decay. Circles and spheres were considered to be perfect geometrical shapes. The planets and stars were fixed on eight crystalline spheres that revolved around an immobile Earth. The spheres were composed of a fifth element called *ether*, that could be neither destroyed nor changed into anything else. Earth was also considered to be a sphere because its shadow could occasionally be seen on the Moon. The incongruence of a perfect spherical shape for an imperfect corruptible Earth apparently caused him no concern. If Earth moved, it would be due to either natural or forced movement. Forced movement caused destruction. Earth has not been destroyed. Therefore any Earth movement would have to be natural movement in a straight line toward the center of Earth. Rotation of Earth would presuppose two natural movements. We don't see anything wrong with that idea today, but it was not part of Aristotelian cosmology. Thus, Earth was supposedly "proven" to be immobile. The Church later came to endorse Aristotle's cosmology because it did not conflict with the Bible.

The Aristotelian system, however, failed to explain all of the astronomical observations. For example, one could observe with the naked eye that the wandering "stars," called *planets*, would sometimes slow down, stop, and then go backward (retrograde motion). The celestial spheres rotated in only one direction in this system. These observations remained unexplained until the Alexandrian astronomer Claudius Ptolemy proposed his theory in the second century C.E. The Ptolemaic system proposed that each planet turns on a small sphere fixed to a main one, somewhat like gears. The combination of a turning main sphere and turning attached minispheres could generate the kinds of motions that had been observed. These successive forward and retrograde motions were termed "epicycles."

The Christian Church had a great interest in calendar reform because, for one thing, the astronomical systems of Aristotle and Ptolemy were insufficient to precisely predict religious holidays such as Easter. The pope, in 1514, asked a Polish priest and mathematician to try to develop a better predictive system. The man thus chosen was Nicklaus Kopernig (better known today as Copernicus). Copernicus was able to do the pope's bidding, but at the terrible price of scrapping the entire Aristotelian system. The heliocentric theory of Copernicus proposed that the center of the universe was somewhere near the Sun. Earth orbited the Sun along with the other planets. A manuscript of the rudimentary Copernican theory, entitled *The Little Commentary*, was circulated on May 1, 1514. The full theory was not published until after his death in 1543. This was done that same year in his book *De revolutionibus orbitum coelestium* (*On the Revolutions of the Celestial Spheres*).

Copernicus's theory had been known and discussed all over Europe for twenty years before his book was published. The earliest religious opposition came from the Protestants rather than from the Roman Church. Martin Luther said, "People give ear to an upstart astronomer who strove to show that the Earth revolves, not the heavens or the firmament, the sun and the moon . . . the fool wishes to reverse the entire science of astronomy."[2] Calvin quoted from the Bible, "the world is also stabilized that it cannot be moved." The Roman Church, on the other hand, initially accepted the Copernican theory for its mathematical elegance and its ability to predict the motions of the heavenly bodies with greater accuracy than the older geocentric (Earth-centered) theories. The Roman Church was willing to use the Copernican system as a "model" (as it would now be called) for reforming the calendar (which was done in 1582). One does not necessarily have to believe that a model represents reality in order for it to be useful in this and other ways. No one in their right mind, even Copernicus, could possibly believe that the heliocentric theory was actually true; or at least that was the assumption of the Roman Church.

There were also some immediate objections to the Copernican system based on scientific grounds. For example, if Earth orbits the Sun, then the apparent position of a star should shift when viewed from opposite positions of its orbit. This is referred to as *parallax*. Copernicus explained that no parallax had been observed because the distance of stars from Earth was so great that parallax was too small to be measured with the instruments then available. From this he concluded, against the prevailing doctrine of a closed universe, that the universe was infinite. His theory also proposed that Earth rotated on its axis once a day. Because he avoided the question in his book, Copernicus apparently could not answer why objects thrown into the air on a revolving Earth did not fall to the ground to the west. About the only thing that he saved from the older cosmologies was circular motion. By removing Earth from the center of the universe he had also removed it from the focus of God's purpose. Man was no longer the center of all things. Copernicus had placed Earth in the heavens and by so doing he had removed the barrier separating the corruptible from the incorruptible.[3] But if Earth was incorruptible, why did Earthly things decay? On the other hand, if the heavens were corruptible, imperfect, and changeable, why hadn't anyone observed these changes?

Then in late 1572, a nova burst forth in the constellation Cassiopeia that was so bright it could be seen during daytime. It lasted two years before fading. The heavens were changeable after all. The new star punched a hole in Aristotelian cosmology. Didn't God complete his creation on the seventh day? If so, where did the new star come from? According to Aristotle, things that changed belonged to the Earthly sphere. But the nova exhibited no parallax, so it must belong to the outer sphere with the rest of the stars. An abortive attempt was made to save the old view by postulating that the nova was a temporary meteorological phenomenon, like a rainbow.

The Danish astronomer Tycho Brahe (1546–1601) had observed the nova and published a book about it in 1573 called *The New Star*. Brahe used massive quadrants that allowed him to measure celestial positions with greater accuracy than any others at that time. His measurements provided the scientific evidence that the heavens had changed and that at least one part of Aristotelian doctrine was in error. The book so impressed the king of Denmark that he awarded Brahe with a feudal lordship of the island of Hven, located between Sweden and Denmark. Brahe built a castle on the island and continued to study astronomy there. In 1577, he observed a comet. Comets had been seen before but had been regarded as belonging to the changeable Earthly sphere. Brahe's careful parallax measurements indicated that it was further from Earth than the Moon, but not as far as the stars. The tails of comets were seen to point away from the Sun by Peter Bieniwitz, indicating that they were under solar influence and therefore in the sphere of the Sun. According to Aristotle, only one object was carried by each planetary sphere. Did each comet have its own sphere? That was impossible because Brahe's measurements proved that the comet had an oval orbit, which required it to move through the planetary spheres. Tycho concluded that there were no crystalline spheres in the heavens. But he could not bring himself to accept the full Copernican system. So he devised a compromise model in which the planets orbited around the Sun, with the group of the Sun and the planets revolving around the Earth-Moon system. The Tychonic model had several problems. It failed to explain how an elliptical orbit could remain regular and not become unstable. If there were no crystalline spheres to hold the planets aloft, through what medium were they traveling and why did they not fall?

It was against this backdrop that Galileo came on the scene. He worked at Padua on Brahe's questions for eighteen years. He initiated an intellectual revolution by proposing that the only way to learn about nature is to experiment, make careful observations, and if possible reduce everything to mathematics. Galileo used common experience as a metaphor to explain why falling objects on a turning Earth did not fall to the west of their release point. He argued that as Earth turned everything on it turned at the same speed, so that any falling object would move not only eastward with Earth, but also move toward the center of Earth. The combination of these two movements results in the object striking Earth directly below its release point. It was analogous to dropping an object from a ship's mast. The object hits the deck directly below its release point because both the ship and the object are initially moving together along the surface of Earth at the same speed. Galileo had thus destroyed the Aristotelian distinction between natural and violent movements. Aristotle had endowed every object with its own "essence" and "quality" that gave it its desires and tendencies. Galileo's new worldview swept away these old ideas by asking *how* things happen, rather than *why* they happen.

A new "looker" (telescope) was invented in 1609 by the Dutchman Hans Lippershey. Galileo had gone to Florence in 1610 and there he improved the instrument (calling it a *perspicillum*) so that it could magnify distant objects a thousand times and make them appear thirty times closer. The Moon had been thought to be a perfect heavenly body without irregularities. This is hard for me to believe because anyone with a healthy set of eyes has seen the "man in the Moon." Such is the power of authority. When Galileo looked at the Moon through his telescope he saw mountains and flat "seas." Because the stars are so far distant, the telescope made them appear brighter, but no closer. He could see many more stars than had Aristotle. The Milky Way appeared to consist of millions of stars in clusters that Galileo called "clouds" (nebulae). On a January night in 1610 he noticed three new shining bodies (he called them "stars") near Jupiter, one to the west and two to the east of the planet. The next night all three bodies were in a line to the west. At that time, Jupiter's movement relative to those bodies should have put them all to the east of that planet. Further observations during that winter convinced Galileo that these bodies were satellites of Jupiter, just like the Moon is a satellite of Earth. Galileo published his observations that spring in a little book called *Sidereus nuncius* (The Starry Messenger). Anyone who wanted to could look through a telescope and verify Galileo's observations for himself. Galileo was an ambitious man, anxious to make a name for himself and improve his lot in life. With the hope of securing a government-supported position in Tuscany at the court of Grand Duke Cosimo II de' Medici, he decided to name the Jovian satellites Medicean stars. So here we see an "unholy" conjunction of science and politics.

There was no unambiguous evidence in *The Starry Messenger* that Galileo espoused the Copernican system. But he had reasoned that if Jupiter had satellites revolving around it while it orbited the Sun, why couldn't the Earth-Moon system be doing the same? One of his former students, Benedetto Castelli, had predicted to Galileo that, in the Copernican system, Venus should show the entire range of phases, as does our moon. Conversely, in the Ptolemaic system, the epicycle of Venus occupies a position between Earth and the Sun, so Venus could only show crescent phases because it never passes behind the Sun for full illumination. Galileo began observations of Venus in October 1610, and over time observed the full range of its phases, just as Castelli had predicted. Galileo had halfheartedly supported the Copernican system up until this time. But with the Ptolemaic system now discredited, he ignored the Tychonic scheme and gave his entire support to the Copernican model.

Galileo succeeded in obtaining a new post as mathematician to the Medici. This gave him the fame and power that he sought. He was egotistical enough to think that this power might allow him to persuade the Catholic Church to accept the Copernican system as reality. Then Galileo turned his attention to the Sun. He found that the Sun also rotates on its axis and that sunspots were imperfections on its surface. He published these observations and conclusions in 1613. Aristotelian doctrine absorbed two more blows. However, Galileo got into a squabble with the Jesuit Christoph Scheiner over who had been the first to discover sunspots. Scheiner, of course, believed that the Sun was unblemished, and that the spots were made by intervening clouds. In his book entitled *The Crime of Galileo*, Giorgio de Santillana suggests that Scheiner held a grudge against Galileo and was active in leading the Jesuits in a vendetta against him, but science historian Owen Gingerich finds no real evidence to support it.[4]

With Galileo's help, Castelli had become a professor of mathematics at Pisa. Soon thereafter, Castelli had breakfast with Cosimo de' Medici and his mother, the Dowager Grand Duchess Cristina. The conversation got around to the question of the reality of the Jovian (Jove is the poetic name for the planet Jupiter) satellites and whether there was any conflict between the heliocentric theory and the Bible. As a consequence of that debate, Galileo was challenged to show that such a conflict does not exist. Galileo wrote his defense in 1616, including an epigram that he had borrowed from Caesar Cardinal Baronius, the librarian of the Vatican. Galileo argued that whereas the Bible could be ambiguous, God's Book of Nature could be investigated and tested. Both the Bible and science have their respective functions. "The Bible tells us how to go to heaven, not how the heavens go."[5] It may be argued that the heliocentric system is compatible with the Bible, but it is another thing to prove that the Book of Nature clearly favors the Copernican explanation.

There are two forms of Aristotelian logic. *Inductive reasoning* draws general conclusions from particular instances. This is the way by which major scientific knowledge is generated, but truth cannot be derived in this way. On the other hand, *deductive reasoning* starts with a generalization and derives a specific conclusion. Given that the major and minor premises are correct, the proper use of logic can lead to a valid conclusion. Gingerich uses the following syllogism:

A. If it is raining, the streets are wet.
B. It is raining.
C. Therefore the streets are wet.

The converse of this is as follows:

A. If it is raining, the streets are wet.
B. The streets are wet.
C. Therefore it is raining.

This last syllogism contains a logical error known as "confirming the consequent." The streets could be wet for a number of other reasons. For example, the winter snow may have melted, the street-cleaning department may have washed the street, or a parade of horses may have just passed through the street and left a memento. Now let's see how these points of logical analysis were applied in Galileo's defense of the Copernican theory.

A. If the planetary system is heliocentric, Venus will show phases.
B. The system is heliocentric.
C. Therefore Venus will show phases.

There is nothing logically wrong with this syllogism. But this is not the form of Galileo's argument. He had interchanged the second premise and the conclusion, as follows.

A. If the planetary system is heliocentric, Venus will show phases.
B. Venus shows phases.
C. Therefore the planetary system is heliocentric.

Galileo had committed an elementary mistake of logic, and even Kepler pointed it out to him. The phases of Venus could have other explanations, such as those of the Tychonic system. He was aware that he could not use deduction to establish the Copernican system. His process of reasoning was what we would now call the hypothetico-deductive method. This involves testing a hypothetical model, which gains ever more credibility as it successfully passes each test. The Copernican model had great explanatory power. It predicted the phases of Venus. It arranged the planets by period. The moons of Jupiter were observed to also be arranged by period, like a miniature solar system. If Earth was a planet, it would be likely that other planets would be Earthlike; the Moon could be seen to be Earthlike with mountains and plains. Thus the Copernican model was gaining in credibility as it passed each test.

Galileo's method was essentially inductive and therefore potentially fallacious. The Bible, however, held the truth. The battle that eventually erupted between the theologians and Galileo was not over the Copernican system, but rather over the route by which true knowledge of the world could be obtained.[6] Because the public might become confused about this fundamental difference, the Church thought it prudent to condemn the teaching of the Copernican system. Galileo was enjoined to desist his propagandizing of the Copernican system. The Holy Congregation of the Index would have preferred to put *De revolutionibus* on their Index of prohibited books were it not for the fact that calendar reform and the date of Easter were dependent upon it. So it was decided to sanitize and edit it rather to proscribe it. Cardinal Bonifacio Caetani suggested to Pope Paul V that the Copernican model was false and opposed to Scripture, but it was not heretical. It would be heretical if it was believed to be true. Thus, the sanitization instructions were as follows: "If certain of Copernicus' passages on the motion of the Earth are not hypothetical, make them hypothetical; then they will not be against either the truth or the Holy Writ. On the contrary, in a certain sense they will be in agreement with them, on account of the false nature of suppositions, which the study of astronomy is accustomed to use as its special right."[7]

While Galileo was in Rome attempting to influence the Catholic leaders to accept his teachings, he was called before Cardinal Bellarmino and was cautioned against any further proselytizing on behalf of the Copernican system. In modern vernacular, he was told to "cool it." Galileo accepted and complied with Bellarmino's warning for seven years as he lived in Florence. After that meeting, however, rumors began to circulate in Rome that Galileo had been forbidden to teach the Copernican doctrine. Galileo was disturbed by the rumors, so he requested clarification from Bellarmino, who wrote him the following response.

> We [*sic*], Roberto Cardinal Bellarmino, having heard that it is calumniously reported that Signor Galileo Galilei has in our hand abjured and has also been punished . . . declare that the said Signor Galileo has not abjured . . . any opinion or doctrine held by him; neither has any salutary penance been imposed on him; but that only the declaration made by the Holy Father and published by the Sacred Congregation of the Index has been notified to him, wherein it is set forth that the doctrine attributed to Copernicus . . . is contrary to the Holy Scriptures and therefore cannot be defended or held.[8]

The injunction did not stop Galileo from his research. In 1618, he published his book on comets called *Il Saggiatore* (*The Assayer*). He avoided discussing the Copernican system, but had so much to say about the nature of science that *Il Saggiatore* has been considered to be his scientific manifesto.

> Philosophy is written in this grand book, the universe, which stands continually open to our gaze. But the book cannot be understood unless one

first learns to comprehend the language and read the letters in which it is composed. It is written in the language of mathematics, and its characters are triangles, circles and other geometric figures. . . . Without these one wanders about in a dark labyrinth.[9]

As *Il Saggiatore* was about to go to press, a new, more liberal pope was elected. Maffeo Barberini took the papal name Urban VIII. He was one of the cardinals who had argued against proscription of *De revolutionibus*. He was, along with Galileo, a member of one of the earliest scientific societies, the Academy of Lynxes. The academy rushed to change the title page on *Il Saggiatore* so that it could be dedicated to Pope Urban VIII. Within a year, Galileo went to Rome for a series of audiences with the new pontiff. Urban was pleased with *Il Saggiatore*, and granted Galileo the freedom in his next book to discuss the Copernican system.

Galileo wrote the *Dialogue on the Two Chief Systems of the World* using three fictitious speakers. One was a sixth-century commentator on Aristotle named Simplicio. Salviati was the one who usually spoke for Galileo. Sagredo was a freethinker who asked intelligent questions and usually was convinced by Salviati's reasoning. The *Dialogue* presented three major lines of evidence that supported the Copernican doctrine. One was the phases of Venus, a second was periodicity of the planetary arrangements, and the third was the existence of tides. Apparently, Galileo had been warned by Urban that God could cause the tides by any way he chose, and not necessarily by moving Earth. This inference is made because of Simplicio's words in the *Dialogue*:

I confess that your hypothesis on the flux and reflux of the sea is far more ingenious than any of those I have ever heard; still I esteem it neither true nor conclusive, but keeping always in mind a most solid doctrine I once received from a most eminent person. I know if you were asked whether God in his infinite power and wisdom might confer upon the element of water the reciprocal motion in any other way, both of you would answer that he could, and in many ways, some beyond the reach of our intellect.[10]

As soon as the book was published, the conservatives were angered and quickly tattled to the pope that the book heavily favored the Copernican system, and that the pope had been made to seem foolish by having his position argued by Simplicio, whose name was a derivative of "simpleton." Urban concurred that Galileo had overstepped his bounds. The Inquisition was activated.

The prosecution of Galileo was hindered by two factors. First, the *Dialogue* had received a license from the censors. Secondly, the Copernican doctrine had never been publicly declared heretical. But then the Inquisitors produced a report of the 1616 meeting between Galileo and Cardinal Bellarmino. According to this report, an official injunction had been served on Galileo, and

Galileo had promised that he would henceforth neither teach nor defend the Copernican doctrine. When the pope learned of this, he was furious. It seemed to him that not only had Galileo made Urban a fool in his *Dialogue*, but he had also deceived him about the results of the 1616 meeting with Bellarmino. Galileo was ordered to immediately come to Rome even though it was winter and he was almost seventy years old. There he was accused of disobedience before a tribunal of ten cardinals. The legality of the report from the Vatican Archives was, however, questionable because it was neither signed nor notarized. Bellarmino had died, so the authenticity of the report could not be verified. The Inquisitors tried to get Galileo to admit that he had been served with an injunction. But instead Galileo produced his copy of Bellarmino's letter exonerating him from such accusations. The Inquisition, it seemed, had no airtight case against Galileo, so the cardinals adjourned. To avoid embarrassment to the pope for having brought him to Rome for nothing, Galileo agreed to settle the matter out of court. He would confess that he had been too zealous in his support of the Copernican doctrine, he would repent, and be sent home and prohibited from writing about cosmology.

But Galileo was not to get off that easy. On June 16, 1633, he learned that the agreement had been overturned and the following mandate was written into the Book of Decrees: "Galileo Galilei . . . is to be interrogated concerning the accusation, even threatened with torture, and if he sustains it, proceeding to an abjuration of the vehement [suspicion of heresy] before the full Congregation of the Holy Office, sentenced to imprisonment. . . ."[11] He was forbidden to write anything more about the mobility of Earth, and his *Dialogue* was banned. Galileo's recantation is recorded in the Book of Decrees as follows: "I do not hold and have not held this opinion of Copernicus since the command was intimated to me that I must abandon it."[12] The document was signed in Galileo's hand. Galileo returned to his house at Arcetri, near Florence. He lived there under house arrest until his death in 1642. Partly as a result of Galileo's persecution, the center of progressive science moved northward to the Protestant countries, especially to England and the Netherlands. The *Dialogue* remained on the Index of Prohibited Books until 1835.

THE LYSENKO AFFAIR

Galileo was silenced by the authority of the Church. Secular authority has similarly been used to repress science. The prime example of political repression in science occurred during the 1940s in Russia.[13] For hundreds of years Russia was ruled by all-powerful czars who kept that country isolated from the progress being made in western Europe and the United States. These isolationist policies kept agriculture and industry in a primitive state. The peasant farmers were poor, uneducated, and still using the hand tools of their ancestors.

The philosophical foundations of communism were established in 1867 in a book by Karl Marx (1818–1883) called *Das Kapital*. One of the basic tenets of communism is the concept of *dialectical materialism*. This concept is based on the historical dialectic, modified from the philosopher Hegel, who proposed that one set of ideas tends to call into being an opposite set, and that the two then interact to form a new synthesis. This, in its turn, forms the thesis for a new antithesis; synthesis occurs again and so the historical dialectic unfolds. Moreover, dialectical materialism assumes that everything that exists is material (no spirits, gods, vitalistic forces) and matter is eternal. Nature is in a constant state of motion, change, and development, but material change is historical; matter can be understood only in terms of its history. Matter consists of opposing elements whose interaction is the cause of change. The duration of relatively unchanging elements (such as the gene) is therefore undialectical. Man, knowing the laws of nature, can influence nature and alter it for his own benefit. The reader will notice how close many of the tenets of dialectical materialism are to those of modern science. Marxism has attracted the minds of many scientists, both in and out of Russia, because the democratic form of communism he proposed was to be based on exploitation of the scientific method.

Russia fought Germany in World War I and suffered greatly. Due largely to a shortage of food and coal, a revolution overthrew Nicholas II in March 1917, and thus put an end to czarist rule. In October of 1917 a second revolt overthrew the provisional government. The Bolsheviks (later called the *Communists*) seized the government and set up a dictatorship under Vladimir Ilyich Ulyanov (Lenin). The Bolsheviks organized the Red Army. Anti-Communists were called "Whites." The struggling country was further impoverished by civil war between Communist and anti-Communist factions from 1918 to 1920. The Red Army was eventually victorious. A thousand years ago, Russia was a small region in Europe. Under the czars it gradually grew by conquering and adding territory on all sides. The Union of Soviet Socialist Republics (USSR) was established in 1922. Now, under Lenin, the new Russia was once again ready to expand its territory and defeat the anti-Communistic "Western imperialists," if not by inciting revolution, then economically, politically, and scientifically in what came (in the late 1940s) to be called the "Cold War." Thereafter, Russia was cut off from contacting non-Communist countries by an "Iron Curtain." It had no need for foreign influences. Russia planned to go it alone and would eventually triumph over bourgeois capitalism. Since its inception, the USSR was completely controlled by a single political party—the Communists. No other political party was permitted.

Joseph Vissarionovich Stalin became general secretary of the Communist Party in 1922, about the time that Lenin became seriously ill. Lenin died in

1924. Meanwhile, Stalin's power in the Communist Party had grown considerably and one by one he defeated his rivals. In December 1927 Stalin won a sweeping victory and instituted full socialism. Stalin did not share the liberal views of his predecessors. He was mainly interested in securing his own political position and concentrating ever more power to that end. A personality cult developed around Stalin by like-minded bureaucrats who were interested in their own self-preservation. By faithfully serving Stalin, they hoped to secure and perhaps advance their own welfare. Stalin did not tolerate views contrary to his own. Argumentation was viewed as "reactionary" and was suppressed. With this cadre of "yes-men" at Stalin's command, the government began taking over private businesses, farms, factories, and other means of production. Small peasant farms were combined into large collective farms. The collective farmers were forced to contribute most of their products to the state so it could feed the growing numbers of people working in the factories of major industrial centers. The peasants rebelled against being forced to join the collectives and destroyed much of their crops and livestock in protest. In reprisal, Stalin sent several million peasants to prison labor camps during the 1930s. Stalin instituted a program of terror, called the Great Purge, against any who opposed his policies. Several million more people were either imprisoned at labor camps or shot by his secret police. Fear spread throughout the USSR as neighbor spied upon neighbor. By 1939 Stalin controlled everything that was published, taught in the schools, or publicly spoken. Nevertheless, Russia's economy expanded rapidly under Stalin's iron fist, and the USSR began to emerge as a powerful nation. In 1941 Russia was invaded by the Germans. Germany surrendered to the Allies on May 7, 1945. But Russia had been devastated by the war from without, and by Stalin's purge from within. It was against this historical backdrop that the Lysenko affair was allowed to develop.

Scientific "truth" is born from argumentation (dialectics). However, in the environment of the massive repressions of the 1930s and the spy hunts that rooted out all "enemies of the people" (read that, "those who dared to disagree with Stalin"), any scientific debate tended to be viewed as a struggle with political overtones. Those who wanted to win such arguments by any means could simply accuse the opposition of being idealistic (antipractical), anti-Marxist, or of practicing "bourgeois" science (i.e., of an upper class or capitalistic nature). The Communist state, of course, was originally designed to be a classless society of workers and peasants. Everyone was supposed to work for the good of the Soviet Union, and the Union in turn would take care of its people; from each according to his abilities, to each according to his needs. The real "enemies of the people" were actually those who used Stalin's power to repress independent thinking. Scientific debates were not the only ones that suffered under Stalin. Progressive ideas in the fields of economics, education, philosophy, history, jurisprudence, literature, and technology were declared to be the work of saboteurs and "wreckers of the system."

The first scientific theory concerning inheritance and evolution was proposed by the French naturalist Jean Baptiste Pierre Antoine de Monet, Chevalier de Lamarck (1744-1829) early in the nineteenth century. Lamarck suggested that specific traits could be acquired by exposure of an organism to a particular environment, and that such "acquired characteristics" could be passed on to its offspring (inherited, hereditary). Lamarck's concept of the phenomenon included traits that become heritable as a result of needs, effort, or experience. For example, the modern giraffe's long neck evolved because their ancestors had a need for longer necks to reach food in the tops of trees and they stretched their necks to reach such food. These needs and/or activities of ancestors changed their heredity slightly so that their offspring inherited what we would today call a "genetic constitution" or genotype that gave them slightly longer necks than they would have had otherwise. Generation after generation of such needs and activities of ancestors gradually accumulated these heritable changes so that today all giraffes have much longer necks than their ancient ancestors. Most modern textbooks of genetics either fail to mention the theory of inheritance of acquired characteristics (IAC) or state that the theory has no scientific support. In reality, there are several well-documented instances of this phenomenon with at least four different mechanisms responsible.[14] Examples of IAC gained as a result of effort or needs (like the evolution of the giraffe's long neck), however, have not yet been reported in the scientific literature.

A belief in this idea of the inheritance of acquired characteristics is often called *Lamarckism*. About sixty years later, in 1859, Charles Darwin published his "natural selection" theory of evolution. Although he disavowed belief in the inheritance of acquired characteristics, some of his writings are unmistakably Lamarckian. Gregor Mendel published his famous work on the inheritance of traits in crosses between different varieties of peas in 1866, but it was ignored and/or misunderstood until 1900. Then Mendel's work was rediscovered and with it the idea that at least some traits were individually controlled by single hereditary elements (later to be called *genes*). The argument for the existence of genes was a statistical one based upon the results of controlled breeding experiments. Beginning about 1880, studies of the cell revealed that it had a membrane-bound nucleus with a collection of chromosomes, the number of which is typical for each species. The chromosomes were discovered to replicate themselves in preparation for each cell division. Sex cells or gametes were observed to undergo a special kind of reductional division by a process called *meiosis*. Gametes contain only half the number of chromosomes found in all other cells of the body (somatic cells). When an egg and a sperm unite in fertilization, a single diploid cell (the zygote) is produced. Thereafter, a different process called *mitosis* ensures that all somatic cells of the embryo and adult normally have the same chromosomal composition as was present in the zygote.

Soon after the rediscovery of Mendel's laws of heredity, it was proposed that the hereditary factors (genes) were on these chromosomes. Some mechanism had to exist that would regularly separate each paternally derived gene from its maternally derived counterpart in the formation of gametes. Meiosis was hypothesized to perform that task, and the "chromosome theory of heredity" soon became a widely tested and verified scientific proposition, at least in the Western world.

In the USSR, a modern champion of Lamarckism arose with the name Ivan Vladimirovich Michurin (1855–1935). He was a railroad man turned nurseryman, with no scientific training. About 1900 he revived Lamarck's idea that an appropriate environment could produce specific heritable changes in plants, especially if their heredity had been "shattered" by hybridization. He coupled the Lamarckian concept to that of dialectical materialism, and about 1915 began to attack Mendelian genetics because its hereditary factors lacked the characteristics of compromisable opposites demanded by dialectical materialism. Michurin put forth his "mentor theory," claiming that plants to be hybridized must first be grafted together and the stock (root and stem) will then train the scion (the foreign shoot or twig) to accept the pollen of the stock plant in subsequent attempts at cross-fertilization. He became as famous for his plant-breeding skills in the Soviet Union as Luther Burbank did in the United States, but this was a case of parallel and independent development of ideas.

Despite the fact that some very progressive work had been done in the young Soviet science of genetics by Agol, Levit, Filipchenko, Serebrovsky, Zavadovsky, and others, a sharp dispute arose during 1929 to 1932 regarding the inheritance of acquired characteristics and the reality of the gene as a hereditary unit. The bulk of experimental evidence supporting the gene concept at that time had come from abroad. It was during this period that Stalin himself coined and applied the absurd term "menshevizing idealism" to the genetics school of philosophy. Political pressure soon resulted in the removal of the majority of geneticists from the Natural Sciences Section of the Communist Academy. Most of these deposed scientists did not suffer personally, but such was not the case for a founder of modern population genetics, S. S. Chetverikov. He was exiled from Moscow to Syerdlovsk and then to Vadimir, and for many years was unable to continue his genetic studies. He died in obscurity, but today is recognized worldwide for his pioneering work.[15]

A partnership was established in 1933–1934 between two men who were soon to dominate Soviet biology and agriculture. One of these men was Trofim Denisovich Lysenko. He was born in 1898 into a peasant family and received a practical education in horticultural schools. Upon graduation from college, he was given a job at an agricultural experiment station. He first rose to prominence in August 1927, when his agronomic work with peas was published in the popular press *Pravda* by the well-known journalist V. Fedorovich. Lysenko had claimed success in winter planting of peas to precede the cotton crop in the Azerbaijani Republic. It seems likely from a practical standpoint that, under mild winter conditions, the green cover of the fields permitted their utilization as pasture even in winter. One wonders why he didn't publish his "research" in a refereed scientific journal. The Russian Zhores Medvedev, author of *The Rise and Fall of T. D. Lysenko,* does not elaborate on this, but I think that it might be because Lysenko's work had been rejected by reputable scientific journals. So he decided to bypass the normal process of science and go to the popular press. (This perversion of scientific methodology is still being used today even in the United States, as witnessed by the cold-fusion fiasco described elsewhere in this book.) Fedorovich describes Lysenko:

> If one is to judge a man by first impression, Lysenko gives one the feeling of a toothache; God gave him health, he has a dejected mien. Stingy of words and insignificant of face is he; all one remembers is his sullen look creeping along the earth as if, at very least, he were ready to do someone in. Only once did this barefoot scientist let a smile pass, and that was at mention of Poltava cherry dumplings with sugar and sour cream. . . . The barefoot Professor [*sic*] Lysenko now has followers, pupils, an experimental field. He is visited in the winter by agronomic luminaries who stand before the green fields of the experiment station, gratefully shaking his hand. . . .[16]

The major crop that sustains Russia is wheat. Wheats are classified into two groups (winter and spring) based upon the normal time of planting. Winter wheats require a longer growing season and are usually planted in autumn. They germinate, establish roots, and begin to tiller (develop new shoots from the base of the original plant) before the freezing temperatures of winter arrive. These plants are remarkably winter-hardy and go into a kind of suspended animation until spring. As soon as the snow melts, the plants begin to grow again; they usually mature and are ready for harvest in early summer before soil moisture becomes critical. In the United States, winter wheat is well adapted to the southern part of the wheat belt from Texas to South Dakota. Spring wheats are usually grown north of this zone (as far north as the Arctic Circle) because they are better adapted to a short growing season. They can yield an adequate crop in a period as short as ninety days because of the longer periods of daylight in these more northerly regions during the summer months. Winter wheats are not normally hardy enough for these harsh regions and the autumn weather would not be long enough to give them a good start before freezing temperatures return. Winter killing would be excessive and crop yields would be lower than those of spring wheats planted as soon as the ground has thawed. As normally planted throughout the world, however, winter wheats give higher yields than spring wheats. It would be desirable, therefore, to combine the high yields of winter wheat with the rapid growth and maturity characteristics of spring wheat.

During 1926 to 1928 Lysenko found that he could treat winter forms of wheat so they could be sown in the spring instead of autumn. He claimed that such treated wheat would grow faster to heading than spring wheat varieties sown in the spring. The first of his experiments was published in 1928 as a monograph of the Azerbaijan Experiment Station. Lysenko presented a paper on this research at the All-Union Congress of Genetics, Selection, Plant and Animal Breeding held in Leningrad in January 1929. The chairman of that meeting was the renowned Soviet geneticist and plant breeder Nikolay Ivanovich Vavilov. Lysenko's paper presented nothing original. "Only the subsequent (and incorrect) interpretation of his experiments and the term 'vernalization' were original."[17] *Vernalization*[18] was Lysenko's term for the process of soaking seed in water to initiate germination, and then arresting further development by refrigeration or burying it in a snowbank until spring, when it is planted. The germinated seed, according to Lysenko, has a head start on untreated seed, and starts growing right away. It reaches maturity and sets seed before soil moisture is depleted, and this gives better yields than untreated spring wheat. In small experimental plots, where hand labor is not a limiting factor, the process might be successful. But it could not be expected to work in large scale farming operations. Vavilov and the other more practical agronomists at that meeting realized this and told Lysenko so. His failure to impress the scientists at that meeting taught Lysenko a lesson. If those high-and-mighty scientists didn't appreciate his work, he would henceforth avoid scientific channels and tell those who mattered—the politicians. His road to political power was through the popular press. That same year (1929) Lysenko began work at the Odessa Institute of Genetics and Breeding where a special department of vernalization had been established.

The second member of the partnership was I. I. Prezent. He was a lawyer by education, but considered himself to be a specialist theoretician on Darwinism and the teaching of natural sciences in secondary schools. He gave Lysenko's theories the persuasive voice they needed to win arguments. His slanderous campaign against B. E. Raikov (an outstanding Darwinian scholar) resulted in the arrest of Raikov and many of his students. Prezent's expertise at scientific-political demagoguery was just what Lysenko needed to launch his campaign to eliminate his detractors. Together, Lysenko's quack theories and Prezent's rhetoric in presenting those theories in ways designed to emasculate opposition made a team that was to eventually destroy the science of genetics in the USSR.

In a speech delivered at the Second All-Union Congress of Shock Collective Farmers in 1935, Lysenko described the vernalization debate:

> In fact, comrades, while vernalization created by Soviet reality could in a relatively short period of some four to five years become a whole branch of science, could fight off all the attacks of the class enemy (and there were more than a few), there still is much to do. Comrades, kulak-wreckers occur not only in your collective farm life. You know them very well. But they are no

less dangerous, no less sworn enemies also in science. No little blood was spilled in the defense of vernalization in the various debates with some so-called scientists, in the struggle for its establishment; not a few blows had to be borne in practice. Tell me, comrades, was there not a class struggle on the vernalization front? In the collective farms there were kulaks and their abettors who kept whispering (and they were not the only ones, every class enemy did) into the peasant's ears: "Don't soak the seeds. It will ruin them." This is the way it was, such were the whispers, such were the kulak and saboteur deceptions, when, instead of helping the collective farmers, they did their destructive business, both in the scientific world and out of it; a class enemy is always an enemy whether he is a scientist or not.[19]

Stalin was at that meeting, and at the end of Lysenko's address he exclaimed, "Bravo, comrade Lysenko, bravo!" Thus, by transforming the debate on ver-nalization into a struggle with class enemies, Lysenko had touched a raw nerve that immediately won Stalin's approval, and whatever Stalin had approved was bound to be approved by his flunky politician "yes-men."

Vavilov was the most eminent plant breeder of his day in the USSR. He was president of the Lenin All-Union Academy of Agricultural Sciences (LAAAS) and of the All-Union Institute of Plant Breeding (AIPB). He was also a member of the government's Central Executive Committee (CEC). On Jan-uary 29, 1931, an article by A. Kol appeared in the newspaper *Ekonomichas-kaya Zhizn* that was the first attack on Vavilov. "Under the cover of Lenin's name a thoroughly reactionary institution, having no relation to Lenin's thoughts or intents, but rather alien in class and inimical to them, has become established and is gaining a monopoly in our agricultural science. It is the Plant Breeding Institute of the Lenin Academy of Agricultural Sci-ences."[20] Vavilov's reply was printed in that same newspaper several months later, "but it was accompanied by unfavorable comments and a hint that pure science served Vavilov as a cover-up for sabotage."[21]

On August 3, 1931, a government decree was published demanding that the LAAAS and the AIPB develop within four years new higher-yielding cereal varieties, each suitably adapted to the regions in which they would be grown. Vavilov had projected that it would take ten to twelve years or per-haps even longer to develop these new varieties. The government's idea was to use hothouses to speed up the process. Vavilov knew that this would not be practical for two major reasons. One reason is that the number of plants that can be grown in hothouses is minuscule compared to the number that can be grown in the fields. Plants bearing more desirable traits (such as higher yield; uniformity; nonlodging; nonshattering; resistance to cold, drought, pests, and diseases; superior baking qualities, etc.) are usually rare. Over a million plants might have to be screened in order to find one that is superior with regard to a single one of these desired traits. Artificial selec-tion would thus be severely restricted by the hothouse method. The second reason that this method is impractical is that the environment of the hot-house does not replicate the natural environment in which the plants are to be cultivated by farmers. Thus selection would again be ineffective in finding the kinds of traits desired. One doesn't have to be much of a biologist to understand these limitations. Lysenko immediately published a claim that he could develop new varieties in 2.5 years.

In 1935 the Odessa Plant Breeding-Genetics Institute, where Lysenko was employed, became incorporated into the LAAAS. Lysenko applied relentless political pressure that forced the dismissal of A. A. Sapegin, the institute's director, and then established himself as its leader. The reports of the effectiveness of vernalization and other Lysenko "innovations" under collective farming were, of course, thereafter inflated by the institute to reflect the progress that Lysenko had promised. These reports were published in a new journal called *Vernalization*, edited by Lysenko and Prezent.

At a special session of the LAAAS on December 19–27, 1936, "Soviet biological and agronomic sciences began their divided course, heading in two theoretically contrasting directions."[22] Lysenko felt that the mantle of Michurin (who died in 1935) had fallen upon him. The Michurinists naturally gravitated to the Lysenko camp, if they had not already done so. Lysenko and Prezent rejected the chromosome theory of heredity. Their idea was that heredity is a general property of a cell, and cannot be compartmentalized into discrete units such as genes or chromosomes. Hereditary elements were just figments of imagination in the minds of formalistic, idealistic, bourgeois, and metaphysical geneticists. Those who had contributed so much to the development of a particulate concept of genetics became denigrated by Lysenko's terms (Mendelism, Weismannism, Morganism).[23] Genes were considered to be very stable elements (Lysenko considered them to be immutable), and they did not adaptively change in response to specific environmental factors. Thus the gene concept did not accommodate the theories of either Lamarck, Michurin, or Lysenko. To them, heredity was an integral property of living matter and was capable of being altered to some extent by its surroundings, providing the proper environment would allow the plant to assimilate there from an adaptive hereditary constitution. For Lysenko, heredity could no more be reduced to elementary units than could temperature. The occasional mutations of genes was explained by Lysenko in another analogy. When we eat too much, we belch. When plants assimilate too much from their environment, they change. Mutations are just nature's belching. The gene concept could never fit into Lysenko's theory because gene mutations are of an indeterminate, undirected nature, not necessarily adaptive. Lysenko reasoned that if he could change winter wheat into spring wheat or vice versa by manipulation of its environment then there must not be any immutable genes for seasonal habit. If there were no genes for seasonal habit, then there were no genes of any kind. This illogical argument was supposedly supported by one of his hothouse experiments involving a single plant that could have been a hybrid, a contaminant, or a mutant. The grains that farmers planted in those days were so highly contaminated with seeds of weeds and other plants that it was commonly believed that wheat could be changed into rye or barley or some other species. With that mentality rampant in his country, Lysenko didn't have much trouble selling most folks on his crackpot ideas. Prezent claimed that Lysenko's hothouse plant was a masterpiece of Soviet biology. To him it indicated that Lysenko's methods could revolutionize agricultural practice, and that it could be done quickly and inexpensively. Russia always had trouble feeding its people and the problem was intensifying. Lysenko had "proven" his claim to be able to rapidly create new cereal varieties, whereas all the geneticists could provide was hypothetical results in ten to twelve years or so. Lysenko's methods bore practical results; geneticists just publish papers and spew hot air. This kind of rhetoric usually won over arguments based on scientific facts.

Most of Lysenko's ideas were just plain stupid. Perhaps his dumbest one was his assertion that pure-line varieties of a self-fertilizing plant could be rejuvenated by intravarietal crossing. There is no genetic variation in a pure line. When selfed, they always (barring mutation) produce progeny that are genetically identical to each other and to the parent plants. Any differences among members of a pure line are of a nongenetic (environmental) nature. It had been known for at least twenty years that selection within a pure line is ineffectual, but that crosses between genetically different lines or varieties of a species often produce progeny exhibiting hybrid vigor. Lysenko apparently thought that hybrid vigor could also be obtained by artificially cross-fertilizing members of a pure line that normally fertilize themselves. Since he was convinced that useful plant traits are not hereditarily transmitted, but arise anew every generation in response to their environment, there was no need to introduce living material from other varieties or species.

Vavilov, on the other hand, had realized the importance of genetic variability for effective selection of useful traits. He was keenly interested in the wild relatives of domesticated crops. These "weedy" cousins contained genes that made them more resistant to diseases or pests, and allowed them to grow under much harsher conditions than those to which cultivated crops are subjected. It was that "wildness" that he hoped to hybridize into crops with high-yielding potential so that they could better tolerate the environmental challenges of the regions in which they would be grown. In the mid-1920s, Vavilov had traveled to all the major agricultural regions of his country and made many journeys to the cradles of civilization (where agricultural crops first were developed) in foreign lands in search of new germplasm from which he one day planned to sculpt new plant varieties to benefit Soviet agriculture. Genetics in the USSR was on its way to making significant improvements in agriculture when Lysenko came on the scene.

At the Sixth International Congress held in the United States in 1932, Vavilov had announced that the Seventh Congress was to be held in Moscow in August 1937. But a vigorous campaign against Vavilov and other Russian geneticists began in 1937, using the popular press and the Lysenko-edited journal *Vernalization*. Through bald-faced lies, Soviet genetics was linked with the aims of fascist racists, especially those in Germany. The antigenetics campaign became so intense that it could not pass unnoticed by the international genetics community. About three months before the scheduled Moscow Congress was to convene, the Soviet organizing committee announced that the Congress was postponed until August 1938. Within a year Meister, Levit, and Gorbunov, all members of the organizing committee, as well as its president, Muralov, had been arrested on trumped-up charges. Lysenko became the new president of the LAAAS. The Seventh International Genetics Congress was finally held in Edinburgh, Scotland, in 1939, but Soviet scientists were conspicuously absent. Vavilov had been elected president of the Genetics Congress, but his government denied him permission to attend it.

The AIPB, directed by Vavilov, was accused by members of the Commissariat of Agriculture of giving refuge to people who cannot be politically trusted. Over Vavilov's objections, Lysenko appointed an undercover member of the NKVD (Soviet Secret Police), S. N. Shundenko, as deputy director under Vavilov. Shundenko was to spy on Vavilov and report directly to Lysenko. In March 1939 a meeting was held at the AIPB, at which Vavilov directly confronted some of his Lysenkoite adversaries with these words:

> . . . this is a long-term operation, especially in our field of plant breeding . . . there are two positions, that of the Odessa Institute and that of the AIPB. It should be noted that the AIPB position is also that of contemporary world science, and was without doubt developed not by fascists, but by ordinary progressive toilers. We shall go to the pyre, we shall burn, but we shall not retreat from our convictions. . . . This is a fact, and to retreat from it simply because some occupying high posts desire it, is impossible. . . . The situation is such that whatever foreign book you pick up, it goes contrary to the teachings of the Odessa Institute. Would you order that these books be burned? We shall not stand for this.[24]

Vavilov's report of the March meeting at the AIPB was received on May 25, 1939, by the LAAAS presidium, chaired by Lysenko. Among the discourse of that group, the following words of Lysenko were of greatest significance: (speaking to Vavilov) ". . . your [act of] being subordinate toward me—and this means AIPB is being insubordinate to me . . . I say now that some kind of measures must be taken."[25] These "measures" turned out to involve the intervention of the NKVD. On August 6, 1940, Vavilov was arrested while on an expedition in search of new germplasm in the foothills near Chernovitsky. Later that night, "three men in civilian clothes" (undoubtedly members of the NKVD) came to the room where he was lodging to get his belongings. In his bag they found a sheaf of spelt, a half-wild local type of wheat. It later proved to be a new species. This was to be Vavilov's last botanical-geographic discovery.

Vavilov's trial took place on July 9, 1941. He was found guilty of belonging to a rightist conspiracy, spying for England, sabotage in agriculture, and other charges, all of which were denied by Vavilov. He was sentenced to death, but the order was not executed immediately, and after several months it was commuted to a ten-year imprisonment. Moscow prisoners were evacuated to the interior during October 1941 because of the advance of the German army. Vavilov was moved to a prison in Saratov and placed in a windowless, underground death cell. He was denied outdoor exercise and received a starvation ration. He spent several months under these conditions until the summer of 1942. After the commutation of his sentence, he was moved to a general cell block in a weakened, malnutritional state. He introduced himself to his fellow prisoners thusly: "You see before you, talking of the past, the Academician Vavilov, but now according to the opinion of the investigators, nothing but dung."[26] He died on January 26, 1943, from (according to the death certificate) pneumonia. Vavilov's wife and son had been evacuated from Leningrad in 1942 and were living in Saratov only two or three kilometers from the prison in which he was dying. His family was never informed of this fact. Instead, the NKVD lied that he was in Moscow. To this date, no one knows where Vavilov was buried.

The entire antigenetics program in Russia had been based on lies, misinformation, false accusations, subversion, and intimidation. The Soviet public was purposefully kept unaware of the reaction of foreign scientists to the arrest and death of Vavilov because such information was not allowed to be printed in the Soviet press and such articles were cut from foreign journals on their arrival.

By the end of World War II, it had become obvious that Lysenko's highly touted agricultural methods had failed to live up to their expectations. Under Lysenko's heavy hand, there had been no elections to the LAAAS since 1935. He had kept the vacancies open, awaiting an increase in the numbers of his supporters before holding elections. Lysenko needed to insure that new members of the LAAAS were his supporters. The best way to do that was to get Stalin to make appointments from a list supplied to him by Lysenko. Lysenko needed to impress Stalin with the need to maintain his supporters in the LAAAS. So he decided to use another of his crazy analogies. Lysenko had championed the idea that intraspecific competition did not occur in nature. For example, one wheat plant does not compete with another wheat plant, but a weedy plant does compete with wheat. Bourgeois biology (practiced by capitalistic countries) held the opposite, truly

Darwinian position that competition can occur between genetically different members of the same species as well as between members of different species. According to Lysenko,

> That is why American scientists could not adopt the practice of cluster sowing. They, servants of capitalism, need not struggle with the elements, with nature; they need an invented struggle between two kinds of wheat belonging to the same species. By means of the fabricated intraspecific competition, "the eternal laws of nature," they are attempting to justify the class struggle and the oppression, by white Americans, of Negroes. How can they admit absence of competition within a species?[27]

In this way, Lysenko was able to show how his philosophy of biology was related to the irreconcilable political and social differences between communistic and capitalistic countries. Unsophisticated in science, Stalin was convinced by such rhetoric and appointed thirty-five new members to the LAAAS from Lysenko's list. Furthermore, Lysenko had prematurely drafted a report on the outcome of the 1948 session of the LAAAS to be held in August, submitted it to Stalin, and was able to obtain its prior approval by the Politburo and by Stalin personally. When Lysenko revealed that Stalin had already approved his report at the LAAAS meeting, it effectively shut off all debate. It was obvious that only complete compliance with Lysenko's philosophy would henceforth be acceptable. Those who held contrary views would be dismissed, or if they wished to remain members of the LAAAS, they would have to undergo the humiliation of public recantation. "Nearly every scientist [in the USSR] had to appear before such a commission and declare his attitude toward the new faith."[28] Within two days of Lysenko's coup d'etat, directives were issued to the universities to dismiss those who had actively fought against Michurinism (Lysenkoism), whether they be geneticists, physiologists, botanists, or other disciplinary specialists. The same order abolished courses and books of heredity based on Mendelism-Morganism. All non-Michurinist research projects were to be eliminated. All stocks of the fruit fly *Drosophila* (from which the genetics school had learned so much) were to be destroyed. All genetic literature was removed from the public libraries. Some geneticists were deprived of their academic degrees. Medical investigations of hereditary diseases were outlawed. The science of genetics in the Soviet Union had been eradicated.

From 1952 to 1964, Lysenkoism dominated Soviet biology and agronomy. Stalin died in 1953. Nikita S. Khrushchev became head of the Communist Party. The new leader favored a policy of peaceful coexistence with the West. But this policy led to a bitter dispute with China over the methods by which the goals of Communism could best be reached. Khrushchev was critical of Stalin's rule by terror, and Russian life became somewhat freer. *Sputnik I*, the first spacecraft to circle Earth, was launched by the USSR in 1957. The United States had fallen behind in this respect. All of a sudden, science education in the United States became a priority. Khrushchev became premier of the USSR in 1958. Yuri A. Gagarin, a Russian air force officer, became the first person to orbit Earth in 1961. Despite the Russian successes in space, Khrushchev was an even more ardent supporter of Lysenko than was Stalin. However, by 1963, many events both inside and outside the USSR had weakened support for Michurinist biology. One of the most important events was the work of a multifaceted Soviet commission representing basic administrative agencies such as education, agriculture, medicine, and the LAAAS. The recommendations of this committee opened the door for consideration of anti-Lysenkoite concepts, along with the continued emphasis on Michurinist biology. These compromises were accepted by the Council of Ministers and gave legal standing for a return of real genetics. The first Soviet texts in medical genetics since 1948 were published in 1963. Also during 1963, the USSR had to purchase large amounts of grain from foreign countries (Canada, Australia, the German Federal Republic, and the United States) for the first time in its history. Lysenko's agricultural policies had shamefully failed to meet the needs of his country. Nevertheless, Lysenko remained in power as long as his unswerving supporter Khrushchev held the top political post. But in 1962 Russia set up missile bases in Cuba. President John F. Kennedy ordered a naval blockade of Cuba and demanded that the Russians remove their missiles. In order to avoid war with the United States, Khrushchev complied. From that moment on, Khrushchev began loosing support from his countrymen. In 1964 the Central Executive Committee (CEC) of the Communist Party unanimously voted to relieve Khrushchev of his positions as Chairman of the Council of Ministers and First Secretary of the CEC. He was replaced by Leonid I. Brezhnev as head of the Communist Party, and Aleksei N. Kosygin as premier. Khrushchev's downfall was due not only to his bumbling foreign policies, but also to his amateurism in dealing with agriculture and industry, his petty tyranny and subjectivism, his endless reorganizations, his nepotism in ruling the country, his unconditional support of Lysenko, and many other factors.

The popular press was almost immediately filled with articles condemning Michurinist biology under Lysenko. The scientific fraud of Lysenko was revealed in various articles, one of which stated: "As may be seen from reports of accountants, Lysenko exaggerated the butterfat percentage figures by at least 0.29–0.45. More than that, compared with 1954, the milk yield per cow dropped by 2660 kilograms."[29] Curiously, neither Lysenko nor his allies, who had acted so tempestuously in the past to even indirect criticism, now remained silent. Although massive reforms were needed to bring Russian biology and agriculture up to the level of modern science elsewhere in the world, a purge of Lysenkoites was not immediately instituted, and many of them remained at their posts. Lysenko, however, was soon dismissed from his post as director of the Academy of Sciences (AS) Institute of Genetics following the annual meeting of the AS on February 1-2, 1965. A new monthly genetics journal, *Genetika*, appeared in September 1965 with P. M. Zhukovsky as editor in chief. "The hundredth anniversary of Mendel's laws of heredity could not have come at a more opportune time."[30] Real science had triumphed over pseudoscience. Lysenko remained as codirector of an experimental farm at Gorki Leninskie, but his authority was gradually eroded. He died in obscurity in 1976. The damage that he did to Russian science and agriculture has yet to be fully repaired.

Analysis

It is ironical that a country whose prosperity was planned to be derived from exploitation of science would come to symbolize the antithesis of science by destruction of an entire branch of science; viz., genetics. Religion suffered a similar fate in the USSR. The Communists looked upon religion as an anti-Communist force, calling it "the opium of the people." Under godless (atheistic) Communism, many churches were destroyed or made into museums. Church leaders who refused to follow Communism were arrested or killed.

The USSR was the first Communist nation the world had ever known. It was built with the intent of creating a "classless society." There would be no private ownership of land, industry, or financial institutions—no wealthy capitalists or poor laborers. The government took over all privately owned factories, farms, and other means of production. All Soviet citizens would serve the government and "Mother Russia" would care for her citizens. The Communist slogan was: "From each according to his ability, to each according to his needs." These ideals were eventually realized to a marked degree. Every child received an education, medical care became available to all citizens, factory productivity markedly increased (especially in the production of military equipment during World War II), and some scientific achievements (such as *Sputnik*, the first artificial satellite, launched on October 4, 1957) earned Russia international respect. Indeed, the USSR deserved this respect because its citizens are so culturally diverse, consisting of over a hundred ethnic groups identified mainly by the different languages they speak. However, the ideal of a classless society eventually faded as special rights were granted to top officials of the Communist Party and the federal government, and to some professionals (engineers, scientists, artists) who conformed to party rules. As news of the affluence of Western nations (especially the United States) filtered through the "iron curtain" of isolationism into the USSR, its citizens became progressively more discontented with the lack of affordable housing and the dearth of consumable goods (such as automobiles, washing machines, and television sets). Following the breakup of the Soviet Union, which began in 1990 when Lithuania declared its independence, the restrictions on privatization began to erode, allowing economic-class distinctions to proliferate. Even the open practice of various religious rights was tolerated in the new republics.

The federal government of the USSR had a legislative branch, called the Supreme Soviet of the USSR, consisting of two houses: the Soviet of Nationalities and the Soviet of the Union. The highest executive branch was the Council of Ministers. The constitution of the USSR gave all political power to the people through their elected representatives. However, the Soviet Union allowed only one political party—the Communist Party. Obviously dialectical materialism, one of the cornerstones of Marxist ideology, could not function in the absence of at least one other opposing political party. Both the federal government and the Communist Party were structured like a pyramid, with each level of representation responsible to the one above it. The Supreme Soviet elected its own Presidium, a thirty-three-member body that handled legislation between sessions of the Supreme Soviet. The chairman of the Presidium was considered to be the head of state, but had little real power. The Supreme Soviet also elected a Council of Ministers. The premier, or chairman of the Council of Ministers, was the political head of the government of the USSR.

The head of the Communist Party was the most powerful person in the USSR. The policy-making body of the Communist Party was an eleven-member Politburo of the Central Committee. This body met in secret and revealed only its decisions. The managerial branch of the Communist Party was an eleven-member Secretariat of the Central Committee. Some members of the Secretariat were also allowed to be members of the Politburo. The Central Committee of the Communist Party consisted of about 350 members who had the responsibility to "elect" a Politburo and a Secretariat. The Central Committee was, in turn, "elected" by an All-Union Party Congress, consisting of about five thousand delegates from lower party organizations throughout the country. In actual practice, however, the Politburo and the Secretariat selected their own members and those of the Central Committee, while the lower levels of the party simply "rubber-stamped" their approval of these selections. Thus, the entire political system was controlled by the highest levels of the Communist Party. While in power, both Joseph Stalin and Nikita Khrushchev were simultaneously head of the federal government (as premier or chairman of the Council of Ministers) and head of the Communist Party. This type of political structure contributed to the development of a dictatorship that (under Stalin's Great Purge) was responsible for the deaths of thousands of Soviet citizens. Everyone lived in fear of being accused of subversive activities and of being sentenced to labor camps in Siberia and/or to death. Science cannot function properly under such tyranny. Scientists must be free to challenge the validity of any research by their peers in the same or foreign countries. Otherwise the self-correcting attributes of the scientific method cannot function.

A similar dictatorship was established in China in 1966 when Communist Party chairman Mao Tse-tung instituted his Cultural Revolution. Scientists and intellectuals were removed from the universities and made to perform manual labor and farm work. Scientists could not publish in China and were forbidden to send their papers outside China for publication. Unlike the Russian story, however, Chinese universities were allowed to continue to receive scientific journals from other countries. So science could be read about, but not practiced during the Cultural Revolution. Only after Mao died in 1976 did the Cultural Revolution end and Chinese science begin again.

Granted, dictatorships are at one end of a spectrum of political systems and therefore are atypical of the environments in which most science is conducted throughout the world. The "democracy" of the United States is a more moderate form of government that should not be as open to the abuse of science as occurred in Communist countries. Our form of government has "checks and balances" of power built into it by our constitution. The Congress (legislative branch) has the power to enact laws and appropriate funds. The judicial branch with its Supreme Court rules on matters of constitutionality of laws enacted by the legislature and on the decisions by lower courts. The executive branch sets foreign and domestic policies and has the power to enforce the laws of the land. Each of these three branches have some powers that affect the other two. For example, the president has the power to veto bills passed by Congress and can influence the federal courts by the kinds of judges he appoints to them. There is no limit on the number of political parties in the United States. A member of one branch of its government cannot simultaneously be a member of any other branch. These conditions help assure that a dictatorship cannot develop here as long as the country remains true to its constitution.

Any form of government can affect the conduct of science and/or the teaching thereof in either positive or negative ways. Conceptually, these effects may be the result of actions or inactions taken by governmental agencies. The Lysenko affair is an obvious example of how direct action by the government of the USSR negatively affected the advancement of genetics and agriculture in that country. The conduct of science in the United States, throughout its history, has also been negatively affected by direct actions of its government. As a recent example, laws or policies were adopted in the early 1990s that prohibited federally funded research involving aborted human fetuses. This topic is discussed in section "Fetal-Cell Research" of chapter 1. Probably the greatest impediment to the advancement of science under most moderate forms of government, however, is the failure (inaction) of the government to adequately fund research. The reason is obvious. There simply isn't enough money to support all of the projects that scientists wish to pursue. Which proposals receive federal funding (and the amount thereof) in any given year may depend largely upon the value to society that politicians place on science and on their inclination to favor either basic or applied research. Unfortunately, the decision (action) to support some projects unavoidably means that other projects cannot be funded by default (inaction). It's a catch-22. These and other monetary aspects of science are also discussed in section "Financing" of chapter 7.

Science is often asked to provide solutions to various problems of major concern to society (e.g., global warming, ozone depletion, environmental impact of a new dam or pipeline, safety and effectiveness of a new insecticide, development of vaccines for emergent diseases). If a society (through its government) provides the money to support the necessary research, it usually wants a definitive answer immediately, but science can seldom work that fast. The best it can do in the short term is to provide periodic progress reports based on preliminary data. Unfortunately, the government may take the tentative inferences of these early studies as definitive conclusions and enact legislation despite the inconclusiveness of the data, as occurred in the case of saccharin's carcinogenicity studies. Both parties are forced into uncomfortable positions; scientists are required to draw premature conclusions and Congress must delay action until more definitive evidence is produced. Science and politics are indeed an "odd couple," but our standard of living has been largely dependent upon their mutual willingness to compromise their operational preferences for the good of society.

CREATION SCIENCE

The strange bedfellows of politics and religion are alive and well today under the banner of "creation science." This is a name given to an ultraconservative political movement by certain fundamentalist Protestant Christian groups to discredit the teaching of evolution in public schools and, failing that, to have their dogma given at least equal time with the teaching of evolution in public schools. These "creationists" are committed to the belief of a literal interpretation of the Bible. The world was created in seven days, Earth is only a few thousand years old, and all creatures were created essentially as we see them now. The Fundamentalists have been around for a long time, but have only recently learned how to flex their political muscle. In this century, "Christian fundamentalism has been a peculiarly American movement; nowhere in Western Europe, in fact, has creationism been an issue."[31] After World War I, these conservative Christians succeeded in establishing Prohibition (the outlawing of alcoholic beverages). Eventually, however, Prohibition was repealed. Nonetheless, that initial success spurred their attempts to prohibit the teaching of evolution in the schools. Due to their political pressure, several states passed antievolution laws. Tennessee was one of the states that knuckled under to their demands. A group of townsfolk in the little town of Dayton, Tennessee, decided to test the constitutionality of that law. In 1925, they persuaded a high school science teacher named John T. Scopes to become the guinea pig in this test case.[32] Scopes was young and unmarried at the time, and didn't have much to lose. Besides, the American Civil Liberties Union (ACLU) was picking up the tab for all court costs. The brilliant lawyer Clarence Darrow was hired to defend Scopes. Darrow consented when he learned that the prosecution was to be handled by William Jennings Bryan, thrice a presidential candidate, a renowned orator, and a pompous fundamentalist. Darrow was itching to sink his legal teeth into Bryan. Many people have been made aware of this famous case through the 1960 film *Inherit the Wind*, starring Spencer Tracy as Darrow and Frederic March as Bryan. This film was so successful that a second version was produced in 1988 starring Jason Robards as Darrow and Kirk Douglas as Bryan.

The Scopes case was called the "monkey trial" because one of the implications of the evolution theory is that modern humans descended from another primate that was a common ancestor of modern apes and humans. The little town was turned into a circus, with entrepreneurs of every description trying to make a buck out of the notoriety given the trial by the press. Caustic H. L. Mencken was the most famous reporter at the trial. His rendition of the monkey trial drama held Tennessee up to world ridicule. There were monkeys on chains begging for money for their masters. Zealots thumped their Bibles and shouted out threats of fire and brimstone. Some walked about carrying placards warning that the end of the world was near. Hucksters sold all kinds of trinkets to memorialize the event. It was a zoo! Many expert witnesses for the defense were not allowed to explain why science accepts the evolution concept. Such testimony was deemed irrelevant. A Tennessee law had been broken and Scopes was judged guilty. His sentence was a $100 fine (a lot of money in 1925), but the ruling was overturned on a technicality. Because the case was dismissed, it could not be appealed to a higher court as the organizers of the trial had wished. It was not until 1968, in the case of *Epperson* v. *Arkansas*, that the Tennessee law and similar laws in Mississippi and Arkansas were declared unconstitutional by the U.S. Supreme Court.

The results of the Scopes trial were such an embarrassment to Tennessee that they caused many other states to indefinitely postpone or dismiss their plans for similar "monkey laws." The creationists then decided to change their tactics and work at the grass-roots level of politics, gaining entrance into and putting pressure on local school boards. In this way they could screen prospective teachers, outline the curriculum to be taught, have input into the selection of textbooks, and so forth. Because of its population size, California has long been a key state in determining what gets into textbooks. Publishers of school texts are in business to make money; they couldn't care less about science per se. They have historically wavered this way or that depending upon their perceived demands of the majority of textbook purchasers. "As California goes, so goes the nation." Little wonder, then, that the headquarters of the creationist movement is located in California. Creationists have been very successful in emasculating evolution from school biology textbooks. In addition, the creationists decided to wrap themselves in a more acceptable cover, which they call "scientific creationism." This term is an oxymoron (self-contradictory), but that doesn't stop them. They have always counted on exploiting the scientific naiveté of the general public. Instead of demanding the exclusion of evolution from the biology curriculum (which their efforts had failed to accomplish in the past), they decided to fight for equal time with evolution in the classroom.

One of the most successful creation research organizations established in the United States is the Creation Research Society (CRS) in California and its current incarnation as the Institute for Creation Research (ICR). The members of CRS must have an advanced degree in some field of science and sign a statement of faith that begins as follows: "The Bible is the written word of God, and we believe it to be inspired throughout, all of its assertions are historically and scientifically true in all the original autographs. To the students of nature, this means that the account of origins in Genesis is a factual presentation of simple historical truths."[33] The statement of faith also pledges belief in God's direct creation of Earth and all things in six days, Noah's flood, Adam and Eve, sin, and salvation through Christ. Thus, no amount of scientific evidence can persuade them to deviate from their dogmatic beliefs. All attempts at logical analysis of the facts will fail to move creationists from their chosen path. I used to participate in debates with creationists, but I have given that up because it is futile to try to use logical persuasion with them. They either do not understand the nature of scientific arguments or they simply don't care. They have the TRUTH. Scientists will never have absolute proof of anything, so why should creationists change their beliefs? How can you argue with that kind of mindset? Many creationists are highly educated, some even with advanced degrees. ICR's main spokesman in public debates is Dr. Duane Gish. He has had experience in industry as a biochemist. Dr. Henry M. Morris, director of the institute, has a degree in hydraulic engineering. According to Dr. Morris,

> It is precisely because Biblical revelation is absolutely authoritative and perspicuous that the scientific facts rightly interpreted will give the same testimony as that of Scripture. There is not the slightest possibility that the *facts* [emphasis his] of science can contradict the Bible and, therefore, there is no need to fear that a truly scientific comparison of any aspect of the two models of origins can ever yield a verdict in favor of evolution.[34]

I do not mean to imply that people without an advanced degree in biology cannot understand the evolution theory, but rather that even people with great talents can close their minds to scientific arguments because of their religious teachings and beliefs. They have every right to do so in a free and democratic society, but science is not advanced by such folks. Their actions also threaten to retard the scientific education of every child in this country. Their use of the word "model" should not be confused with a scientific model because their "model," being based on the supernatural, cannot be falsified, and thus (according to philosopher Karl Popper) is outside the realm of science.

One of the things that has made the United States of America such a progressive country is its educational system. There are still many things that could stand improvement in this system, but one of the most important principles that has made it great is the separation of church and state. Public schools should be free of indoctrination into any particular religious beliefs. This country was founded to allow people to pursue their religious interests as they choose. The best way to do this is to prohibit the teaching of sectarian religious beliefs in a proselytizing manner in schools supported by public funds. Of course our children should be made aware of the importance of religion in history and in shaping the moral and philosophical attitudes of nations. They should understand the basic tenets of the great religions of the world and be aware that there are so many "minor" religions that if one searches, one is likely to find just the right belief system to suit each individual's personal needs. And failing that, one is free to start his own religion. What a country! Unfortunately, comparative religions is not a subject taught in most primary and secondary public schools. Rather than risk the accusation of proselytizing while teaching *about* religions, most school boards discourage the teaching of anything that smacks of religion. Instead, we leave it up to parents to instruct their children about religion. This is unlikely to be an unbiased education. Parents usually want their children to follow the teachings of the religion to which they themselves currently subscribe. Furthermore, many parents do not know much, if anything, about other religions and so cannot inform their children about the wealth of choices that exist. Few churches (or other religious institutions) are openminded enough to teach anything but their own system of beliefs. So most people go through life without an understanding of different religions, and thus are unable to comprehend different attitudes toward solving world problems by peoples of different nations with different moral standards and philosophical beliefs. If ever a case could be made for teaching *about* religions in the public schools, I believe that this could be the core of it.

The "creation scientists," however, want their way of thinking to be given equal treatment (at least with regard to time) along with the teaching of evolution in the biology classrooms of public schools. Now this may sound like a fair way to handle the problem, but it violates the principle of separation of church and state. What about a religion that believes the world was hatched from a giant turtle egg? Is it not entitled to "equal treatment" along with beliefs of the multitude of other religions of the world? Furthermore, biology is a scientific discipline where dogmatic beliefs have no place. Presenting religious arguments in a science class is like trying to mix oil and water. It makes no sense philosophically and is pedagogically indefensible. The proper place for a discussion of "creation science" is in a course of comparative religions, which essentially doesn't exist in the primary and secondary grades of public schools, although in August 1999 such a course was proposed by the school board in Altoona, Pennsylvania. Such courses are available in many public colleges and universities. But the "creation scientists" would not want it taught there even if such a class was offered because students would then be told that many religions allow that it is possible to believe in evolution without becoming excommunicated either officially or otherwise.

In his book, *Evolution and the Myth of Creationism,*[35] Tim Berra lists and discusses sixteen claims of the creationists that, to their minds, justify a rejection of the evolution theory. These are not the only objections they raise, but merely a sample. I shall review a few of them here.

1. Evolution violates the second law of thermodynamics

This law of physics states that complexity (order) does not arise from disorder (chaos), and entropy (disorder) is always increasing in the universe. What the creationists fail to mention, and what they are counting on their audience to be unaware of, is the fact that order can be generated from disorder if energy is added. This law certainly applies to a closed system, but Earth is not such a system; it receives energy from the Sun. Because the Sun is losing more energy than Earth is gaining, entropy is increasing in our solar system. Another way to look at the entropy of a system is by the amount of information necessary to define it. It doesn't take much information (a simple mathematical formula for example) to define a crystal of salt. That is because every part of the crystal has essentially the same structure. There isn't much entropy there. Life, as we know it today, has a lot of entropy because (among other factors) it consists of millions of species produced by the natural processes of evolution. It would require much more information to adequately define the living world than to describe a crystal.

2. The rate of decay of Earth's magnetic field indicates that Earth was created about 10,000 years ago

How many people are aware that Earth's magnetic field waxes and wanes irregularly and repeatedly throughout geological time, along with polar reversals when the north magnetic pole becomes the south pole and vice versa? Any good earth-science course would certainly address this phenomenon.

But how many people have had a "good earth-science course"? As a matter of fact, this information has become known only within the last thirty years. As molten rock solidifies, magnetic material in the rock acquires the polarity of Earth's magnetic field at that time. That polarity will not change until the rock again becomes molten. Thus, by mapping the polar reversals of sea-floor rocks, geologists came to the conclusion that new sea floor is being created and is spreading out from midoceanic ridges. Earth's crust is made up of several large plates that move slowly (perhaps only a few centimeters per year) over the face of the earth. Old sea floor sinks back into Earth's mantle at subduction zones near some of these plate boundaries. The old idea of continental drift has finally gained respectability with the development of the plate tectonics theory. But the creationists don't want to share this information because it would weaken their case for a young Earth.

3. *If gradual evolution were true, then there should be numerous transitional fossils. But (according to the creationists) there are none; therefore evolution has not occurred*

While it is true that there are not nearly as many transitional fossils (those showing characters intermediate between major groups of organisms) in the record as paleontologists would like, it is not true that there are none. Perhaps the best known of such transitional fossils is *Archaeopteryx*, a bird-like dinosaur that lived some 200 million years ago. The skeleton is definitely reptilian, with a long tail, clawed forelimbs, and teeth in its mouth. Were it not for the impressions of feathers in some of these fossils, it would be classified as a reptile. But because today we classify any feathered animal as a bird, we would have to classify *Archaeopteryx* as a bird too, albeit a very primitive one. Now, what better transitional form could one ask for? The creationists, however, refuse to accept it as a transitional form. Instead they simply state that it was a uniquely different creation that owes nothing in its ancestry to other forms of life. Thus, the observations have again been made to conform to their belief system. Other examples of transitional fossils are known. The fossil called *Eusthenopteron* exhibits several unmistakable intermediate characteristics between those of the lobe-finned fishes and primitive amphibians. There are numerous examples in the mammallike reptiles. One can almost trace the gradual changes in certain bones supporting the jaw in reptiles to the bones that now form the apparatus for sound amplification in the middle ear of mammals. Even in humans, there is a marvelous intermediate between small-brained, apelike ancestors and fully bipedal modern forms in *Australopithecus*. It is obvious that no series of intermediate fossils, regardless of how complete the record, would sway creationists from their belief in each form representing a unique creation event.

4. *Since fossils of complex organisms first occurred about 600 million years ago without any recognizable ancestors, they must have been created*

The earliest microfossils go back almost 3.5 billion years. A variety of multicellular life is found in the fossil record as old as 670 million years. This is about 80 million years before the beginning of the Cambrian period, when a multitude of shelled animals seems to burst upon the scene. This explosion of fossil forms is attributed by science to the fact that once shells had evolved, it greatly enhanced the chance of their preservation in the fossil record. Many people do not understand what special conditions must prevail in order for a dead organism to be preserved as a fossil. It certainly is not true that there were no forms of life before the Cambrian explosion. Unfortunately, most Precambrian rocks have been either destroyed or so badly deformed that it is surprising any of these earlier forms of life have been discovered. This does not fully address the creationists' argument, but the problem remains an open one for science to continue its investigations. Who knows what new discoveries may soon be found to shed more light on the controversy? Creationists don't need to search for fossils. They already have the answer.

5. *All fossils were formed at the time of the Noachian flood*

This argument continues to be offered by creationists despite the fact that there is no evidence for a worldwide flood at any time in the geological record. Why they continue to insist that one occurred in the face of so much negative evidence is a testimony to their faith. Their idea of time goes back only about 10,000 years because Archbishop Ussher, in the seventeenth century, calculated from the ages of the patriarchs in the Bible that Earth was created in 4004 B.C.E. They refuse to believe that the estimates of rocks in the billions or even millions of years made by radiometric analysis of long-lived isotopes could be true. Again if the data do not fit the model, then the data cannot be trusted. This is the exact opposite of the scientific method. In science we let the data generate the hypothesis or model, and if the hypothesis cannot accommodate the data, then the hypothesis should be rejected. Thus, creationists refuse to believe that calculations of half-lives for radioactive isotopes can be trusted to behave consistently through long periods of geological time. Notice that such beliefs strike at the heart of nuclear physics. Astronomy relies on the red-shift principle to estimate the speed at which galaxies are receding from us. This is analogous to the Doppler effect observed when a train, traveling at high speed, passes us with its horn blaring. The pitch of the horn drops abruptly as it passes us because the sound waves become longer. In a similar way, the faster a galaxy recedes, the more its light is shifted toward the red end of the spectrum. The most distant galaxies are estimated to be millions of light-years away; i.e., their light that we see today has been traveling for millions of years to reach us. How do the creationists account for this? They propose that the universe was recently created with the light from distant galaxies in the location it would have been if it had been traveling for millions of years. To those who balk at such concepts, the creationists remind us that there is no limit to what God can do. Our job is not to question how God made the universe, but simply believe that it was done according to the plan set forth in the Book of Genesis in the Bible. Thus, scientific creationism is not only an attack on biological evolution. It threatens all of science.

6. *Evolution is also a religion and requires faith*

Scientists may harbor a set of basic beliefs accepted on faith, such as the concepts that the physical universe is real, that it behaves in regular ways that can be explained by what we call natural laws, that these laws operate uniformly in all parts of the universe throughout time, etc. However, they do not have to believe that the solar system is about 4.5 billion years old. This proposition can be tested by applying various methods of dating to rocks of Earth, the Moon, and meteorites. Scientists do not have to believe that transitional forms exist in the fossil record. We can search that record in attempts to find them. Scientists do not have to believe in natural selection as an evolutionary mechanism. We can observe that light-colored moths changed over a period of about one hundred years to a predominantly black form as a consequence of sooty pollution from industrial factories and differential predation by birds. Herein lies the major difference between science and religion. Science exists *because* of the evidence, whereas religion exists upon faith apart from the evidence, or in the case of religious fundamentalism and creationism, *contrary* to the evidence.

Our children should be given a science education that stresses analytical abilities and critical thinking, not blind acceptance of religious or any other dogmas. Both the powers and limitations of science should be clearly understood by all people. It is important that the history of science should be an integral part of every science course, so that students can appreciate how science has evolved from primitive superstition, magic, witchcraft, mysticism, religion, and other metaphysical beginnings to its present state of understanding the universe. Students should be made aware that many people do not believe in even the most firmly established scientific principles, such as the theory of evolution. This does not necessarily make these nonbelievers ignorant or evil. It may simply be a matter of upbringing (education) and/or free choice. In teaching my evolution course, I have found it very useful to examine some of the so-called evidence that the creation scientists offer in support of their position. However, I do not devote half of my class time to the creationists' arguments because the bulk of them have been so thoroughly discredited by scientific research. Of course, the origin of the universe and of life itself can never be proven by scientific investigation. These topics will forever be a matter of which philosophical position one chooses to follow. Aside from their historical interests, why should any science course be obliged to give equal treatment to subjects that have long since been abandoned in favor of more holistic scientific explanations? Should astronomy courses devote equal time to astrology and the idea that the Sun goes around Earth? Should geology courses give equal emphasis to the "flat Earth theory"? Should chemistry courses discuss the "phlogiston theory"[36] on the same level as oxidation/reduction reactions? Then why should biology courses have to share equal time with the religious dogmas of "creation science" whenever evolution is taught? Yet this is precisely what the "creation science" movement wants to accomplish.

Perhaps if these "creationists" reflected more on the history of science, they might adopt a more flexible philosophy by which to run their lives. The Catholic Church had for centuries supported the notion that Earth was at the center of the universe. When Copernicus proposed in 1543 that the Sun was the center of the solar system, the Church was shocked and labeled his idea a blasphemy. Giordano Bruno, a disciple of Copernicus, was burned alive at the stake in 1600 for this and other heresies. This was about twenty years before the first settlers left Europe, to escape religious persecution, and came to America. Galileo suffered house arrest during the last eight years of his life for his support of the Copernican system. Only within the last dozen years has the Church admitted its errors in this respect. But the lesson is clear. Religious convictions may be falsely placed. Just as the Copernican revolution displaced Earth from the center of the universe, Darwin's theory of evolution removed humans from their supposed central position in the world of life. This may not be what most people would prefer to believe, but reality oftentimes goes counter to our wishes. History has shown that the best way to understand our physical universe is by scientific investigation of the facts of nature rather than by trying to force nature to conform to our preconceived (or learned) system of dogmatic beliefs.

NOTES

1. Owen Gingerich, "The Galileo Affair," *Scientific American* 247 (August 1982): 133–43.

2. James Burke, *The Day the Universe Changed* (Boston: Little, Brown and Company, 1985), 136.

3. Ibid., 135.

4. Gingerich, "The Galileo Affair," 137.

5. Ibid., 143.

6. Ibid., 138.

7. Ibid.

8. Ibid., 140.

9. Ibid.

10. Ibid., 142.

11. Ibid., 143.

12. Ibid.

13. Until recently, "Russia" and "the USSR" have generally (but improperly) been used as synonyms, and are used in this way in this report. In reality, however, Russia became an independent republic within the Soviet Union. The Union of Soviet Socialist Republics (USSR) was dissolved in 1991. At this writing, Russia remains the largest and most influential member of a loose commonwealth or confederation of independent states.

14. Otto E. Landman, "The Inheritance of Acquired Characteristics," *Annual Reviews of Genetics* 25 (1991): 1–20.

15. Zhores A. Medvedev, *The Rise and Fall of T. D. Lysenko* (New York: Doubleday & Company, 1971), 9.

16. Ibid., 11–12.

17. Ibid., 13.

18. Today, the term "vernalization" represents the process by which floral induction in some plants is promoted by exposing them to cold temperatures for a certain period of time before planting.

19. Medvedev, *The Rise and Fall of T. D. Lysenko*, 17.

20. Ibid., 18.

21. Ibid.

22. Ibid., 22.

23. Gregor Mendel (1822–1884) is known as the father of genetics. August Weismann (1834–1915) provided experimental evidence that nongenetic factors (such as mutilation of body parts) were not heritable, thus arguing against Lamarckism. The genetic studies of Thomas Hunt Morgan (1866–1945) supported the chromosome theory of heredity.

24. Medvedev, *The Rise and Fall of T. D. Lysenko*, 59.

25. Ibid., 64.

26. Ibid., 73.

27. Ibid., 107–108.

28. Ibid., 122.

29. Ibid., 231.

30. Ibid., 239.

31. Tim M. Berra, *Evolution and the Myth of Creationism* (Stanford, Calif.: Stanford University Press, 1990), 142.

32. William D. Stansfield, *The Science of Evolution* (New York: Macmillan Publishing Company, 1977), 3–6.

33. Berra, *Evolution and the Myth of Creationism*, 125. In May 2000, the ICR Web site (www.icr.org) presented a unique statement of faith for its faculty and students, excerpts from which follow here. "The administration and faculty of ICR are

committed to the tenets of both scientific creationism and Biblical creationism" as specified in their Web page. "All genuine facts of science support the Bible. Teleological considerations are appropriate in scientific studies. There are many scientific evidences for a relatively recent creation of the earth and the universe, in addition to strong scientific evidence that most of the earth's fossiliferous sedimentary rocks were formed in an even more recent global hydraulic cataclysm. All theories of origins or development which involve evolution in any form are false." Listed among the ICR resident faculty are eight Ph.D. and one M.D. professors. There are eight Ph.D, five Ed.D., and one M.S. degrees among the ICR's adjunct faculty members.

34. Ibid., 4-5.

35. Ibid., 126-32.

36. Phlogiston was a hypothetical substance that was thought to be a volatile component of all combustible substances released as flame in combustion.

CHAPTER 6

LADY LUCK

A popular misconception about scientific discoveries is that they are always made by a straightforward application of scientific methodology. A little reflection on the history of science, however, reveals that many scientific problems of long standing have failed to be solved and that many discoveries have been made by accident rather than by design. The most extensive collection of such lucky events of which I am aware is in a book entitled *Serendipity:Accidental Discoveries in Science*, by Royston M. Roberts.[1] He defines *serendipity* as "the gift of finding valuable or agreeable things not sought for" or "the faculty of making fortunate and unexpected discoveries by accident."[2] Horace Walpole coined the term "serendipity" in a letter to his friend Sir Horace Mann in 1754. Serendip (or Serendib) is an ancient name for the island nation of Ceylon, now known as Sri Lanka. Walpole had read a fairy tale about the accidental discoveries made by the three princes of Serendip, so Walpole used the term to describe some of his own accidental discoveries. The term first appeared in English dictionaries in 1974, so it has only recently come into popular use. Most of those who have benefited from serendipity have been eager to admit the role that it played in their discoveries because they know that being lucky is only part of discovery. One must also be sagacious (of sound perception, keen judgment, wise) enough to recognize how to put the accident to good use. The famous French scientist Louis Pasteur appreciated the twin factors of scientific discoveries (luck and sagacity) when he said, "In the fields of observation, chance favors only the prepared mind."[3] Paul Howard Florey, a Nobel laureate and member of the team that brought penicillin to commercial production, reiterated Pasteur's thoughts when he received the Perkin Medal (the highest honor given by the American Chemical Society):

> Significant inventions are not mere accidents. The erroneous view [that they are] is widely held, and it is one that the scientific and technical community, unfortunately, has done little to dispel. Happenstance usually plays a part, to be sure, but there is much more to invention than the popular notion of a bolt out of the blue. Knowledge in depth and in breadth are virtually prerequisites. Unless the mind is thoroughly charged beforehand, the proverbial spark of genius, if it should manifest itself, probably will find nothing to ignite.[4]

Roberts coined the word *pseudoserendipity* for cases of "accidental discoveries of ways to achieve an end sought for, in contrast to the meaning of (true) 'serendipity' which describes accidental discoveries of things not sought for." For example, Charles Goodyear had been trying without success to find a way to make rubber useful. Then one day he accidentally dropped a piece of rubber that had been mixed with sulfur onto a hot stove, and voilà!—he discovered the vulcanization process. That was pseudo-serendipity. But George deMestral was not trying to invent the Velcro fastener when he investigated why some burs had stuck so tightly to his clothing. That was serendipity and sagacity in operation.

PENICILLIN

One of the best-known examples of serendipity in science is the discovery of the antibiotic drug penicillin by the British bacteriologist Alexander Fleming. He was born in Scotland in 1881 and gained his medical education at St. Mary's Hospital (a branch of the University of London). Fleming had planned to become a surgeon, but he fell under the influence of his bacteriology teacher, Sir Almroth Wright. Following graduation, Fleming was offered a job working in Wright's laboratory where he remained for the rest of his life. He became professor of bacteriology in 1929. Fleming and Wright were sent to France during the First World War as physicians to wounded soldiers. Phenol (carbolic acid) was the first chemical antiseptic (discovered by Joseph Lister in 1871), but it could only be used topically (on the body's surface) because it was too caustic to be used systemically (inside the body). Paul Ehrlich had discovered the first specific chemotherapeutic substance (a "magic bullet") near the end of the nineteenth century. His search for such a substance was essentially a random one. After testing 606 substances, he finally found an arsenical compound that was effective in fighting the bacterium causing syphilis. Ehrlich had sent his "compound 606" (named *salvarsan*) to Wright. Wright, however, believed that stimulating the body's immunity, not drugs, was the key to prevention and recovery from microbial infections. So he passed salvarsan on to Fleming, who had such remarkable success with it that he was nicknamed "Private 606." Many chemicals were known to kill microbes, but virtually all of them were harmful (if not lethal) to animal cells. Only 606 had specificity for a particular kind of bacteria, but even it had undesirable side effects in the animal body.

Fleming caught a cold in 1922 and while examining a culture dish containing bacteria, a tear fell from his eye onto the plate. The next day he noticed a clear area of lysed (disintegrated) bacteria on the plate where his tear had landed. The tear dropping on the plate was an accident (pure serendipity), but without Fleming's keen observation and intuition, it would never have become a scientific discovery. He reasoned that tears contain an antibiotic enzyme that could lyse bacterial cells but was harmless to human cells. Wright named the enzyme *lysozyme*. Fleming had hoped that lysozyme would be an important medical tool, but it was later shown to be active only on relatively harmless bacteria. His attempts to extract the substance in pure form were unsuccessful. Nevertheless, Fleming's experience with lysozyme demonstrated that natural substances other than antibodies produced by living organisms could be active against some microbes, and this idea prepared his mind for the later discovery of penicillin.

During the summer of 1928, Fleming was working on influenza. He had looked at many petri plates (the flat covered glass dishes containing nutrient agar for growing bacteria) and had stacked them in a tub of antiseptic fluid. A few of the top plates, however, had not been touched by the fluid. For no particular reason, Fleming took a last look at a few of these old plates. He noticed that one of them had been contaminated with a mold. Around the mold was a clear area of dead bacteria. He remembered the lysozyme incident and concluded that the mold was producing something that was killing the bacteria. He wrote,

> But for the previous experience [with lysozyme], I would have thrown the plate away, as many bacteriologists must have done before. . . . It is also probable that some bacteriologists have noticed similar changes to those noticed [by me] . . . but in the absence of any interest in naturally occurring antibacterial substances, the cultures have simply been discarded. . . . Instead of casting out the contaminated culture with appropriate language, I made some investigations.[5]

Fleming identified the mold as belonging to the genus *Penicillium* (later identified as *P. notatum*) and he named the antibacterial substance *penicillin*. But would penicillin be effective on pathogenic bacteria without harming human cells; i.e., was it a true (medically useful) antibiotic? Additional research by Fleming proved that penicillin was effective on many (but not all) such bacteria in doses that were not toxic to animals or humans.

> It was this nontoxicity to leukocytes [white blood cells] that convinced me that some day it would come into its own as a therapeutic agent. . . . The crude penicillin would completely inhibit the growth of staphylococci in a dilution of up to 1 in 1000 when tested in human blood, but it had no more toxic effect on the leukocytes than the original culture medium. . . . I also injected it into animals and it had apparently no toxicity. . . . A few tentative trials [on hospital patients] gave favorable results but nothing miraculous and I was convinced that . . . it would have to be concentrated. . . . We tried to concentrate penicillin, but we discovered . . . that penicillin is easily destroyed. . . . and our relatively simple procedures were unavailing.[6]

Actually, Fleming's experiments on the antibacterial action of penicillin with blood cells in vitro found that, while it was able to kill bacteria, the penicillin itself was rapidly degraded. He leaped to the unwarranted conclusion that it would therefore also be quickly destroyed in the animal body (in vivo). Thus, he demonstrated that penicillin was not toxic to animals, but apparently these in vivo trials were not designed to test the effectiveness of the antibiotic. Fleming thought that penicillin might prove to be beneficial as a local antiseptic. He continued to work with penicillin as an aid to isolation and identification of certain kinds of bacteria and published his results in 1929. His paper had little impact at the time and it was soon forgotten. However, "The myth that Fleming abandoned penicillin after 1929 has no factual foundation. Though his paper in 1929 attracted little attention, Fleming continued to experiment with the substance, to talk about penicillin, to write papers on it, and to endeavor to interest chemists in extracting it."[7]

While it was fortunate that the plate (on which Fleming discovered the mold) had not been touched by the antiseptic solution, it was even more fortunate that there had been an exceptionally cold spell of weather at that time of year. Molds usually grow well at temperatures lower than that at which most bacteria are cultured in the laboratory. So even if a plate had become accidentally contaminated with such a mold in the past, its growth in a warm bacterial incubator would have been inhibited; after inspection of plates, they would normally soon be chemically or thermally sterilized. But this batch of plates had been neglected in the cold for several days, thus allowing the mold to grow and release its antibiotic to kill the surrounding bacteria. More serendipity!

The first of the sulfa drugs (prontosil) shown to have medical potential in killing infectious bacteria was discovered by the German bacteriologist Gerhard Domagh in 1935. Domagh demonstrated that prontosil killed streptococcus infections in mice. The sulfonamides comprise a large class of synthetic chemicals that starve bacteria to death. These sulfa drugs were widely used during World War II, but were later largely displaced by more effective natural products (called *antibiotics*) with fewer side effects, such as penicillin. We now know that penicillin kills bacteria by interfering with the development of their cell walls; animal cells do not have cell walls outside their limiting plasma membranes. Other kinds of antibiotics exploit more subtle differences between bacterial cells and animal cells (such as the nature of the cellular machinery by which proteins are made). Domagh was offered a Nobel Prize in medicine in 1939, but the Nazi regime in Germany would not allow him to accept it.

In the late 1930s a professor of pathology at Oxford by the name of Howard W. Florey began a research collaboration with the biochemist Ernst Boris Chain, a Jewish refugee from Hitler's Germany. Using much more sophisticated chemical techniques, they were able to concentrate and eventually purify enough penicillin to begin clinical trials. Penicillin originally grew best on shallow liquid culture medium as a mat of fungus on its surface. Large shallow containers (providing much surface area for fungal growth) were not common laboratory vessels in those days. A biochemist named Norman Heatley was brought into the Oxford team to try to produce sufficient penicillin for the first clinical trials. Heatley was a master of making do with what was available (which, during World War II, was very little in Britain). He initially used bedpans from the hospital, but later on had flat ceramic culture pans made that could stack on top of one another. Heatley also found a way to extract the notoriously unstable penicillin into a brown powder and preserved it by freeze-drying in a vacuum. The powder contained only about 1 percent penicillin, but it proved to be effective in a dilution of one part in a million. The first protection experiment with mice was conducted on May 25, 1940. Eight mice were infected intraperitoneally (within the body cavity) with a virulent streptococcus. An hour later, four of them were injected subcutaneously (under the skin) with penicillin; the other four remained untreated as controls. Heatley watched over the experiment until 3:45 A.M., by which time all four of the controls had died. The results of the Oxford team were published in the journal *Lancet* on August 24, 1940.

They then began treating human patients, but the drug was so scarce that they had to try to recover and reuse as much of it as they could from the patient's own urine. Penicillin would never be widely used unless they could find some way to mass-produce it. The urgency of doing this became evident during World War II when soldiers were dying from diseases and wounds. Britain was under attack and had lost much of its industrial capacity. So Florey and Heatley went to the United States to gain help in scaling up the production of penicillin. They eventually arrived at the Northern Regional Laboratory of the U.S. Department of Agriculture in Peoria, Illinois. This institution had for some time been trying to find some use for "steep" (a dark, viscous extract by-product of the corn-milling process). Adding steep to the culture medium unexpectedly increased the yield of penicillin tenfold over the previously used yeast extract medium. At the same time, the world was being scoured for new strains of penicillin-producing molds. Samples came into the Peoria laboratory from all corners of the globe. But the best strain was, unbelievably, discovered locally in Peoria by a woman named Mary Hunt. She had earned the name "Moldy Mary" because of her enthusiasm for finding new sources of penicillin. She found a moldy cantaloupe in a Peoria fruit market. This fruit was infected with a fungus later identified as *Penicillium chrysogenum*. It had a "pretty golden look" and this new strain of mold produced twice as much penicillin as any of the others. The combination of corn steep culture and the new species of mold increased production of penicillin twentyfold. The further development of a deep-tank fermentation system, in which air was bubbled through the tank and its contents constantly stirred, put penicillin production in high gear. In 1944, the use of mutagenic agents on the molds produced genetic variants (mutants) that boosted penicillin production a hundredfold.

The Nobel Prize in physiology and medicine for 1945 went to Fleming, Florey, and Chain. All three men were later knighted for their lifesaving research. Each Nobel Prize is awarded to no more than three people. Many people believe that Heatley has been the "forgotten man" of the penicillin story. In October 1990, Heatley was awarded an honorary doctorate of medicine from Oxford, the only person to receive that award in the university's eight-hundred-year history.

Fleming acknowledged the importance of serendipity thus: "The story of penicillin has a certain romance in it and helps to illustrate the amount of chance, or fortune, or fate, or destiny, call it what you will, in anybody's career."[8] Fortunate accidents, however, require an intelligence or sagacity to make them into useful discoveries.

Most people thought the penicillin story was over, but those familiar with history know that it has a record of periodically being revised and reinterpreted. So I was not surprised when a quite different story emerged in 1989. "Question: What was the first antibiotic put to clinical use, and who discovered it? If you answered penicillin and Alexander Fleming, you would be wrong on both counts. The correct answers are gramicidin and René Dubos."[9] Carol Moberg, a research associate at Rockefeller University, even called a conference in October 1989 to set the record straight. René Dubos was born in a little town outside Paris, France, but came to the United States in 1924. While visiting a fellow French scientist at Rockefeller, he happened to be seated at lunch next to the noted bacteriologist Oswald Avery. Avery told Dubos of his quest to find a microbe that could dissolve the gummy layer (consisting mainly of multiple sugar molecules; polysascharides) surrounding the cell wall of the deadly Type III pneumococcus. Animals infected with this germ cannot disrupt this capsular coating, allowing the bacteria to proliferate. Dubos reasoned that since the polysaccharide has not accumulated in nature, there must be some microbe that can digest it. He began a search, and within three years, Dubos discovered such a microbe in a New Jersey cranberry bog and extracted from it the enzyme that digests the polysaccharide. Since it only removed the pneumococcus capsule, but did not kill the bacterial cell, this enzyme did not qualify as an antibiotic. Host immune defenses could, however, dispose of the uncoated germ. It proved effective in curing pneumonia-infected mice in 1930. George B. Mackanness, president emeritus of the Squibb Institute for Medical Research, claimed that, "Dubos showed how you could look for antibacterial products in nature by defining what you are hoping to discover and devising the test systems to reveal their presence."[10]

By 1939, Dubos had discovered that the antibacterial properties of the microbe *Bacillus brevis* lay in a substance called *tyrothricin*; this substance later was shown to consist of the polypeptide tyrocidene and gramicidin. The antimicrobial action resided in the gramicidin component. According to Rollin Hotchkiss, an early associate of Dubos at Rockefeller, "It was the first antibacterial principle that resulted from a deliberate, systematic search for antagonistic principles among soil microorganisms."[11] Gramicidin was used successfully on animals to protect them from otherwise fatal doses of Gram-positive pathogenic microorganisms such as pneumococci, streptococci, and staphylococci. Unfortunately, gramicidin proved too toxic to be used systemically (throughout the body) on humans.

Sir Edward P. Abraham, a member of Florey's group that helped purify and determine the structure of penicillin, seemed to think that Florey and Chain's early penicillin work was initially of purely academic interest because he quoted Florey as saying, "I don't think the idea of helping suffering humanity ever entered our minds."[12] Norman Heatley pointed out that the rush to commercial production of penicillin was based on very slim evidence (a toxicity test in rodents and a clinical trial in six subjects, two of whom died). Had the regulations currently promulgated by the British Committee on the Safety of Medicine been in effect at the time, penicillin would undoubtedly have been rejected. Asks Heatley, "Is it too whimsical to suggest that our greatest piece of luck so far might be the fact that this worthy body did not exist in 1941?"[13]

Robert Crease, who wrote the 1989 report for the journal *Science*, seems to see British national pride at stake in the outcome of the Moberg symposium, for he states,

> Characteristically, the British speakers at the symposium came prepared with carefully crafted papers available for distribution, while the U.S. participants tended toward anecdotal speeches delivered from recently com-

posed notes—reflecting, perhaps, the national differences in the valuation of the history of science that had conspired to disguise Dubos's role in the first place.[14]

When Joshua Lederberg, an eminent bacterial geneticist, was asked at that symposium about the value of the history of science to ongoing research, he replied:

> Well, that raises the question of history itself. Science is often thought of as an automatic process, whereas in fact it profits from an immense diversity and richness of styles, personalities, approaches, and so forth. Histories such as that of the development of antibiotics—and of Dubos in particular—are valuable because they remind us of that.[15]

In recounting the "classic" story of penicillin, Crease reports that Heatley told the symposium that Fleming originally misidentified penicillin, misunderstood its properties, never succeeded in isolating it, and published nothing on it after 1931. H. T. Swan, in a letter to the editor of *Science*, took exception to the Crease report in which

> we are told that Dubos was the first to put an antibiotic to clinical use, having discovered gramicidin around 1939. This misrepresents the published facts concerning penicillin. Fleming's penicillin had been used on patients in 1930 by G. C. Paine, a former student of Fleming. Paine used crude penicillin successfully on patients by local application in the treatment of particularly severe eye infections in Sheffield, U.K.[16]

Swan also objects to the remark about Fleming having lost interest in penicillin. Penicillin needed to be purified and mass-produced before it could be of much benefit to medical practice, but these chores were beyond Fleming's abilities. Furthermore,

> Fleming did not "misunderstand" the properties of penicillin. Indeed, working as a bacteriologist but not as a chemist, he characterized them so well, even in his initial publication of 1929, that when Ernst Chain read Fleming's publication nine years later he was immediately attracted by the potential of penicillin for what was to become his subsequent research with Howard Florey.[17]

History has a way of being periodically revised, so I doubt that we have heard the last word on the penicillin story.

ANIMAL ELECTRICITY

Is there anyone who hasn't heard about Galvani and his frog's legs? In 1786 the Italian physiologist Luigi Galvani (1731–1798) observed a dissected frog leg twitch as it lay on a table close to an electrostatic generator. Although a similar observation had been made about thirty years earlier by Floriano Caldani, he did not make any subsequent experiments or propose any theory to explain his observations. But Galvani, in the spirit of scientific inquiry, followed up his initial observations with experiments to determine if the frog leg was actually responding to an external source of electricity. In one of these experiments he hung a frog leg from an iron railing by a brass hook. The leg was serendipitously set to swaying in the breeze. When the lower leg touched the railing it contracted or jumped. This had been observed about a century earlier by the Dutchman Jan Swammerdam, but Galvani apparently was unaware of it. This time there was no external source of electricity nearby, so he hypothesized that the frog leg must have made its own electricity; he called it "animal electricity."

When Galvani published his research it caught the attention of the Italian Alessandro Volta, professor of physics at the University of Pavia. Volta proposed that the frog leg twitched, not because of electricity generated within the animal, but rather because of the difference in electric potential between the two dissimilar metals (the hook of brass, which is mainly copper, and the iron of the railing) which the frog leg accidentally connected. Volta viewed the frog muscle as a very sensitive electroscope, able to detect currents much weaker than could be detected by any apparatus at that time. Galvani answered Volta's challenge with an experiment in which he used a drawing compass, made entirely of iron, to connect the crural nerve of the frog with the leg muscle. The twitch thus produced demolished, at least in Galvani's mind, Volta's bimetallic hypothesis. However, the controversy continued about another twenty-five years. Galvani's work helped to establish the new discipline of electrophysiology. We now know that animals do make their own electricity. The conduction of a nerve impulse involves a wave of depolarization of the nerve-cell membrane as sodium ions rush into the cell. The electric potential of the membrane becomes reestablished by a "sodium pump" mechanism that trades potassium ions for sodium ions, and the cell is then again ready to transmit another impulse. All of this active transport requires the expenditure of energy supplied by the food that the animal eats.

Meanwhile, Volta continued to experiment with dissimilar metals and eventually produced the first electric battery, which he described in a letter to the Royal Society of London in 1800. His battery ("pile") consisted of a series of "cells" composed of silver and zinc plates with moistened pasteboard discs between the two metals. His battery provided the first useful source of continuous electrical current. The electrostatic generators of his day could produce discharges of high voltage, but could not deliver a constant current. Volta found that the power ("voltage") of his battery depended on the number of cells connected in the series. Sir Humphry Davy soon put the voltaic pile to use in the discovery of the elemental metals sodium and potassium. In 1820 the Danish physicist Hans Christian Oersted noticed that a directional compass needle could be deflected when an electric current was sent through a wire passing over the compass. This discovery led to the invention of the electromagnet by William Sturgeon in 1825. Today's technology heavily depends on electromagnets for everything from doorbells to electric motors.

To understand the story of animal electricity more fully, it is necessary to review the history of this phenomenon before Galvani's time. The phenomenon of animal electricity was debated long before Galvani and Volta. The animals that provoked these studies were not frogs but electric fish such as the ray (torpedo), eel, and catfish. The earliest records of the "shocking" ability of such fish go back to about 2750 B.C.E. for the Nile catfish. Aristotle understood the function of the electrical discharge (although he did not know its nature) in capturing food. The Roman physician Galen (131–200 C.E.) thought that the shock was due to the action of extreme coldness generated by the animal. This idea held sway for a thousand years. Francesco Redi (1671), and later his pupil Stephano Lorenzini (1678), were the first to dissect the torpedo. They concluded that what we now call the electric organ was modified muscle tissue. Lorenzini thought that microscopic effluvia or corpuscles darted forth from these muscle fibers to produce the shock. Giovanni Borelli (1680) proposed that the electric organ contracts rapidly, producing a numbing sensation like a blow to the elbow. This theory went unchallenged for almost a hundred years.

Then in 1745 Pieter van Musschenbroek invented an instrument that was destined to change the course of experimental research into the phenomenon of animal electricity. It was the Leyden jar. This apparatus is formed by lining the inside and outside of a glass jar with metal foil; it is able to function as a capacitor for storing an electrical charge. The electrostatic generator had been invented about 1672 by von Magdeberg. When used in conjunction with the electrostatic generator, the Leyden jar could store a much greater amount of electricity for use in animal experiments. Van Musschenbroek inferred from his experiments with the electric eel that his Leyden jar was an inanimate model of the eel's electric organ. The Leyden jar discharges all of its stored electricity in one burst. The electrostatic generator, however, could produce high voltages, depending on how fast the wheel was turned, but could not store a charge. John Walsh, a fellow of the Royal Society, performed some experiments on the European torpedo that had been suggested by Benjamin Franklin. In 1773 he found that the two sides of the torpedo's electric organ were in opposite electrical states. In that same year, the great surgeon and anatomist John Hunter did his own dissections of the torpedo, paying particular attention to the unusually large nerve trunks that innervate its electric organ. Hunter speculated that the nerves play a role in the animal's shocking ability, but he was careful to avoid the use of the word *electric*, aware that the nature of the discharge had not yet been conclusively proven. In 1775 Jan Ingenhousz challenged Walsh's theory that the torpedo's shock is an electrical discharge on the grounds that the fish failed to attract light bodies or to electrify a coated bottle or even the experimenter. Furthermore, when the fish gave a shock in the dark, no cracking noise was heard and no spark could be seen. That same year, Hugh Williamson in the United States succeeded in passing a shock across an air gap, but no spark was observed. The production of a spark was to be the sine qua non (Latin, "without which not," an essential element or condition) for full acceptance of the electrical nature of the fish's shock.

Henry Cavendish published a paper in 1776 outlining three principal difficulties with Walsh's electrical theory that prevented its full acceptance: (1) If the torpedo is electric like a Leyden jar, it should be short-circuited by the sea water in which the beast lives; (2) A large battery of Leyden jars weakly charged could not discharge across an air gap, although it was capable of giving a severe shock; and (3) Cavendish attributed the absence of attraction or repulsion, commonly associated with static electricity, to the transient nature of the shock. No electrometer then available was sensitive enough to measure such a brief discharge. In order to produce a spark, Walsh turned from the torpedo to the electric eel because of its higher degree of electrification. He succeeded in 1776 in doing so, and demonstrated it about a dozen times before members of the Royal Society of London, but never published the results of these experiments himself. Fortunately, Tiberius Cavallo (1786) was there to give us an eyewitness account:

> The strongest shocks of the gymnotus [electric eel] will pass a very short interruption in the circuit. . . . When the interruption is formed by the incision made by a pen-knife on a slip of tin foil that is pasted on glass, and that slip is put into the circuit, the shock in passing through the interruption, will show a small but vivid spark, plainly distinguishable in a dark room. . . . The night that Walsh first demonstrated the spark to his friends could well be said to mark the birth of electrophysiology. Before this, the agent of nervous and muscular activity had been thought of as a mysterious animal spirit, but Walsh proved that at least some living animals produce their own electricity. This was generalized about fifteen years later to the physiological principle that all living matter produces its own electricity and that electricity is the agent of nervous and muscular activity.[18]

Now Galvani enters the picture with his theory of inherent animal electricity which he thought was synthesized in the brain and stored in the muscle via the nerve. When he touched the nerve-muscle preparation of a frog with the metal arc, he thought the muscle was discharging its stored electricity, thereby contracting the muscle. Galvani sent a copy of his paper to Volta. Initially Volta had nothing but praise for Galvani's work. Soon, however, using a sensitive condensing electroscope that he had invented earlier, Volta repeated Galvani's experiments and came to the conclusion that the electric currents were generated in the contact of dissimilar metals in the metal arc touching the muscle. So began one of the great debates in science. After about twenty-five years, Galvani lost the debate essentially by default. By 1794 he was plagued by personal and health problems that caused him to withdraw from the arena. Napoleon invaded Italy in 1796. Galvani lost his professorship in the University of Bologna because he refused to pledge allegiance to Napoleon's puppet regime, the Cisalpine Republic. Volta was well aware of the challenge that electric fish made to his denial of intrinsic animal electricity. He had even done detailed anatomical dissections of the torpedo's electric organ. Volta only acknowledged the existence of intrinsic animal electricity after he had invented the first electric battery. He even called it an "artificial electric organ" because he saw it as a mimic of "the natural electric organ of the torpedo or electric eel."[19] He thought that the electric organ was constructed like his "pile." The organ was able to discharge by bringing some of its parts together and causing some moisture to flow between its normally separated elements. This, of course, is not how animal electricity is generated.

Both Galvani and Volta were at least partly right and partly wrong. Living cells do produce electricity, and electrical forces are generated at the junction of dissimilar metals in an electrolyte solution. Galvani erred in proposing that animal electricity is produced in the brain and delivered by a nerve to a muscle where it is stored (analogous to a Leyden jar). Animal electricity is not stored, but generated as needed both in all nerves (not just those in the brain) and in the modified muscle of electric organs. Volta failed to recognize that the electrical current generated at the junction between dissimilar metals serves only to excite the nerve, which in turn propagates a wave of depolarization along its length (an action potential) that leads to muscle contraction. Serendipity played a major role for Volta too. It was in his attempt to refute Galvani's theory of animal electricity and to justify his contact theory that he stumbled upon the fact that alternating dissimilar metals, separated by pasteboards soaked in salt water, could produce a steady current of electricity. Next time we turn on a flashlight or any electrical device we'll be in a better position to appreciate how it all began.

WHAT A GAS!

Nitrous oxide was discovered by Joseph Priestly in 1772, just two years before he discovered oxygen. The ability of nitrous oxide to induce a variety of mental states in those who breathed this gas was soon discovered. Some people responded pugnaciously, some became fearful, still others became euphoric. Because of this latter property, nitrous oxide became popularly known as "laughing gas." A whole field of pneumatic medicine began to grow around the inhaling of medicines. In 1798, Humphrey Davy became the director of England's Pneumatic Institute. The next year he discovered that prolonged inhalation of nitrous oxide produces temporary unconsciousness. Davy suggested that nitrous oxide might be useful as an anesthetic during surgical operations, but no one immediately put his suggestion to practical use. The gas was, however, widely used (and abused) as a source of entertainment. Nevertheless, Davy's exploits got him promoted in 1801 to a position in the Royal Institution of London, where he went on to discover many chemical elements. Davy recognized the abilities of the electrical wizard Michael Faraday, and made him his assistant. Faraday succeeded Davy as professor at the Royal Institute, and was knighted at age thirty-four. In 1818 Faraday noted: "When the vapor of ether mixed with common air is inhaled, it produces effects very similar to those occasioned by nitrous oxide."[20] Why the medical profession did not seize upon these early suggestions remains a mystery. Over a quarter century would elapse before these gases would be put to medical use.

"Professor" Gardner O. Colton (a title he had bestowed upon himself) gave an entertaining public lecture on nitrous oxide in Hartford, Connecticut in 1844. Colton called for several volunteers from his audience to come to the stage for a demonstration of the effects of the gas. A Hartford dentist named Horace Wells was in the audience with his friend Samuel Cooley. Cooley became one of the volunteers. After inhaling nitrous oxide, he became violent and had to be physically subdued. During the scuffle, he fell down and inflicted a deep gash in his leg, but he was not aware of it at the time. After he calmed down and returned to his seat, Wells saw blood trickling down his friend's leg. Wells questioned Cooley about the pain, but he said he felt none. Wells immediately thought of applying nitrous oxide as an anesthetic for suppressing the pain of tooth extractions. Wells had a dentist friend pull one of his own diseased molars after breathing nitrous oxide. When Wells regained consciousness, he confessed that he had felt no pain during the tooth extraction. Flushed with this success, Wells then arranged to perform a demonstration before a group of medical students in the class of the famous surgeon Dr. John Collins Warren at the Harvard Medical School in Boston. His patient was a man who volunteered to have his tooth

DEATH OF A RAT

extracted after inhaling nitrous oxide. Unfortunately, Wells had not learned how to control the dosage of his anesthetic, and he gave the order to have the tooth pulled before his patient had become completely unconscious. The man howled in pain as his tooth was pulled. Wells was hissed from the room in disgrace. Undaunted, Wells returned to his dental practice and successfully used nitrous oxide on several of his patients. But he was unaware of the addictive nature of the gas and his mental faculties began to erode as a consequence of his own abuse of it. Eventually he was arrested for throwing acid at a prostitute, and committed suicide while in prison in 1848.

Another dentist, and former partner of Wells, William T. G. Morton, had observed Wells's successful use of nitrous oxide on his patients. In 1846, Morton decided to try nitrous oxide on his own patients. He asked his mentor, Charles T. Jackson, how to procure nitrous oxide. Jackson suggested that he use sulfuric ether (now correctly known as diethyl ether) rather than nitrous oxide. Morton extracted a tooth without pain from a patient under ether anesthesia on September 30, 1846. Claiming to have discovered a new anesthetic, he then quickly arranged with Warren for a demonstration of its use during surgery at the Massachusetts General Hospital in Boston on October 16, 1846. Realizing the potential lucrative rewards for such an anesthetic, Morton tried to disguise the volatile liquid ether by adding a red dye to it and calling it "letheon." The patient had a tumor removed from his jaw without experiencing pain while under Morton's "new" anesthetic. The report of this successful use of sulfuric ether anesthesia appeared in the *Boston Medical and Surgical Journal* on November 18, 1846. This was the article that introduced the use of ether anesthesia to the world.

But his plans for instant wealth were soon shattered when Jackson, on learning of Morton's famous demonstration, began claiming that he had told Morton to use ether because he had personally experienced its pain-suppressing properties by accident. According to Jackson, it was during the winter of 1841–1842 that he was conducting an experiment with the deadly gas chlorine. The container broke and Jackson inhaled some of the chlorine which irritated his throat. He was able to soothe his throat pain by alternatively inhaling ether and ammonia. The next day he tried ether alone and found it more effective in relieving pain even to the point of losing consciousness. Jackson stated, "The idea flashed into my mind that I had made the discovery I had for so long a time been in quest of—a means of rendering nerves of sensation temporarily insensible, so as to admit of the performance of a surgical operation on an individual without his suffering pain therefrom."[21] Morton and Jackson temporarily settled their differences by jointly applying for and receiving a patent on their anesthetizer (containing letheon). Very soon, however, surgeons at the Massachusetts General Hospital barred Morton and his anesthetic until they learned what kind of substance it was. Morton broke his secrecy by confessing that it was highly purified sulfuric ether. From that point on, Morton could not protect his patent because ether was readily available from commercial sources.

Jackson was not satisfied to share the glory with Morton. He started claiming full credit for using sulfuric ether as an anesthetic; Morton was just a person who delivered it. This was not the first time that Jackson had displayed this kind of behavior. He had previously tried to take credit from Samuel F. B. Morse for the invention of the telegraph, and also from Dr. William Beaumont for his discoveries of gastric functions (discussed in chapter 1). Thus there were three contenders for the title of discoverer of anesthesia: Wells, Morton, and Jackson. The dispute was waged by their supporters even after Wells died and it was eventually brought to the United States Congress for settlement.

By this time, a fourth candidate appeared. Dr. Crawford W. Long was a country doctor practicing in Jefferson, Georgia. Some of Long's friends asked him for some nitrous oxide for their party. Like elsewhere, it was a popular pastime in the early 1840s to "get high" on nitrous oxide. Long didn't have any nitrous oxide, but he did have ether and he knew from his own experience that it would produce "equally exhilarating effects." So they took the ether, and after "getting high" they thought it might be fun to try it on the male black slave who was serving refreshments. The slave did not want to participate, and during the ensuing scuffle, he inhaled so much ether that he became unconscious. Fearing that the slave was dead, the revelers notified Long. After examining the slave, Long found that he was breathing regularly and had a regular pulse. When the man revived over an hour later, he felt normal but had no memory of what had happened. Long immediately saw the possibility of using ether as an anesthetic for surgical operations. On March 30, 1842, he removed two tumors from the neck of a patient under ether anesthesia with the same positive results. By chance, the type of operation (tumor extraction) performed by Long was the same kind in which Morton would demonstrate ether anesthesia to the medical world in Boston four years later. Morton knew nothing of Long's experiments, however, because Long had not published his findings at the time. Some historians say that Long was too busy with his practice to publish his findings, but this makes one wonder what could be more important than sharing this knowledge with the rest of his professional colleagues. At any rate, after the controversy between Wells, Morton, and Jackson arose, friends of Long made the U.S. Congress aware of Long's priority in the discovery of ether anesthesia. Congress refused to settle the argument. Posthumously, Wells was acknowledged as the discoverer of anesthesia in the United States by the American Dental Association in 1864, and by the American Medical Association in 1870. The French Academy, unable to resolve the dispute, split the Prix Montyon between Jackson and Morton.

In those early days of ether anesthesia, no way had been devised to adequately control the dosage of ether that a patient received. Furthermore, there was no knowledge that different patients could respond in various ways to the same dose of ether. Some patients died of overdose. Conservative dentists and doctors began to turn away from the use of ether. Morton, who continued to use ether on his patients, was harassed by these objectors until he was forced to abandon his profession. In 1868, he was found unconscious in New York's Central Park and he subsequently died from injuries thought to have been inflicted by an unknown assailant(s). Jackson received no credit from any association in the United States, but was awarded the Order of the Red Eagle of Prussia. He spent the last seven years of his life in the McLean Asylum, a department within the Massachusetts General Hospital, where he died in 1880. Only Crawford Long remained aloof from the controversy. Long died in 1878 of natural causes while assisting in the delivery of a new life into the world. A statue of him was placed in the Hall of Fame in Washington, D.C., by the state of Georgia. Its inscription reads: "Georgia Tribute, Crawford W. Long, discoverer of the use of sulphuric ether as an anesthesia in surgery, on March 30, 1842, at Jefferson, Jackson County, Georgia, U.S.A."

In Edinburgh, Scotland, Dr. James Young Simpson began using ether on his obstetrical patients but found it to have undesirable side effects such as odor and a tendency to irritate the bronchi. So he began a random search among the then known volatile chemicals for a substitute. Whenever he found time, he would sniff aromatic substances such as acetone and benzene. One day he inhaled some chloride of hydrocarbon or "Dutch oil" that caused him to vomit and put him in bed for a month. But even this narrow brush with death did not stop him from further self-experimentation. Eventually his trial-and-error method led him to discover the anesthetic action of chloroform. This substance worked much faster than ether and at lower dosages, and had a less unpleasant odor. After using it successfully on his patients in 1847, he campaigned vigorously for its use in obstetrics and surgery. "Men of the cloth" objected to the use of chloroform for women in labor, claiming that it was God's will that women should bear the pain of childbirth with patience and fortitude. (It is too bad that there were no "women of the cloth" in those days to offer a different interpretation of Scriptures.) These objections, however, were silenced when Queen Victoria used chloroform during the delivery of her eighth child. For a time thereafter, chloroform was widely used until reports of deaths due to overdose began to appear. Chloroform proved to be more dangerous than ether, and its dosage had to be carefully regulated. Ether again became the anesthetic of first choice. Thus one of the greatest advances in medical history has been difficult to assign to any one person. But the role of serendipity in the discovery of anesthetic gases is central and secure.

DYNAMITE–A NOBEL (*SIC*) SUBSTANCE

Alfred Bernhard Nobel was born in Stockholm, Sweden, in 1833. His health was poor from birth and this condition persisted throughout his life. His father, Immanuel Nobel, was a self-taught inventor and engineer. His business in Stockholm went bankrupt in the same year that Alfred was born. Immanuel left his family to escape debtors' prison and to try to make enough money so that he could pay off his debts and return home. Several years later, Immanuel moved his family to St. Petersburg where he had a good job and was part owner in a factory making military explosives for the Russian government. But Immanuel went bankrupt a second time at the end of the Crimean War in 1856. The family moved back to Sweden and Alfred helped his father manufacture nitroglycerine, an explosive that had been discovered by an Italian chemist, Ascanio Sobrero, in 1846 or 1847. In 1863, at the age of thirty, Alfred made and patented his first major invention, a blasting cap of mercury fulminate used to ignite nitroglycerine. Alfred had made more than fifty experiments at his father's workshop in Heleneborg before he perfected his blasting cap, so this invention was not accidental.

His brother Emil and five other people died in an explosion at his father's factory in September 1864. Immanuel suffered a stroke that physically incapacitated him until he died eight years later. Alfred had to take over the family business after the accident. He became determined to find a safe way to make and transport nitroglycerin. A new factory was built in an isolated area near Stockholm with the financial help of wealthy merchant J. W. Smitt. Another one was later built near Hamburg, Germany. Despite the danger of using this unstable liquid explosive, nitroglycerin saved so much time and money in blasting railroad tunnels and mining operations that it soon replaced black powder explosives, not only in Europe but also in the United States. Without nitroglycerin, the building of the Central Pacific railroad across the Sierra Nevada Mountains from California to Nevada might not have been completed. Failure to handle nitroglycerin with extra care caused many deaths, but probably even more loss of life was caused by the inherent instability of the substance. Sometimes it could be handled roughly without blowing up, but at other times the least little jarring would set it off. Alfred's factory in Germany was destroyed by an explosion in 1866, but he was not there at that time. Among the many attempts to reduce the capricious nature of nitroglycerin, Alfred tried mixing it with various solid powders and fibers such as sawdust, paper, brick dust, and charcoal. None of these treatments were without drawbacks. Although Alfred denied the role of chance observation in the discovery of a solution to the problem, a popular account persists. According to this legend, a metal canister of nitroglycerin sprung a leak and the fluid soaked into the surrounding packing material called *kieselguht*, a porous mineral common in northern Germany. Alfred noticed the pasty mixture formed by the leak and discovered that it could be molded into a compact solid. Tests proved that it had just as much blasting power as liquid "nitro," and was safe to handle; he called his new explosive "dynamite."

One day in 1875, Alfred was working in his laboratory and cut his finger on a piece of glass. He applied collodion, a common treatment of the time, to the injury. Collodion is a viscous solution of cellulose nitrate in alcohol and ether. When the solvents evaporate, it forms a highly flammable film as a temporary covering for wounds. Guncotton was already known to be a powerful explosive consisting of a more highly nitrated form of cellulose. Alfred had previously tried, without success, to combine nitrocellulose with nitroglycerin in the hope of producing an explosive more powerful than either one, but as safe as dynamite. Now unable to sleep because of his aching finger, he thought about those experiments and wondered if a lower amount of nitration, as existed in the collodion on his finger, would allow it to become better mixed with nitroglycerin. He jumped out of bed at 4 A.M. and hurried to his laboratory to try out his idea. Within a few hours he had made a clear jellylike combination that had the very properties he had hoped for. Alfred patented this "blasting gelatin," as he called it, in 1875 in England and in 1876 in the United States. The cut on his finger was an accident, but such an event would never have produced a discovery unless it occurred to a person whose mind had been prepared, by previous experience with collodion, to envision its possible use in the solution to one of his unsolved problems. He also immediately followed up his idea with action. How many discoveries do not get made because people with good ideas never get around to testing them?

The inventions of Alfred Nobel made him a very wealthy man. In addition to his poor health, he was often mentally depressed throughout his life. Because he never married, some rumors exist that he may have been homosexual. But the truth is that, besides his mother to whom he was very devoted, at least three women were important in his life. At age eighteen, Alfred had a romance with a girl he met in Paris. He wrote poems to her and she gave him great happiness. She died very young and Alfred grieved deeply over her passing. Alfred was forty-three and in Paris again when he needed a secretary and housekeeper. The Countess Berta Kinsky saw his advertisement and applied for the job. She was a beautiful and talented woman from an impoverished Austrian family. She was in love with a young Viennese aristocrat named Arthur von Suttner and had come to Paris to be with him. Alfred fell in love with her almost at first sight, but her heart belonged to von Suttner. She worked only briefly for Nobel and then returned to Austria to marry von Suttner. She and Alfred remained lifelong friends, however. Berta von Suttner became actively involved in promoting international peace, an interest she shared with Nobel. Even before Berta left her employment with Alfred, he met a beautiful twenty-year-old woman named Sofie Hess in a Parisian flower shop. It was a mismatch from the start. Nobel was an older, wiser, wealthy man who mainly wanted companionship; she was a young, uneducated, and undisciplined vamp who just wanted to have fun. Alfred placed her in a nice apartment in Paris with servants and later in a villa near Vienna. Alfred addressed her as Madame Sofie Nobel in many letters to her, but their marriage was never made official. During the next eighteen years, Alfred often was away on business trips around the world. Sofie's attention gradually turned toward her many young admirers in the fashionable resorts around Europe. When she became pregnant by a young Hungarian officer, Nobel gave up his attempts to reform her, gave her a liberal annuity, and vowed to see her no more. Sofie married the Hungarian officer but did not live with him. Both of them tried to extort money from Nobel for the rest of his life.

Nobel had frequently suffered from angina pectoris (chest pains) and had several heart attacks in later life. He wrote in a letter shortly before his death, "It sounds like the irony of fate that I should be ordered to take nitroglycerine internally. They call it Trinitrin so as not to scare pharmacists and public."[22] By that time his parents and brothers had all died. In the autumn of 1895, he wrote into his will the guidelines that would be used to establish the Nobel Foundation and its prizes. One part of the will says,

> The capital shall be invested by my executors in safe securities and shall constitute a fund, the interest on which shall be annually distributed in the form of prizes to those who, during the preceding year, shall have conferred the greatest benefit on mankind. The said interest shall be divided into five equal parts, which shall be apportioned as follows: one part to the person who shall have made the most important discovery or invention within the field of physics; one part to the person who shall have made the most important chemical discovery or improvement; one part to the person who shall have made the most important discovery within the domain of physiology or medicine; one part to the person who shall have produced in the field of literature the most outstanding work of an idealistic tendency; and one part to the person who shall have done the most or the best work to promote fraternity between nations, for the abolition or reduction of standing armies and for the holding and promotion of peace congresses.[23]

Although Nobel and Berta von Suttner shared concerns over world peace, they disagreed on how to attain it. In 1892, she asked Alfred to join her at a Peace Congress in Switzerland. He declined, stating that "My factories may end war sooner than your Congresses. The day when two army corps will be able to destroy each other in one second, all civilized nations will recoil from war in horror and disband their armies."[24] Alfred Nobel died in 1896 at his villa in San Remo, Italy. The first year in which the Nobel Prizes were awarded was 1901. Berta von Suttner was the fifth recipient of the Peace Prize in 1905.

RAYS OF HOPE?

Wilhelm Konrad Röntgen (Röentgen) was born in 1845 in Lennep, Prussia (now Remscheid, Germany). He received a diploma in mechanical engineering from the Polytechnic School in Zürich, but soon became more interested in science than in engineering. He began to study mathematics and physics and earned a doctorate from the University of Zürich through his studies of gases. He became professor of physics and director of the Physical Institute at the University of Würzburg in 1888. It was there that he discovered X rays by accident in 1895.

It had been known for almost forty years that the walls of a glass tube become phosphorescent when electricity at high voltage is discharged through air or other gases in a tube with most of the air removed (Crookes tube). Today, neon signs, television tubes, and fluorescent light tubes are the outcomes of these early experiments. The inner surface of these tubes can be coated with highly fluorescent substances to produce different colors when excited by electrons. By 1892, Heinrich Hertz had discovered that cathode rays could penetrate thin metal foils. The first discharge tubes having thin aluminum windows were made in 1894 by Philipp Lenard. The window allowed the cathode rays to pass out of the tube where they could be detected by aiming them at a fluorescent screen, causing it to glow. No one at that time, however, had been able to detect cathode rays emanating from an evacuated all-glass tube. Röntgen had a theory that the strong phosphorescence of the cathode tube obscured the weak fluorescence of the detecting screen. To test his theory, Röntgen covered the tube with a black cardboard shield, darkened the room, and turned on the high-voltage coil to energize the tube. He did not observe any light escaping from the tube. He was just about to shut off the coil when he noticed a weak glow from a spot about a yard from the tube. He struck a match to see what might be located at the glowing spot, and there lying on a bench was a little fluorescent screen that he had planned to use as a detector near the blacked-out cathode-ray tube. Cathode rays (streams of electrons, equivalent to beta rays from radium and other radioactive elements) were known to travel only a few centimeters in air at ordinary pressure outside the evacuated tube. Yet here was evidence that something was traveling at least a yard away from the tube. Röntgen knew that he had discovered a different form of radiation. After several weeks were spent investigating this new phenomenon, he reported his findings in a paper dated December 28, 1895, and titled "A New

Kind of Ray: A Preliminary Communication." He acknowledged that his understanding of these new rays was rudimentary by naming them *X rays* (*x* being the symbol for an unknown in mathematics). He found that X rays could not be deflected by a magnet the way that cathode rays can. He further reported that

> All bodies are transparent to this agent, though in very different degrees. . . . Paper is very transparent; behind a bound book of about one thousand pages I saw the fluorescent screen light up brightly. . . . In the same way the fluorescence appeared behind a double pack of cards. . . . Thick blocks of wood are also transparent, pine boards two or three centimeters thick absorbing only slightly. A plate of aluminum about fifteen millimeters thick, though it enfeebled the action seriously, did not cause the fluorescence to disappear entirely. . . . If the hand be held between the discharge tube and the screen, the darker shadow of the bones is seen within the slightly dark shadow image of the hand itself.[25]

These images could be captured on a photographic film, and in a very short time X rays were being routinely used in medical practice worldwide. The first Nobel Prize in physics was awarded to Röntgen in 1901. He had been called by the government of Bavaria to head the Physical Institute at Munich in 1900, and he stayed there for the remainder of his life. He died in 1923 at the age of seventy-eight. The damaging effects of X rays were not fully appreciated during his lifetime. Fortunately, Röntgen had built a booth in his laboratory for the development of photographs. It is thought that his life was not severely shortened because he was shielded behind this booth during most of his experiments with X rays.

It is now known that when the negative electrode (cathode) of a vacuum tube is heated to white-hot intensity by an electric current, a stream of electrons is released. These cathode rays travel toward the positive electrode (anode) at speeds proportional to the difference in electric potential (voltage) between the two electrodes. In modern X-ray tubes, the anode (target) is usually made of tungsten, partly because that metal will not melt from the heat of the electrons that smash into it. Electrons close to the center of atoms in the target are displaced from their normal positions by the energy of the collision with high-speed cathode rays. When these electrons fall back into place, energy is released in the form of X rays. The higher the electric voltage, the faster the cathode rays strike the target, and the shorter the wavelengths of the resulting X rays. "Soft" X rays, produced at lower voltages, have longer wavelengths and less penetrating power than the "hard" X rays that are produced by higher voltages. X rays are equivalent to gamma rays emitted by radioactive elements. They both are *photons* (particles of electromagnetic radiation) that travel with the speed of light. When an unstable radioactive atom emits an alpha particle (helium atom) or a beta

DEATH OF A RAT

particle (electron), it is *transmuted* (changed) into an atom of another element. After the atom has become transmuted, it has more energy than in its stable state. It becomes stable by emitting gamma rays. No further transmutation of that atom occurs as a result of its gamma-ray emission.

Antoine Henri Becquerel (1852–1908), a French professor of physics at the École Polytechnique, had read Röntgen's paper on X rays. Becquerel wondered if certain substances made phosphorescent by visible light might emit a penetrating radiation similar to X rays. To test his theory, he wrapped a photographic plate in black paper, placed a crystal of a uranium compound on the paper-wrapped plate, and set it out in the bright sunlight. After the plate was developed, an image of the crystal was visible. Becquerel reasoned incorrectly that since the black paper protected the photographic plate from the sunlight, the crystal had, as a consequence of absorbing sunlight, given off rays that could penetrate the black paper. Then an accident of nature occurred that was to set the stage for the development of atomic energy. The sun failed to shine in Paris for several days. Because Becquerel believed that sunlight was necessary to activate the phosphorescence of the uranium crystal, he put the crystal in a drawer on top of a black-paper-wrapped photographic plate to await the return of sunny weather. After several days in the drawer, Becquerel developed the plate, expecting to see only a weak image from residual phosphorescence in the crystal. To his surprise, the image was just as bright as the experiments done in full sunlight. Becquerel abandoned his earlier idea, and came to the correct conclusion that the uranium crystal was spontaneously producing its own radiation whether it was in the sunlight or not. He found that all uranium ores emitted this kind of radiation. An electroscope could be used to measure these radiations because they ionized the air through which they passed. The amount of radiation was found to be directly proportional to the percentage of uranium in the ore. The only exception he found was in an ore called *pitchblende*, that had a much higher radiation than pure uranium. Becquerel concluded that another kind of radioactive substance besides uranium was in that ore.

At the suggestion of Professor Becquerel, Marie Sklodowska Curie chose as the subject of her doctoral research the identification of the radioactive impurity in the uranium ore pitchblende. With the help of her physicist husband, Pierre, they were able to isolate and characterize two new elements that were more radioactive than uranium. They called one of them *radium* for obvious reasons; the second one was named *polonium* after Marie's native Poland (she was born in Warsaw on November 7, 1867). Radium was sixty times and polonium four hundred times more radioactive than uranium. The yield from pitchblende, however, was very low; only about one part of radium per ten million parts of ore. The Curies eventually worked their way through fifty cubic feet of pitchblende ore, placing up to forty pounds at a time into boiling mixtures in large cast-iron basins and stirring it with iron bars. They announced their discoveries of the two new elements in 1898, just two years after Becquerel discovered natural radioactivity. In 1903, the Nobel Prize in physics was shared, half of it going to Becquerel for discovering natural (spontaneous) radioactivity, and half to the Curies for their discoveries of two new radioactive elements. Pierre Curie died in a traffic accident in 1906. Otherwise he would have shared the 1911 Nobel Prize in chemistry with his wife for her work on isolating radium and polonium and for investigation of their chemical properties. Marie died in 1934 of leukemia, a cancer of the blood, very likely due to the radiation she received while experimenting with these unstable elements.

Irene Curie (1897–1956), daughter of Marie and Pierre, and her husband, Frederic Joliot (1900–1958), discovered artificial radioactivity in 1934. They found that alpha-particles ejected from atomic nuclei could be used to bombard certain stable (nonradioactive) elements and induce those elements to become unstable and emit their own radiations. They were awarded the Nobel Prize in chemistry for this work in 1935. Irene and Frederic directed the project that put the first French atomic reactor into operation in 1948. Irene died in 1956 of leukemia, the same disease that killed her mother twenty-two years earlier.

Shortly after the discovery of artificial radioactivity, Enrico Fermi blasted a stable isotope of uranium with neutrons. He expected to produce nuclei of elements close in weight to uranium. But Otto Hahn and Fritz Strassman, at the Kaiser Wilhelm Institute in Berlin, found some barium among the products of neutron bombardment of a stable uranium isotope. Barium is an element about half the size of uranium. Since there was no barium in the sample bombarded, the barium must have came from splitting the uranium atom in two. In December 1938, Hahn wrote a letter to Lise Meitner, who had worked with Hahn for thirty years. She had to flee Germany to escape the Nazi purge of Jews. He explained to her the unexpected results of his experiment. Lise's nephew, Otto R. Frisch, visited her at Christmas. Frisch was also a physicist and had worked with the famous Danish physicist Niels Bohr in Copenhagen. Lise and Otto discussed Hahn's letter. They recalled that Bohr had theorized in 1936 that nuclear bombardment might cause the atom to split in two with a tremendous release of energy. Frisch named the atom-splitting process *nuclear fission*, as an analogy to cell division in biology. A few days later, Frisch caught up with Bohr in Copenhagen and told him of Hahn's results. Bohr was on his way to a conference on theoretical physics in Washington, D.C., and he carried the Meitner/Frisch interpretation of Hahn's experiments to that meeting. Bohr calculated that only a rare form of uranium, the isotope U-235, was capable of sustaining a chain reaction. Leo Szilard soon made the first chain reaction. The work of Szilard and Enrico Fermi prompted Albert Einstein to write a letter to President Franklin D. Roosevelt, warning him that manufacture of an extremely powerful bomb might now be possible. The United States scrambled to beat Hitler in the race to make an atomic bomb. The rest is well-known history. The long "cold war" between Russia and the United States, both of which possessed nuclear weapons, did not escalate into a shooting war perhaps because, as Alfred Nobel responded to Bertha von Suttner and her pacifist conferences, "My factories may make an end of war sooner than your congresses."[26]

NOTES

1. Royston M. Roberts, *Serendipity: Accidental Discoveries in Science* (New York: John Wiley & Sons, 1989).

2. Ibid., ix.

3. Ibid., x.

4. Ibid.

5. Ibid., 161.

6. Ibid., 162.

7. Parke, Davis & Company, "The Era of Antibiotics," narrative accompanying *A Pictorial History of Medicine* (1964).

8. Roberts, *Serendipity*, 164.

9. R. P. Crease, "Righting the Antibiotic Record," *Science* 246 (November 17, 1989): 883. Gramacidin proved to be too toxic to be used systemically on humans.

10. Ibid., 884.

11. H. T. Swan, "The Antibiotic Record," *Science* 247 (March 23, 1990): 1387.

12. Ibid.

13. Ibid.

14. Ibid.

15. Ibid.

16. Ibid.

17. Ibid., 1387–88.

18. C. H. Wu, "Electric Fish and the Discovery of Animal Electricity," *American Scientist* 72 (September–December 1984): 603.

19. Ibid., 606.

20. Parke, Davis & Company, "The Era of Antibiotics."

21. Roberts, *Serendipity*, 39.

22. Ibid., 89.

23. Ibid., 90.

24. Ibid., 91.

25. Ibid., 141.

26. Ibid., 91.

CHAPTER 7

STONES
IN THE
ROAD

Most nonscientists have a warped view of how science is conducted. It is popularly believed that when a new problem arises a scientist or team of scientists goes to work on it, and by use of "the scientific method" they should be able to solve the problem. Simple problems should be solved quickly; more difficult problems may take longer. If the project requires a lot of money to pay salaries; build facilities; buy equipment, supplies, technical help, computer time, whatever—there is no reason why the problem can't be solved by simply throwing enough money at the project. In my opinion, it is this type of mentality that has done more damage to science's reputation among the general public than any other factor. The news media doesn't help in this regard either. Journalists love to find a preliminary scientific report and sensationalize its tentative conclusions. How often have we seen headlines declaring "New Cure for Cancer Discovered" or "AIDS Vaccine Imminent"? The results of one study, especially a preliminary one based on small sample size and not yet subject to independent confirmation, should not be blown out of perspective as a widely accepted scientific breakthrough. The public has grown to expect great things from science, but after being repetitively bombarded by journalistic hype, having our hopes and expectations raised only to be dashed by long periods of stagnation and failure to deliver what was promised, it is understandable why the public's esteem for the merits of scientific work would be diminished. Look what happened after the 1988 announcement of "cold fusion." In this instance, however, it was the scientists themselves that had announced prematurely to the media that they had discovered a way to fuse atoms in a test tube with the release of more heat than the electrical energy necessary to cause the "fusion reaction" to occur. Although numerous attempts at independent confirmation of their experimental results have been attempted, few of them to date have claimed success. The whole idea was initially contrary to the predictions of current nuclear physics theory, and was met with a great deal of scientific skepticism. This in itself is not unusual, for most great scientific advances have met with the same uphill road to travel before gaining wide acceptance. Very quickly, however, "cold fusion" fell into the scrap heap of useless concepts because the initial experiments could not be replicated. Scientists should never use the media to announce their results and conclusions. The general public is not the appropriate forum for testing the merits of a scientific theory. Works of science should be published in a refereed scientific journal that will be distributed to other experts in the appropriate discipline(s) and provide the necessary information to repeat the observations or experiments by which confirmations or refutations may be obtained. Unfortunately, when financial gain and scientific prestige are at stake, there are powerful pressures to sidestep the normal scientific procedures of publication and go directly to the news media. The result of failing to follow conventional procedures for reporting science in this case caused diversion of much-needed money and a waste of "manpower" (throughout this book "man" is generally meant in the broad sense of "humanity") from other more promising lines of work. This is just one reason why science does not progress as rapidly as it might.

There are many other "stones in the road" to the advancement of scientific knowledge, most of which are less familiar to the general public. The majority of scientific work today requires rather large sums of money. Government agencies (such as the National Institutes of Health) provide a large portion of the money for scientific research in this country. Our government was set up to serve the people, and as such it should be sensitive to public opinion. If the public views science as less valuable than highway construction or other worthwhile projects, then science winds up getting less financial support from the government. Of course this does not stop science dead in its tracks, because industries and private donations are also important sources of funding scientific projects. However, because the government is such a large contributor, it is important that the functions and limitations of science be well understood by the public, so that fiascoes such as "cold fusion" do not sour public opinion in such a way as to reduce funding of science in general.

The power of the people in this regard, if they understand it, is one reason why science education is so important for an "enlightened electorate." Science is doomed to suffer in countries where their citizens have not received a good science education. Most educational institutions in the United States, from grade school through college or university, depend on public funds. The voice of the people, especially at the level of local school boards, can either favor or disfavor the teaching of science. Witness the disruption to the teaching of evolution (the unifying concept of all the biological sciences) by fundamentalist groups that have pressed for teaching of special creation in the classroom and in textbooks along with the teaching of evolution. No wonder our students are confused about science! They don't know where religion leaves off and science begins, and apparently neither do those educators who have allowed the idea of "equal time" for these two ways of "knowing" to become a part of the science curriculum. Each generation that misses out on a good science education can adversely affect the progress of science in two major ways. One of these, I have just said, is public opinion regarding science funding and science education. The other way is, by failing to stimulate students' interest in the sciences, we wind up with fewer well-trained scientists and science teachers than are required to serve our national needs. Already, the United States has had to hire many foreign scientists and engineers for our space program and to staff unfilled university teaching/research positions in math, science, and engineering because we are not able to generate enough of our own citizens trained in these disciplines. Students who get a poor science education and who decide to go into teaching are then likely to do a poor job educating yet another generation of students. Thus, poor science education has a way of perpetuating itself. Perhaps more than we care to recognize, science fails to progress as rapidly as it might because our educational system has let us down.

There are still other stumbling blocks in the road to scientific progress. Many studies, especially those in the biological sciences, are based on sample sizes that are too small to yield statistically significant results. Of course it is desirable to have scientific studies based on large samples that adequately reflect the populations from which they were drawn. But it is often not possible to have large samples for a variety of reasons. Perhaps the subjects are rare humans as in "orphan diseases" that do not receive adequate funding because they affect so few people. Or perhaps the subjects are members of an endangered species, none of which can afford to be lost through experimentation. Alternatively, the measurements or observations that need to be made are so costly or time-consuming that large samples are not feasible. Thus, many experiments would like to be carried out in space aboard a satellite or the space shuttle. But even small loads are extremely costly cargo and this severely restricts the kinds of experiments that are likely to be funded. During 1990, a high school class received fairly large numbers of tomato seeds that had been in space for some time. The students then planted the seeds and, when the plants bore fruit, compared the results with a comparable control group of seeds that had not been in space. What a marvelous opportunity to introduce students to scientific methodology using "exotic" materials. Where experimental animals are involved, the investigator needs to choose a minimal sample size that is likely to provide the needed data without the expenditure of lives, pain, or anguish inflicted on surplus animals. This is not an easy determination to make. Biological materials are fraught with more potential difficulties than nonbiological materials. All of the seeds planted may not germinate. Not all of those that germinate may grow to maturity. Some may be eaten by insects, or stunted by disease. Similar difficulties are met in animal research. It is unrealistic to expect that all the animals at the inception of an experiment are likely to be present at its conclusion. Some may get sick and/or die. Individual animals are often much more expensive than individual plants, and the expense of rearing or maintaining animals (e.g., feeding mice and cleaning their cages) can be much greater than that needed to care for the same number of plants. When two or more groups of individuals are to be compared, the statistical analyses are usually enhanced if the groups are of equal size. There are methods for analyzing groups of unequal size, but the results may be less useful. Hence, it is desirable to avoid any loss of individuals, or differential losses in the various groups being compared. If the researcher is overly conservative and uses too few animals, the study may be inconclusive and will need to be repeated. On the other hand, if the researcher chooses too large a sample, it may be clear in retrospect that fewer animals could have been used and still obtained the data needed to answer the question. Thus one is left with feelings of guilt that some animals, time, and money could all have been spared. Hindsight is always better than foresight! The animal-rights activists love to jump on such cases to prove their point that animals are suffering needlessly in the cause of science.

THE HELA DEBACLE

One of the alternatives to animal testing is cell or tissue cultures. For some purposes, such as toxicity testing, it is not necessary, as a first step, to assay a chemical in an intact animal. Individual cells or a tissue sample can be assayed in isolation within a suitable container. These kinds of experiments have historically been referred to as *in vitro* (literally "in glass," although the container need not be of glass), in contrast with *in vivo* experiments that are carried out in a living organism. Until recently, cell or tissue culture was a black art. Many people had tried to keep cells alive, growing, and reproducing in vitro without success or with only sporadic success. In the 1950s human tissue culture was still in its infancy. At that time it was known that the vessels must be exceptionally clean, both biologically and chemically. Vital growth factors (some still not adequately known) in such extracts as blood from human placentas, beef embryos, and chicken plasma had to be added to the cultures to keep them alive. The nutrients must be changed regularly and the waste products promptly removed. Constant attention must be given tissue cultures if they are to last any useful length of time. Even with the best of care, however, most animal cells in tissue cultures can only go through less than about fifty generations before they die. Leonard Hayflick discovered this principle, now known as the *Hayflick limit*. Biologists are still searching for all the reasons that cells normally undergo this programmed death known as *apoptosis*.

One of the groups pioneering in tissue-culture techniques was the husband-and-wife team of George and Mary Gey at the Johns Hopkins Medical School. By early 1951 they had managed to keep human cancer cells alive in tissue culture for a few months. No continuously growing or permanent cell lines had yet been established (such cells are said to have become *immortalized*). On October 4, 1951, a thirty-one-year-old African American woman named Henrietta Lacks died at the Johns Hopkins Hospital from metastasis of a cervical cancer. Metastasis is the spreading of tumor cells from their original site to various sites throughout the body. When her cancer was first detected in February 1951, it was an unusually bright red (highly vascularized or supplied with blood), puffy lobe of tissue on the cervix (the opening into the uterus or womb). Most cervical tumors are pale. A biopsy specimen of Lacks's tumor was taken to the Geys' lab and put into tissue culture. It started growing immediately and continued to grow past the normal Hayflick limit. The culture was code-named HeLa (for Henrietta Lacks) and became one of the most famous cell lines in tissue-culture history. HeLa cells were found to be so rugged that they could withstand shipping through the mail. Soon laboratories around the world had HeLa cells to work with.

Shortly thereafter, other labs reported being able to establish permanent cell lines from a wide variety of tissues. With so many cells now in circulation, mix-ups were inevitable, and there was no means for properly identifying them. Of course it was possible to distinguish mouse from human chromosomes or perhaps use antibodies to distinguish certain molecular cell markers, but these were very crude methods. As expected, various labs began finding that mouse cells had invaded their human cultures, while human cells were growing in cultures of pig, duck, and mouse cells. Faulty technique was usually to blame, for example, mislabeling of cultures, using the same pipette to feed different cultures ("changing their diapers"), or taking it on faith that newly acquired cells were from the species and tissue type specified by the sender. The American Type Culture Collection (ATCC) in Washington, D.C., was established in 1962 as a private supplier of tissue-culture cells and it became the nation's leading repository of reference cells.

In 1966 a Seattle geneticist named Stanley Gartler announced at a scientific meeting in Bedford, Pennsylvania, that he could identify types of human tissue by the isoenzymes that they produced. *Isoenzymes* or *isozymes* are various forms of an enzyme that perform the same function, but are constructed differently so that they can be detected by such techniques as electrophoresis or amino-acid-sequence analysis.[1] One of these isoenzymes is named G6PD (glucose-6-phosphate dehydrogenase) and exists in two forms, A and B. Type A occurs almost exclusively in blacks (Negroes). Another isoenzyme was PGM (phosphoglucomutase) that normally occurs in types 1, 2, and 1-2. Type 1 occurs only in about one-third of an average population.[2] Gartler had examined eighteen long-lasting cell lines, supposedly from Caucasian patients, and found all of them had type A G6PD. Only the HeLa cell line was at that time known to have come from a black patient. The tissue-culture specialists at that meeting were disturbed by and skeptical of Gartler's report. Leonard Hayflick was at the meeting, and he was an officer of the Tissue Culture Association that sponsored the conference. He had originated a cell line called WISH, one of the lines that Gartler had discredited. It was taken from the amniotic sac ("bag of water" surrounding the fetus) in which his daughter Susan was born. (WISH stands for Wistar Institute Susan Hayflick.) Both Leonard and his wife were Caucasians. The point that Gartler was trying to make was that at least some of the samples in circulation were not what their labels indicated. It was not known how long these impostors had been used in research, but it was possibly as long as fifteen years. No one wanted to entertain the thought that much of the time and money spent on research using these cells during that interval might be worthless. If it were true, how much of that work would now have to be repeated? Two years elapsed before Gartler's conclusions were confirmed (although why it took so long was not explained in Michael Gold's book *A Conspiracy of Cells*).[3] Twenty-four of the thirty-four cell lines of the ATTC proved to be either contaminated by or totally composed of HeLa cells.

Lewis Coriell of the Institute for Medical Research in Camden, New Jersey, found in 1961 that merely pulling a stopper from a test tube or dispensing liquid from a dropper could create aerosols containing cells that float around just waiting to find their way into a culture bottle. If the floating cells were HeLa, they usually could survive the airborne phase, and upon contaminating a different cell line, could rapidly outgrow the natives. After several transfers to new media, most, if not all, of the native cells would have disappeared leaving a culture of mostly, if not totally, HeLa cells. Until the late 1950s or early 1960s, "spontaneous transformation" was unknown in human cell lines. Most spontaneous transformations of tissue-culture cancer cells seemed to have the same characteristics as follows:

- They undergo rapid transformation to an aggressive growth form.
- They have the same nutritional requirements.
- They seed new tumors when inoculated into a hamster cheek pouch (an immunologically privileged site, where foreign materials are not subject to attack by the immune system).

• Many transformed cells have abnormally shaped chromosomes and most of them carry the same antigens on their surface membranes.

A warning appeared in the 1968 catalog of the ATTC that twenty-four of their cell lines were actually HeLa cells. In 1970 it was discovered from the original biopsy slides that HeLa cells were not epidermal cells (as claimed by the pathologist who in 1951 described Henrietta Lacks's type of cervical cancer), but were immature glandular cells, a tissue type known to be far more aggressive in the body. Another theory suggested that HeLa cells had undergone evolutionary changes in tissue culture that amplified their aggressiveness. Indeed, several subfamilies of HeLa cells did exist.

In 1972 the USSR gave the United States six different cancer-cell cultures in exchange for a set of thirty animal viruses implicated in cancer. President Richard Nixon had negotiated an agreement with Leonid Brezhnev in May 1972 calling for the two nations to cooperate in biomedical research. The cultures found their way to Walter A. Nelson-Rees, director of the Cell Culture Laboratory of the University of California at Berkeley, located at Oakland's Naval Biosciences compound. Nelson-Rees was instructed that no one was to be provided with these materials or with any data acquired using these cultures unless specific authorization had been obtained through approved governmental channels. Nelson-Rees immediately disobeyed the order and sent samples of each culture to Ward Peterson at the Child Research Center in Detroit for biochemical tests to characterize each line. While waiting for the results, Nelson-Rees investigated the chromosomes of these cells and found that each line was missing the Y chromosome. They were therefore all female cells. Nelson-Rees immediately "smelled a rat." About a month later, Peterson's results were received—all six lines were G6PD type A. Nelson-Rees then contacted Wade Parks, a virologist at the National Cancer Institute, and had him identify the virus in each cell line. The answer was that they all contained the same monkey virus even though each culture had come from a different research institute in the Soviet Union. Nelson-Rees and his assistant Bob Flandermeyer then applied a new chromosome-banding technique to these cultures. This technique enabled them to identify each type of human chromosome (there are twenty-three chromosomes common to both males and females, whereas the Y chromosome is found only in males). HeLa cells were found to be characterized by six distinctive chromosomes, including one called Mickey Mouse (with "ears") and one called Zebra (because of its bold stripes). All six of the Russian cultures had the HeLa cell chromosomal markers.

The Washington bureaucrats feared to make this information known because it might jeopardize the future of the entire biomedical exchange program with the Russians. Nevertheless, the decision was made to inform a Russian delegation on November 12, 1973, at the National Cancer Institute. The Russians were upset at the revelation, but they did not want to break off the exchange program because of this "bad news." After the meeting with the Russians, Nelson-Rees was free to publish his findings. He sent his manuscript to the *Journal of Virology*. One of the journal's referees (judges) said it was an "attack on the Russians." Another said it's just a footnote to history and not a new study with new scientific implications. His paper was rejected. Livid with rage, Nelson-Rees withdrew the paper and sent it to the *Journal of the National Cancer Institute*. It was accepted in June 1974 and published the following September, almost two years after the Russian cells arrived in the United States.

In the meantime, Nelson-Rees had found five American cultures to be HeLa cells, including the widely used human breast-tumor lines HBT3, HBT39B, and a human embryonic-kidney line HEK. He submitted a paper to the journal *Science*, but it was rejected. Apparently *Science* was not interested in alerting the scientific world to the fact that some major cell lines used in cancer research were not what the researchers thought them to be. Bob Bassin, who originated HBT3 in 1972, wrote to twenty researchers around the world whom he knew were using his cell line. He warned them that they may be HeLa cells and asked each one to inform any others with whom they had shared these cells. Bassin sent a copy of the letter to Nelson-Rees who forwarded it with his manuscript again to *Science*. This time it was accepted and rushed into print on June 7, 1974. Of course when the press heard of the scandal it made headlines across the nation.[4]

CANCER WAR SET BACK: GOOF COSTS 20 YEARS OF RESEARCH

A line of human tumor cells used by laboratories around the world for more than 20 years may have invalidated millions of dollars worth of cancer research, according to a scientist's report. As a result, says the author, Dr. Walter A. Nelson-Rees, checks are in order for dozens of laboratories engaged in cancer research.
—*Los Angeles Herald Examiner*

DEAD WOMAN'S CANCER CELLS SPREADING

Dr. Walter Nelson-Rees, one of the most experienced cell biologists in the world . . . has reported that many cell lines are by no means what they are thought to be by the laboratories handling them.
—*Miami Herald*

A SHOCKER FOR SCIENTISTS

"The main situation has probably existed for years," said the main author of the report, Walter A. Nelson-Rees, a highly respected researcher. . . . Nelson-Rees said the contaminating potential of the HeLa cells is well known, but that sufficient precautions against it have apparently not been taken.
—*San Francisco Chronicle*

Nelson-Rees was made a member of the advisory board to the cell-banking program of the National Cancer Institute (NCI) by its director, Robert Stevenson. Both men launched a campaign to persuade journal editors to require complete, authentic descriptions of every cell line used in a published research report. Even the Tissue Culture Association's own journal, *In Vitro*, had no such requirement. Relda Cailleau worked at the M. D. Anderson Hospital and Tumor Institute in Houston. She and Nelson-Rees knew each other professionally. In June 1975, Nelson-Rees asked some potentially embarrassing questions of Cailleau about her lung-cell culture MAC-21 that had been established sixteen years earlier. She became furious and accused him of sabotaging other peoples' work and ruining their reputations. Nelson-Rees checked into MAC-21 in May 1974 and found it had the A form of G6PD. This cell line had been used extensively for lung-cancer research. When he informed Cailleau of this, she refused to believe it. She did not bother to have an independent test made to settle the case. She simply knew he was wrong and carried on her research as before. Her reaction was typical of many of the other researchers contacted by Nelson-Rees. But Cailleau's response got him all fired up, and spurred him to draft a new "hit list." It was published in *Science* in January 1976. In addition to all the other indicators, his report included the HLA profile of each strain, and the name of the person who supplied the culture for him to analyze. HLA stands for *human leukocyte antigen*. These antigens are immunological molecules found on all nucleated cells of the body, but are most easily detected on white blood cells (leukocytes) by reaction with specific antibodies. Each individual has a virtually unique set of HLA antigens, almost like a fingerprint. Thus, barring identical multiple births such as identical twins, no two unrelated individuals should have the same set of HLA antigens. HLA typing had only recently been developed. Nelson-Rees had been advised to omit the names of suppliers, but he thought to do so would defeat the whole point of the list.

In a 1972 issue of the *Journal of the National Cancer Institute*, Richard Akeson of the UCLA Medical School claimed to have found antigens specific to a lung-tumor-culture designated line 2563. This line had been derived from Cailleau's MAC-21. Nelson-Rees tested 2563 and found it to be HeLa. He then wrote to John Bailar III, editor of the *Journal of the National Cancer Institute*, asking that this information be published quickly as "an important addition to the increasing knowledge of what does and does not constitute organ- or tissue-specific antigens."[5] Bailar sent Nelson-Rees's letter to Akeson and Cailleau for rebuttals. This time, Cailleau had her one remaining ampule of MAC-21 analyzed independently for chromosomes and isoenzymes. The report confirmed it was HeLa. Cailleau then confessed this to Bailer. It took a year from the time that Nelson-Rees notified Bailar until his letter was published in the *Journal of the National Cancer Institute*.

During 1978 Nelson-Rees received a report by a microbiologist at the

NCI and found that it had omitted technical descriptions of certain cell cultures at the request of Cailleau (they were her cultures). Nelson-Rees was amazed that, first of all, Cailleau would have the gall to dictate the handling of someone else's manuscript, and secondly, that the author would knuckle under to such a demand. A few weeks later, the Tissue Culture Association asked Nelson-Rees to evaluate some abstracts. One report claimed that a breast tumor contained cells with HeLa-like chromosomes, but these cells had come directly from a pleural effusion and were analyzed immediately (and therefore, by implication, could not have become contaminated by HeLa cells). The authors were Sen Pathak (a karyologist or specialist in chromosome identification), Michael Siciliano (an enzymologist), and Relda Cailleau. Nelson-Rees asked Pathak for the photographs of the chromosomes. They arrived two months later. Nelson-Rees and Flandermeyer could not find HeLa-banded chromosomes in the photos. Then Bailar asked Nelson-Rees to review a new twenty-five-page paper by these same authors, in which they claimed that isoenzyme tests and chromosomal analyses were of questionable value (i.e., not unequivocal evidence) in assaying for HeLa contamination. It was, in essence, an indictment of Nelson-Rees's indictment. Nelson-Rees studied the breast-cancer cells of Pathak et al. and found that twelve of the fourteen isoenzymes tested were identical to HeLa, including G6PD. However, the patient was a black woman and many of the isoenzymes tested were common to 90 percent of the human population. One of the enzyme differences was PGM3 (a standard "marker" molecule characteristic of HeLa cells). Nelson-Rees and Flandermeyer went carefully over the chromosome "mug shots" and discovered that they were photographic composites from various cells in this breast-cancer culture.

Shortly thereafter, at the 1978 Tissue Culture Association meeting in Denver, Cailleau campaigned for Nelson-Rees's resignation from his newly elected vice presidency of the association, because of his "hit list" and other "paranoid" behaviors. Pathak presented the paper entitled "Fresh Pleural Effusion from a Patient with Breast Cancer Showing Characteristic HeLa Markers" without reference to Nelson-Rees or challenge to the usefulness of his techniques. Before the conference began, Nelson-Rees had told Bob Stevenson, "I'm really going to throw the fat on the fire."[6] The word had gotten around to the chairman of the meeting. So when Pathak had finished delivering his paper, Nelson-Rees raised his hand. The chairman, anxious to avoid an incident, called on someone else. Nelson-Rees waited for that question to be answered, but would not be denied again. He simply stood up and without the aid of a microphone began to read his statement. The chairman tried to cut him off, but Nelson-Rees moved from his seat to the front of the room, saying to the chairman, "I'm going to finish this."[7] He related his discovery of how the Pathak team had "fished about" in several cells from this culture to find their so-called HeLa chromosomes. "We (Nelson-Rees and

Flandermeyer) have carefully studied many different cell lines and observed a good number of marker chromosomes, but we've *never* had to create a collage such as has been done here in order to sensationalize a point."[8] The audience gasped in astonishment. Apparently they were more concerned with the breach of etiquette displayed than with correct scientific method and inference. Nelson-Rees also indicted the authors for plagiarism, accusing them of lifting HeLa marker chromosomes from the work of Heneen as published in the Swedish journal *Hereditas* and presenting them without credit as their own observations, thus: ". . . some normal chromosomes were erroneously classified as markers and vice versa, and at least in one instance a chromosome's image was used twice, albeit after reversal of the negative."[9] In conclusion, Nelson-Rees said (to the Pathak team), "I suggest that you stick to pleural effusions and leave the HeLas to us."[10] He deposited 150 copies of his statement at the back of the room as he left the participants in stunned silence. Suddenly pandemonium broke loose as half of the audience rushed for the exits. Relda Cailleau grabbed a stack of Nelson-Rees's statements and began passing them out as proof of her claim that he was not fit to review manuscripts for technical journals. "Just look at how he violated the reviewers' oath of confidentiality!"[11] He did not deserve to serve as vice president of the Tissue Culture Association because of his rude and unethical behavior. In contrast with Cailleau's behavior, Pathak was so distraught by what he felt was wrongful accusations and public humiliation that he refused to do any research on the chromosomes of tumor cells for an entire year. Nelson-Rees offered a different explanation years later: "A little dog—or even a big one—who has peed on the living room rug and is found out *always* hides underneath the couch for a couple of hours."[12] Despite Nelson-Rees's protest to the *Journal of the National Cancer Institute*, the paper delivered by Pathak at that meeting was published in February 1979 essentially in an unrevised form.

Contaminated cell lines had occurred even in the best of labs. For example, in October 1978, Jonas Salk (creator of the famous polio vaccine) announced that HeLa cells had taken over one of his monkey-heart tissue lines. He had actually injected terminal cancer patients with HeLa cells believing them to be monkey-heart cells![13] A debate had been going on for many years as to the safety of growing viruses in continuous cell lines for the purpose of making vaccines. Salk, like many others, thought that cells might gradually evolve HeLa-like attributes in tissue culture even though there was (and still is) no scientific evidence to support this theory.

In 1957, a paper was published that described the first culturing of cells from an eight-year-old Dutch boy operated on for kidney stones. A scientist named van der Veen started a cell line from the boy's kidney tissue. It only grew for twelve days and then stalled. After many transfers, the culture's growth suddenly took off. He named it T-1. In 1979 a Pennsylvania State University report claimed that small amounts of gamma radiation would kill T-1 cells. Paul Todd, one of the authors of this report, tried to submit their cell line to the ATCC but were rejected because HeLa markers were in it. The authors tried to mention it in previous reports on T-1, but reviewers for the journal *Radiation Oncology* called the information "cell culture folklore" and said it was out of place in their journal. Over twenty years, various T-1 strains had become favorites of radiation-health researchers all over the world. Nelson-Rees published a report in the August 1980 *Science* about the HeLa nature of T-1 cell lines. In that same year, the Oakland lab's budget was cut 20 percent.

Nelson-Rees learned in February 1981 that the genetic engineering company Genentech was using the WISH cell line to make an immunoregulatory protein called *interferon*. Genentech refused to believe that their cell line was actually HeLa. The motivation for the biomedical boom in the 1970s was the government's "war on cancer" program. In the 1980s, the motivation for the biotechnology boom was capitalistic profit. Proprietary information and corporate secrets are not in the best interests of the advancement of science. The budget cutback caused Nelson-Rees to retire in 1981.

In his book *A Conspiracy of Cells*, Michael Gold lists some of the reasons he thinks caused HeLa cells to persist for such a long time. One of the reasons was the overconfidence that tissue-culture workers had in their own abilities to prevent contamination. Even when confronted with evidence of contamination, many of them were reluctant to admit their mistakes. There was an unwillingness to discount perhaps years of research that was thought to be valid. Naturally they were worried that if news of their carelessness became known it would put their grant applications in jeopardy.[14] In other words, they might be out of a job. There was (and still is, in most cases) a tendency to rush into publication so as to be first (prestige factor); success in a recent project usually helps in obtaining a grant for the next one.

Nelson-Rees never received the proper recognition he so justly deserved. "They don't award Nobel Prizes for finding out that things are wrong."[15] His contributions were of a nature that was bound to make a lot of enemies. After Nelson-Rees retired, the Oakland facility was closed and all of its cultures were transferred to the ATTC. Until that time, researchers could have their cultures verified for free. Thereafter they would have to pay. My question is: if this service was available, and especially since it was free, then why wasn't everyone taking advantage of it regularly as a part of their quality-control programs? I can only surmise that there were no such programs in most tissue-culture labs. That, in my opinion, is poor science—really not science at all. Nelson-Rees is gone from the tissue-culture scene now, but he left science with a valuable legacy. During his lifetime, he found that about one-third of the cell lines in use throughout the world were HeLa cells. He showed us that scientists cannot always be trusted to "know what kind of cells they are working with," and to be intellectually honest in their reports. It takes guts to admit mistakes and make that information known to the scientific world so that further use of contaminated lines can be halted. He also showed us that we cannot always trust journal editors and reviewers to act in the best interests of science. Should information concerning contaminated lines be withheld from those who need to know in order to protect someone's reputation? Doing the kind of job that Nelson-Rees did is a thankless task. Scientists of his caliber are rare. Who will carry on the fight now that he is gone?

UP THE RIVER

The Charles River Breeding Laboratories in Wilmington, Massachusetts, is one of the largest suppliers of laboratory animals in the United States. Its stock includes several inbred lines of mice that are widely used in research. By a system of brother-sister matings, a line of mice can be brought to genetic purity in about twenty generations. These highly inbred lines are often of much greater use than non-inbred mice for many purposes. For example, a lot of modern in vivo cancer research relies on tumor transplantation. A tumor transplant will normally be rejected by a recipient unless the donor and recipient are genetically identical. One of the greatest potential sources of variation inherent in any biological experiment is the genetic variation from one individual to another. By using animals from one of these highly inbred lines, the experimenter has eliminated at least one major source of variation from his data. This increases the likelihood of discovering a statistically significant difference between "test" and "control" groups with smaller numbers of animals.

A strain of mice called BALB-c has been inbred since 1913, so that all of the individuals of the same sex in that line should be as identical as identical twins. During 1981, when Brenda Kahan was studying tumor-cell differentiation, she began to get results that were "either very spectacular or a mistake. She was sufficiently cautious to doubt her results."[16] She and her coworker, Robert Auerbach, traced the problem back to the BALB-c mice she was using. They analyzed the mice for two biochemical markers and found that they were not the same ones that characterized the BALB-c strain. "By no means were we convinced, even then, that anything was wrong with the mice at Charles River," said Auerbach. Just to be on the safe side, he and Kahan ordered more mice from three different Charles River breeding units. They repeated the tests and, according to Auerbach, found that at least 75 percent of the mice from two of the three units were not pure BALB-c. At Auerbach's request, the tests were independently repeated and her results were confirmed at the University of Minnesota by Barbara Alter and Fritz Bach. Kahan and Auerbach announced their findings in the July 22, 1982, issue of *Science*, stating that "the seriousness of our findings cannot be overemphasized . . ." because the conclusions of "possibly hundreds" of published experiments that relied on the genetic purity of this strain may be wrong. Henry Foster, executive vice president of Charles River, did not dispute Kahan and Auerbach's charges, but called their actions "unorthodox" because they failed to inform the company as soon as they suspected something was wrong with their mice. Foster claimed that the company had already destroyed the questionable colonies of mice as a result of their own genetic monitoring system by the time the paper was submitted to *Science* in January 1982.

Harold Hoffman, program director for the genetics quality-control program at the National Institutes of Health, revealed that a similar complaint had been previously lodged with him by an NIH immunologist the previous summer. The results were the same, and his agency informed Charles River of the problem in the fall of 1981. Regardless of how or when the company was made aware of the problem, says Auerbach, it is "unfortunate that Charles River did not notify their users." Hoffman agrees that the contamination "has put a lot of research in jeopardy. . . . One Sloan-Kettering investigator lost one whole year of work as a result of this."[17] Moreover, many researchers are literally "up the creek (or "river," if you like) without a paddle," not knowing if their results are valid because the questionable mice have all died and cannot be retested.

LEGALITIES

John Moore was diagnosed in 1976 as having a rare form of blood cancer called hairy-cell leukemia. Seeking a second opinion, he went to a recommended expert in the field, Dr. David W. Golde, chief of hematology/oncology at the University of California at Los Angeles (UCLA) Medical Center. Golde confirmed the diagnosis and recommended that a splenectomy (removal of the spleen) be performed. Another UCLA physician removed Moore's grossly enlarged spleen and, at Golde's request, sent a piece of it to him for further analysis. Golde and his colleague Shirley G. Quan studied the spleen tissue and found that it produced high quantities of a variety of interesting and potentially useful immunological regulator proteins. In January 1981, over four years after Moore's leukemia was diagnosed, Golde and Quan applied for a patent on a cell line named Mo that they developed from Moore's spleen cells. The patent was issued in 1984 to UCLA, but any profits derived from Mo were to be split evenly; one-half going to UCLA and the other half to Golde and Quan. A few months later, the researchers had signed a contract with the Genetics Institute of Cambridge, Massachusetts, worth $400,000 to study the therapeutic properties of products derived from the Mo cell line. By 1985, the Genetics Institute had successfully cloned two important products of the Mo cell line, the first step in commercial production that had the potential to generate millions of dollars in profits. In the meantime, Moore had recovered from his surgery, but was unaware of the research that had gone on and the plans for commercial exploitation of his cells' products. Even though the symptoms of his disease had disappeared, he had agreed to return to Los Angeles from Seattle twice a year for follow-up examinations. Golde took blood samples at each visit over an eight-year period and found that Moore's blood contained over half a dozen valuable components that increased the effectiveness of the immune system. According to Moore, Golde soon offered to pay for his flights and to foot the bill for his stay at the posh Beverly Wilshire Hotel. Moore accepted the free flights, but preferred to stay with relatives rather than at the hotel. Golde's generosity aroused Moore's suspicions. Moore claimed to have asked him several times if there was any commercial value to the material they were taking from him, and was given a negative answer.

Moore had originally signed a consent form that authorized his surgery. He was asked in September 1981 to sign a paper giving the university rights to his spleen cells, but he refused until September 1983. Moore then gave the university the right to conduct research on his tissues, but declined to relinquish his rights to any products that might be developed from his blood. Moore thought there might be some things that the university wasn't telling him about, so he hired Beverly Hills attorney Sanford Gage in April 1984 to investigate the case. Gage soon found that Golde had been granted a patent on the Mo cell line the previous month, and that Golde and UCLA had signed a contract with the Genetics Institute and Sandoz Inc., an international drug firm, for worldwide distribution of future products from the Mo cell line and from Moore's blood cells. Moore was angered when he found all this out. He authorized Gage to file a lawsuit against the Regents of the University of California, UCLA Medical Center, Golde and Quan, requesting the Los Angeles Superior Court to grant him a share of whatever profits might be generated from the use of his cells. Gage argued that, "They're taking something that doesn't belong to them, which is of value, and they're not disclosing that value to the patient."[18] Gage further contended that "If John Moore wanted to make his cell line available to other researchers—to the world—he could not. He would be infringing on their patent on his cell line."[19] Allen B. Wagner, a University of California lawyer, rebutted, "The university's position is that the research was conducted upon a specimen of his therapeutically removed diseased spleen. . . . John Moore came to where he knew the most current research was being conducted. Can he now say, 'if you want to do any research after me, I want to control it when it applies to anyone else'? Moore wants the property rights to his cells—the rights to exclude others from using or enjoying the benefit of them."[20] I disagree with Wagner's summary of what Moore was asking for. All I see Moore asking for was a share (amount unspecified) in the profits from the use of his cells. I am unaware that he made any demands on controlling the use of his cells, although I can see where some people may want that right. For example, if a person knew that a virus to be used in germ warfare was going to be extracted from his cells, he should have the option to prevent this. Thus, Moore's case is not just about money, but represents a pioneering bioethics adjudication.

The university argued that California's health and safety code stipulates that body parts obtained during surgery may be retained for scientific research. The court did not accept this argument because "simple consent to surgery does not imply a consent to medical research unrelated to treatment nor to commercial exploitation of the patient's tissues."[21] The Superior Court of California dismissed the case in 1986, ruling that the complaint was technically defective and did not demonstrate that a taking of property had occurred. The case was appealed, and on July 21, 1988, the California Court of Appeal overturned the lower court's decision, stating: "To our knowledge, no public policy has ever been articulated, nor is there any statutory authority against a property interest in one's own body."[22] The case was to be returned to the Superior Court for a trial unless an appeal to the California State Supreme Court was successfully made (which it eventually was). The university based its appeal at least partly on the dissenting opinion from appeals court judge Ronald M. George: "A patient who consents to surgical removal of his bodily substances has no reasonable expectation as to their subsequent use other than an understanding that licensed medical personnel will comply with applicable medical standards and legal restraints."[23] George maintained that the legislature is the proper body to define the issue of human-tissue property rights, rather than the court. He warned that, if Moore wins his case, a multitude of former patients might file for remuneration for parts removed from them and successfully manipulated to produce valuable products. An enormous record-keeping system would have to be established at each research institute to keep track of each patient's cells as they moved through various research channels and from one institution to another as scientists farmed out or shared various aspects of each research project. It might take many years before anything of commercial value is produced from a given patient's cells. The vast majority of these projects would be failures, and all of that bureaucratic red tape would have been for naught—adding to the burden of "overhead" costs of research. Guidelines would have to be established that addressed the percentage of profits that the patient should receive, the time frame during which the patient's rights existed, the question of heir's rights, etc. "The potential complexity is horrifying to contemplate."[24] Patients might even postpone needed surgery to shop around for deals on their used parts. John Fletcher, former chief of the bioethics program at the National Institutes of Health, predicts that the final ruling in this case will "make investigators think three or four times about the potential use of their research materials."[25] Advocates of patient's rights claim that a victory for the university would allow researchers to mine human bodies for private profit. The ruling of the Supreme Court had not been made at this writing, but its decision will be of great importance to both patients and science for generations to come.

Even before the Moore case, excessive litigation was recognized as thwarting scientific and technological innovation in the United States. For example, most companies have stopped research in the United States on contraceptives and drugs to counteract infertility and morning sickness. The president of one pharmaceutical company asked in 1986, "Who in his right mind would work on a product today that would be used by pregnant women?"[26] Not everyone reacts the same to a given drug. Even drugs that have undergone extensive testing for safety and effectiveness may cause adverse side effects in rare individuals. The costs of such litigation, which can be in the millions of dollars, can nullify all the profits derived from the drug, and in some cases may cause the manufacturer to go bankrupt. In such a litigious climate, pharmaceutical companies may opt to stay with their "tried and true" products, rather than pump any money into research and development of new products. According to Peter Huber,

> The various elements of liability in the courts today all join to thwart inno-vation The most regressive effects are felt precisely where fruitful innova-tion is most urgently needed. Liability today is highly—and often indis-criminately—contagious. Progress is undercut the most in the markets already battered by a hurricane of litigation: contraceptives, vaccines, obstetrical services, and light aircraft, for example.[27]

Huber suggests that our judicial system needs reforms. In cases concerned with private risks (e.g., in transportation or personal consumption), "fair warning and conscious choice by the consumer must be made to count for much more than they do today—not because individual choices will always be wise (they surely will not be) but because such a system at least allows positive choice and the acceptance of change."[28] In cases where the public does not have a choice (e.g., mass vaccination programs, chemical and nuclear waste disposal, and central power generation), the public safety obvi-ously must continue to be delegated to the expert agencies acting for the col-lective good. "At the very least, full compliance with a comprehensive licensing order should provide liability protection against punitive, if not compensatory damages. It has always been true that ignorance of the law is no excuse. Today knowledge of the law is no excuse either. It should be."[29]

Since 1790, the United States Patent Office has issued more than 4.75

million patents on various products and processes. However, it wasn't until 1980 that any biological organism received a patent. In that year, the U. S. Supreme Court ruled that Ananda Chakrabarty, a scientist employed by General Electric Company, was entitled to a patent on a novel strain of bacteria that digests oil slicks. This decision set a legal precedent that was to encourage investments in commercial genetic engineering research. George B. Rathmann, chairman of the Industrial Biotechnology Association and chief executive officer at Amgen in Thousand Oaks, California, said, "The Chakrabarty decision set the stage for Genentech's public offering, and was directly linked to our ability to start this company [Amgen]."[30] In 1985, the Patent Office Board of Appeals and Interferences allowed the patenting of novel plant forms. On April 3, 1987, the same board allowed the patenting of animals (humans excepted) with unique, man-made characteristics that do not occur in nature. This decision was made while considering an application for an oyster (*Crassostrea gigas*) with an unusual chromosome number of thirty instead of the normal number of twenty. The alteration was designed to prevent a souring of the oyster's meat that normally occurs during the summer reproductive period. The oyster patent application was rejected on the basis that it was not unique. Another species of oyster (*C. virginica*) had already been developed that also contained thirty chromosomes. But the way was now open for patenting unique forms of animal life. Animal-rights advocates, a few farm coalitions, and environmentalists petitioned Congress to ban all animal patents. Issues raised included the morality of patenting life, whether farmers would have to pay royalties on the progeny of patented animals, and the risks of ecological disruption involved in releasing new organisms into the environment. To date, these pleas to ban patenting of life-forms have won little support in Congress.

Most of the patents awarded to genetic engineering companies have been for unique processes rather than for products such as animals or other organisms. Since it is far easier to detect and prove infringement on a product than on a process, most companies would prefer to have patents on products. "Process patents have also been a source of international friction. While product patents receive the same protection outside the U.S. as they do domestically, process patents do not." A quirk in U.S. law allows foreign companies, manufacturing in plants outside the United States, to employ a U.S. patent process and sell the resulting product in this country without paying royalties.[31] According to David J. Mugford, associate general counsel at Schering-Plough in Madison, New Jersey, "The issue is extremely important to biotechnology" because of the large number of process patents in the United States.[32] Rathmann (CEO, Amgen) concurs, stating that "what took a U.S. inventor five years can be replicated abroad in a matter of days."[33]

The first U.S. patent on an animal was issued to Harvard University in April 1988.[34] The animal was a genetically engineered mouse, containing a gene that caused a rapid development of tumors when exposed to small amounts of carcinogenic chemicals. The gene was introduced into early stage embryos, so the trait became heritable; i.e., offspring of these mice are also highly susceptible to developing tumors. This animal model was expected to be very useful in testing substances for their carcinogenic (cancer-producing) potential and also for testing potential anticancer drugs. This means that now lower doses and fewer animals will be needed than before.

ACCESS TO DATA

Science cannot advance knowledge rapidly unless researchers have the freedom to investigate wherever they think that the best data can be obtained. This freedom may be severely restricted in totalitarian countries, as exemplified by the Lysenko affair discussed in chapter 5 in this book. Many problems in modern science require international cooperation and the free exchange of ideas and sources of data. All too often, provincialism prevents foreign scholars from studying the problems they wish to tackle. For example, archeologists foreign to Turkey who wish to search for Noah's ark on Mount Ararat have had a tough time obtaining permission to enter that country for that purpose. Geologists who want to study specific geological formations in various countries may find it difficult or impossible to get visas into the respective countries, depending on the political situation at the time. The giant panda is such an endangered species that China is unwilling to release any specimens to other countries for study. This is understandable because the kind of bamboo they prefer to eat grows only in restricted regions of China. The panda's fate probably depends more on China's success in preserving the native stands of bamboo in their natural habitat than any other factor. Foreign zoologists, if allowed visas to study pandas in China, must do so without interfering with China's research plans. This is commonly the rule when scientific specimens are rare or unique. National interests usually take precedence over the free inquiry that science is ideally supposed to represent. In some cases, foreign scientists have been prevented from studying such specimens until "native" scientists have essentially exhausted their interests, as you will soon see when the following case study has been presented.

Amphibians are thought to have evolved from bony fish about 400 million years ago. Bony fishes (Class Osteichthyes) are classified into two major subclasses: the Actinopterygii contains the ray-finned fishes, including the bulk of modern-day fishes, whereas the Choanichthyes (Sarcopterygii) contains the fleshy-finned fishes such as lungfishes (Dipnoi) and the Crossopterygii (fringe-finned fishes such as coelacanths). The amphibians (today represented by salamanders, frogs, and toads) evolved from the fleshy-finned fishes that had a bony structural support in their paired fins. All tetrapods (four-limbed vertebrates, including ourselves) descended from these early amphibians. The lungfishes are not good candidates for amphibian ancestors because their skeletons are quite different (especially in the cranium) from that of the earliest amphibians found in the fossil record. Crossopterygian fishes, however, are better candidates for ancestors, but they were thought to have all perished with the dinosaurs some sixty-five million years ago at the end of the Cretaceous period. A hot debate is currently in progress as to whether the dinosaurs (along with more than half of the other species of plant and animal life at that time) died out at the end of the Cretaceous period because of climatic changes brought on when Earth collided with an extraterrestrial object such as a comet or an asteroid, or whether it was because of massive volcanic outpourings, or both.

In 1938, Margaret Courtnay-Latimer, curator of the local museum in the port town of East London, South Africa, discovered an odd-looking fish among the catch of the day caught in the net of a trawler from the Chalumna River. There before her eyes was this ugly, heavy, dirty, oil-dripping fish, the likes of which she had never seen before. She didn't know exactly what it was, but she knew it was important. After preserving it, she sent a sketch of it to the renowned ichthyologist (fish scientist) James Leonard Brierly Smith, professor of chemistry at Rhodes University (then College) in Grahamstown, South Africa. It was he that made the identification of it as a true coelacanth. But how could this be? All of its near relatives were thought to have become extinct millions of years ago. Wonder of wonders! She had found a "living fossil"; that is, an extant (living) species that has somehow survived through long periods of geologic time with very little evident evolutionary change (evolutionary stasis). It was like finding a dinosaur in your backyard! Smith named it *Latimaeria chalumnae* in her honor and to acknowledge the location from which it was taken. It is now agreed that this species is probably not in the direct line of evolution to the amphibians, but is thought to be closely related to that line. It is the great-grand-cousin rather than the great-grandfather, so to speak. Nevertheless, it offered a marvelous opportunity to test some ideas about the early evolution of tetrapods. Smith immediately set into motion a plan to capture more coelacanths. He plastered the shores of the Mozambique channel with posters offering a reward for the delivery of other coelacanths.

It wasn't until December 1952, fourteen years later, that the second coelacanth was found. This time it was found nearly three thousand kilometers away from East London by a fisherman named Ahmed Hussein Bourou, from the Comores Islands (in the Indian Ocean). He brought it up at night, from the depths, on his two-hundred-meter fishing line. This second coelacanth looked like the first: bony scales of steel-gray flecked with white, powerful jaws and teeth, a double tail, an elastic unsegmented notochord in place of a backbone, and a primitive single vestigial lung characteristic of the sub-class Sarcopterygii.[35] The lung was filled with fat, possibly as an adaptation for buoyancy control. But the fish lacked two fins that later were judged to have been bitten off and the scars healed. "It was a proper prehistoric monster too, huge and primal. Although it was a zoological sensation in the outside world, it was not considered unusual by the natives. A few 'kombessan' ('gombessa') or 'mame' were caught, measuring one to two meters in length, near that emerald archipelago each year and sold in the marketplace, dried and salted."[36] The meat is oily and not very good to eat, so it isn't worth much as a food fish. But what a waste of a scientific resource!!! Coelacanth fossils had been found all over the world, and the most ancient of these goes back some 370 million years. These ancient fossils were found in Scotland, Norway, and Canada, all northern parts of an ancient landmass called Laurasia. This region had many lakes and rivers. The climate was subject to periodic droughts. In such environments, lobe-fins could provide a selective advantage to a fish in a pond that was drying up, because it could use its fins to crawl a short distance overland in search of a more favorable aqueous habitat. Those ancient coelacanths had five-toed fins and other skeletal similarities to the earliest amphibians. Later on, coelacanths migrated from their freshwater habitats to the open ocean where *Latimeria* was found. By 1976, eighty-five coelacanths had been taken from the Indian Ocean, all near two of the four Comores Islands. No one knows why they have been found exclusively there and nowhere else.

When Bourou pulled the coelacanth into his canoe, he bashed it on the head to kill it because these fish can be very dangerous when loose in a small boat. First thing in the morning, he took it to market. Abdellah Houmadi, a local hairdresser, spotted it and told the head of the primary school, Affane Mohamed, about it. Mohamed knew from the posters that it might be a coelacanth and that he should take it to a "responsible person." Captain Eric Hunt used a small schooner to trade between the mainland and the Comores Islands. He had been advertising a reward for a coelacanth. Mohamed was aware of that fact and that Hunt happened to be about forty kilometers away at Matsamudu on the other side of the island. The islands were preparing for a sports festival and Mohamed was the captain of the soccer team from Anjouan Island. He had already made arrangements to travel to the games on Hunt's schooner, so this is how he knew where Hunt was. The games were to be held on the island of Mayotte where the French governor lived. Hunt learned very quickly from the "palm radio or bush telegraph" that a possible coelacanth was on its way to him. Hunt offered to take it to the capital. Mohamed had to agree because none of the natives knew what it was or how to preserve it. According to Mohamed, Hunt had begged him not to ask the governor for compensation for the fisherman Bourou. He would pay the entire 50,000 francs of reward money when they returned to Anjouan.

Hunt had met J. L. B. Smith and his wife earlier in the year at Zanzibar. Smith had asked Hunt to keep his eyes open for a coelacanth and gave him some reward posters for the French authorities who administered the Comores Islands. Hunt was instructed that if no Formalin was available to preserve a specimen, that he use salt. When Smith returned to Durban from his collecting trip about ten days later, he found a cable waiting for him from Hunt. It read, "Have specimen coelacanth five feet injected formalin stop absence Smith advise or send plane immediately." Formalin was not available at Mutsamudu because the local doctor was away. So Hunt had split and salted the fish according to Smith's directions, but did some damage to the specimen in the process. When Hunt and Mohamed arrived at Dzaoudzi, the local doctor supplied Hunt with all the Formalin he had, and they injected the carcass at many sites. The French governor, Pierre Coudert, was apprised of the fish's importance, and he was sympathetic to Hunt's argument that "since Smith was the one who had recognized the fish in the first place, had the posters printed, and was offering the reward, he should have the specimen."[37] Coudert didn't feel comfortable about making such an important decision on his own, so he cabled the nearest French research station on Madagascar. Jacques Millot was in charge of this station and it was he who later would lead the French research program on coelacanths. But it was Christmas Day, and the station was closed. Coudert decided to make a compromise. If Smith claimed the specimen in person within a few days, he could have it. Otherwise it would be sent to the French research station on Madagascar. Hunt fired off another cable to Smith: "Charter plane immediately authorities trying to claim specimen but willing to let you have it if in person stop specimen different yours no front dorsal or tail remnant but definite identification Hunt." After five days of frantic activity, Smith finally got a ride aboard an army plane, thanks to Prime Minister Daniel Malan. The fact that Mrs. Malan just happened to have chosen to read Smith's *Sea Fishes of Southern Africa* during her Christmas vacation might have been instrumental in the prime minister's decision.

Smith arrived in the Comores at daybreak on December 30, 1952. He hurried to Hunt's schooner to see for himself if the fish was really a coelacanth. After all, Hunt was not an ichthyologist and the specimen was reportedly missing two of its fins. It lay in a crate wrapped in cotton. Smith recalled that, "My whole life welled up in a terrible flood of fear and agony, and I could not speak or move. They all stood staring at me, but I could not bring myself to touch (the crate) . . . and . . . after standing as if stricken, motioned them to open it, when Hunt and a sailor jumped as if electrified and peeled away the enveloping shroud. . . . God yes it was true . . . it was a coelacanth alright [*sic*]."[38] But now he had to rush to the governor's reception and the pilot was worried about the weather on the return flight. The fish was carried to the plane while Smith tried to be a polite and grateful guest at the party. He announced that he would guarantee a reward for the next specimen, and that it would go to the French government. He left as soon as he could.

Affane Mohamed had been at the airport to greet Smith and was there as he departed. But no one had spoken to Mohamed. Mohamed felt personally slighted that Smith was in such a hurry to be off with his prize. When French scientists learned that Smith had absconded with one of their precious coelacanths they were furious and the French government barred him from ever returning to the Comores. The French administrators became security conscious, and for thirteen years thereafter claimed the next eighty-three specimens for exclusive study by French scientists. Consequently, most of the definitive work on coelacanths was published by the French. Beginning about 1963, foreign scientists were offered coelacanth specimens at more than $5,000 each. Sometimes these specimens carried unwritten restrictions on competing research, thereby allowing the French to stay in the forefront of that scientific work. The four Comores Islands remained French possessions until 1975. Three of them later gained their independence after much political upheaval.

It had been known that a fossil coelacanth contained tiny youngsters within its body cavity, but they were dismissed as either artifacts or the result of cannibalism. The most popular hypothesis of the day was that coelacanths were oviparous, i.e., they laid large, grapefruit-sized, yolk-filled eggs that allowed them to develop in the sea apart from the mother. Robert W. Griffith, a Yale researcher, knew that no shell glands had been found in the reproductive tract of coelacanths, and that the eggs had a delicate outer membrane that would be likely to rupture easily in the open ocean. Sharks are the only other fishes known to have similar large, fragile eggs. Furthermore, embryo sharks are very late in developing the retention of urea in their blood as an osmotic adaptation to living in sea water. Griffith therefore concluded that either the eggs of the coelacanth must be laid within heavy capsules that isolate them from sea water (as in sharks) or be retained in the mother's body until late in development and delivered as live young (ovoviviparous). Griffith and his Yale associate Keith S. Thomson published these ideas in 1973, but they were not accepted at the time. The American Museum of Natural History in New York obtained a preserved specimen from a French doctor in the Comores in 1962. According to museum ichthyologist C. Lavett Smith, "We were bound by a gentleman's agreement not to publish until after the French had finished their anatomical work. We could have dissected it as long as ten years ago, but it was the only specimen in the Western Hemisphere, and we didn't want to mess around with it. I guess we were treating it as an art object instead of a working scientific specimen. We're kicking ourselves now, though . . . ,"[39] because the coelacanth turned out to be gravid (pregnant), the only one yet caught. When hematologist

Charles S. Rand and Smith cut into the specimen, they found four baby coelacanths, each about one foot (thirty centimeters) long and perfect miniatures of the adult. This finding supported Griffith's ovoviviparous hypothesis. This mode of reproduction requires internal fertilization, but no intromittent (insertion) organ has been found. It has been hypothesized that two pairs of erectile caruncles (claspers) flanking the male's cloaca (external opening for both reproductive and waste products) might serve as an intromittent organ for placing the male's sperm into the female's reproductive tract.

Keith Stewart Thomson, director of Yale's Peabody Museum of Natural History, was the first to request from the Comoresan government a frozen specimen.[40] All previous requests had been for preserved specimens. On Good Friday, April 5, 1966, Thomson received a call from the U.S. State Department in Washington, D.C., informing him that a frozen-food ship reached Marseilles, France, where it was to undergo refitting. The freezers were to be turned off. A box addressed to the Peabody Museum was in one of those freezers. The consul-general, Mr. D. V. Anderson, had the box moved to a commercial storage facility, and had made arrangements for its passage to New York. A few days later, Thomson received a letter from the Comoresan ministry stating that a coelacanth had been caught on March 14, frozen immediately, and shipped on a fruit boat to Marseilles. This letter was sent on the following boat. The specimen arrived at the Hoboken docks on Memorial Day, 1966. The Peabody Museum put it on public display for a few days on a small bier in a borrowed ice-cream freezer case with a glass top. Most of Connecticut filed past the carcass like it was Lenin's Tomb.

Thomson knew that he could only thaw the specimen once. He contacted everyone that he thought might possibly want samples for research. When the fish was thawed and dissected by the museum staff, it was almost an anticlimax. They had hoped to study the gut contents to see what a coelacanth eats, but the gut was empty. They explored the spiral valve of the intestine that is normal for cartilaginous fishes like the shark, but is not common in bony fishes. The heart was surprisingly small. Professor Clem Markert had a taste of its meat and found it to be good but oily. A defining characteristic of crossopterygian fishes is a skull that is hinged on the posterior part, with the cheeks and lower jaws mainly attached to the anterior part. The brain runs through this skull joint. French researchers had already published the conclusion that this joint was totally immobile in the coelacanths. Thomson had been interested in this joint since 1964. It seemed obvious to him (although naive to others) that the complexity of the jointed elements bespeaks of an actual flexure of the skull itself, even though that would involve bending the brain stem. Thomson took the fresh specimen in his hands and lifted the snout. To his great delight, the snout moved easily and at the same time the lower jaws slid forward, exactly as he had predicted they should. It is now known that many types of vertebrates (mammals being one exception) have some sort of intracranial kinesis (movement) so that the mouth can be opened by the simultaneous lowering of the lower jaw and raising of the upper jaw (along with movements in at least some parts of the cranium).

Professor Grace Pickford and Keith Thomson were associates at the Peabody Museum of Yale University. Pickford took a sample of its blood and found that it was full of nonprotein nitrogen, making it slightly hyperosmotic (containing more dissolved substances) than seawater. It was immediately suspected that the nitrogenous substance would prove to be urea, and later studies confirmed that hypothesis, because that is how sharks prevent water loss from their bodies. This is because water tends to flow from regions of higher concentration (seawater in this case) to regions of lower water concentration (higher content of other molecules such as salt and urea) in the tissues of the coelacanth. Many other studies on blood and tissue substances were made from that specimen and subsequent fresh-frozen specimens. An unsuccessful joint expedition to the western Indian Ocean was made in 1969 by the Royal Society and the National Academy of Sciences. In 1972 an expedition by the Royal Society, the National Academy, the National Geographic Society, and the Muséum Nationale d'Histoire Naturelle was able to obtain and briefly film a living coelacanth.[41] After those expeditions, however, the Comores experienced political upheaval that made it difficult to obtain more specimens.

In January 1975 John E. McCosker, superintendent of the Steinhardt Aquarium in San Francisco, mounted a major expedition to bring a coelacanth back alive. He offered a round-trip, two-week pilgrimage to Mecca to anyone who captured a live coelacanth. In response, he had over a hundred black Moslem fishermen working full time in their own interests, but with potential benefit to science. I view this as a delightfully bizarre juxtaposition of religion and science. Despite all his efforts, McCosker was unsuccessful in obtaining a live specimen. He was able to return with two frozen coelacanths, however. His advice to future coelacanth hunters was: "(1) Wait until the political future of the Comores is defined (this may require a year or so). (2) Request permission through the president of the Comores, listing the potential value of such a mission on a trade-off basis, e.g., publicity and tourism. (3) Bring everything that might be needed—make no assumptions about local availability of anything. (4) Bring lots of Kaopectate" (a treatment for diarrhea).[42]

No coelacanth has survived captivity at sea-level pressure for more than twenty hours. Using a two-man submersible vehicle in 1987, Hans Fricke[43] observed that coelacanths swim oddly, sometimes backward, sometimes belly-up. The right front fin works in tandem with the left rear fin, and similarly for the left front and right rear, like the gait of a trotting horse. The flexible front fins are also able to rotate nearly 180 degrees, allowing the fish to "scull" and stabilize its sluggish body as it drifts with the current along the bottom. Although some coelacanths were observed to rest on the bottom, they were not seen to use their fins to walk along the bottom.

Thomson[44] concludes his article by stating "There is no great moral in this story," to which I say, "Au contraire, mon ami." There is a great moral in this story. It is in the importance of maintaining good relationships between nations so that information and access to limited data sources can be freely shared. A scientific problem should never be thought of as the sole property of a single nation. Science ideally knows no boundaries. The isolation of the rest of the world from coelacanths, after the second one was taken back to Rhodes University by Smith, should never have occurred. Such actions can place stumbling blocks in the path of scientific progress. With all their possessiveness, the French never did discover such things as intracranial kinesis or urea retention in coelacanths. Maybe Smith should have been more diplomatic in his acquisition of the second specimen. But the retaliation of French scientists was not just against South Africa, but also against scientists from every other nation. Scientific advances are hard enough to win without adding these kinds of unnecessary restrictions.

The coelacanth story also shows us how great discoveries often depend on chance events. Miss Latimer, who recognized the importance of the original specimen, the alert hairdresser Houmadi, the schoolmaster Affine Mohamid, who knew whom to contact when the second specimen turned up, the accidental meeting of the Smiths with Captain Hunt in a Zanzibar marketplace, Mrs. Malan's choice of reading material and her possible influence on her husband's decision to fly Smith to the Comores, the message that could not be delivered on Christmas Day, the Solomonic decision of the colonial French administrator Coudert, and the single-mindedness of a lone scientist (Smith) all played significant roles in the discovery of the second coelacanth.

Thomson correctly points out that "biology . . . proceeds iteratively using hypotheses, data, inferences from these data, and more refined hypotheses, turn and turn about."[45] There were, for example, two schools of thought on the subject of intracranial kinesis, but which explanation should one choose?

> Obviously this is a problem in all sciences, but because the data biologists (and especially evolutionary biologists) use are almost always incomplete, and multiple possible explanations of them abound, the exercise of discretion in the choice of hypothesis is a major art. . . . The choice of hypothesis can therefore be a controversial business, which some highly productive people avoid whenever possible, although no blame should attach when work honestly done turns out to be superseded. We have to expect it. It has to happen all the time if the field is to advance.[46]

The discovery of living coelacanths gave us a special opportunity to encounter (as T. H. Huxley put it) those "ugly facts that slay beautiful hypotheses," and to develop new and better ones.

BONES OF CONTENTION

In October 1982 the government of Ethiopia stopped all foreign prehistory expeditions in that country which, equivocally, has the richest source of fossils relating to early human origins. Ethiopia's Hadar region was the site where a team of United States anthropologists (including Tim White of the University of California, Berkeley, and Donald Johanson, later to become head of the Institute of Human Origins, Berkeley) had earlier discovered the famous prehistoric human fossil named "Lucy." White, Johanson, and Desmond Clark had planned to return to the Hadar region in the fall of 1982 under a half-million-dollar grant from the National Science Foundation (NSF), but were prevented from doing so by the moratorium. Ethiopia's commissioner of science and technology, Haile Lul Tebike, said, "We are not being anti-American or nationalistic. We are merely trying to establish in Ethiopia the kind of rules and regulations that govern this type of research activity in other countries. I can't say our work will be finished next year. I can say that we have no interest in delaying a decision."[47] Clark and White submitted a report to Steven Brush, program director for anthropology at the NSF, stating, "We were assured by all officials that the action was *not* directed against our expedition and that we would be invited to continue our research once new antiquities regulations were formulated."[48] The NSF report also relates some "deeper reasons" for the temporary ban on prehistory research in Ethiopia.

The history behind that decision goes back at least to 1975 when Jon Kalb, a geologist now at the University of Texas at Austin, formed the Rift Valley Research Mission in Ethiopia. Kalb and several students from the United States and Ethiopia did anthropological, archeological, and geological research in the Middle Awash region. This is the site where the Berkeley team worked in 1981 and from which they were barred in 1982. Kalb had been associated with Johanson in research in that area, but they split up at the end of 1974 because of, according to Kalb, "disputes . . . over scientific and management issues."[49] In August 1978 the Ethiopian government expelled Kalb from that country because there had been persistent rumors that he was associated with the CIA (Central Intelligence Agency). His research project in Ethiopia collapsed and some of the students in the project became tainted with suspicion merely by their association with Kalb. An investigation by the NIH failed to reveal any evidence to substantiate the rumors. Understandably, no foreign country would put much credence in an investigation by the accused's own country. Clark wrote to Bisrat Dilnes-sahu, dean of the faculty of science at Addis Ababa University, rather than to an Ethiopian governmental agency, for help in planning a research trip to the Awash region. Permission was eventually granted, and the Berkeley team made their first expedition there in 1981. Even though Kalb was later barred from doing such research in Ethiopia, he somehow felt "that a legitimate claim on access to the Middle Awash and other localities established by his colleagues in the research mission, both American and Ethiopian, has been unfairly and unethically overridden by the Berkeley team."[50] However, Fred Wendorf (of Southern Methodist University in Dallas, Texas, and a former member of Kalb's research mission) had decided to shift his research to the Nile Valley and invited Clark to take over the research that his mission had started in the Middle Awash.

Clark had attempted to involve other members of Kalb's research mission, but four of Kalb's Ethiopian students felt that they had been deliberately excluded. They submitted a position paper to the Ethiopian government and academic officials claiming their priority in the Middle Awash and stating their belief that French and American researchers had done an unsatisfactory job in training Ethopian students. All four of those students were studying in the United States on Fulbright scholarships. When Clark learned of this petition, he wrote to the Ethiopian Centre for Research and Preservation of Cultural Heritage suggesting that invitations to join his 1981 expedition should be extended to these students. Clark did not know the whereabouts of these students. One of them, Sleshi Tebedge, was at the University of Texas in Austin. Tebedge claimed that neither he nor his colleagues received an invitation. He felt that "in any case, why should we have to be invited by Clark? That is our research area."[51] When Tebedge arrived in 1982 at the Middle Awash region on Wenner Grenn money, he found his access blocked because Clark had the required permit to work there. Another Ethiopian, Tsrha Adefris, was studying anthropology with Clifford Jolly at New York University. She had planned to study fossil-preservation methods in Nairobi so that she could clean and restore the Bodo skull, a 300,000-year-old cranium of *Homo erectus* found in 1986 by Kalb's research mission. When she returned to Addis Ababa to begin a teaching position at the university, she found that the Bodo skull was on an officially approved, nine-month loan to Berkeley where White was doing her planned work. The Bodo skull had been lying around for six years without receiving any serious work, so White decided to take it upon himself to do the restoration and return it to the Ethiopian collection. Kalb, however, wanted it to remain in Ethiopia where trained Ethiopians would eventually prepare it.

Tebedge and Adefris joined forces during the summer of 1982 to lobby Ethiopian officials to block the second stage of the Berkeley expedition. Clark and White wrote in their NSF report, "They [Tebedge and Adefris] contacted governmental and university officials and poisoned the atmosphere with a welter of false and ridiculous accusations that we had stolen fossils from Ethiopia, excluded Ethiopians from the field, failed to train Ethiopians, failed to provide facilities in Ethiopia et cetera et cetera."[52] One of the very incriminating facts in this case was revealed in Johanson's book *Lucy*.[53] There he described how he and his colleague, Tom Gray, removed a leg bone from a recent Ethiopian grave for comparison with a newly discovered fossil knee joint estimated to be about three million years old. The Ethiopian commissioner for science and technology declared, "Desecrating graves is not tolerated in any culture, is it?"[54] Johanson also made some unkind remarks (perhaps true, but undiplomatic) about corruption and ignorance in Ethiopian politics and culture that did not sit well with the Ethiopians. There appear to be legitimate complaints on both sides. But Lewin[55] documents his belief that the central complaint against Clark—that he failed to include Ethiopians in his projects—flies in the face of the facts. Glynn Isaac, a Berkeley archeologist, is quoted as saying, "Clark has worked in many parts of Africa over many decades. He has always been assiduous in helping and training local students and academics."[56] Noel Boaz, a New York University anthropologist, declares that "Desmond [Clark] has always done his best to arrange for foreign students to study at Berkeley. He was one of the first people to do this."[57] Since 1974, all of Clark's expeditions have included one or two local students and one or two faculty members from Addis Ababa University.

The wider problem of how to foster research in Third World countries is of great concern. Richard Leakey, then director of the National Museums of Kenya, commented that, "They [foreigners] have an obligation not to take material out of the country and to develop within the country facilities where they can be studied."[58] A Harvard paleontologist, David Pilbeam, stated that, "It's not so much the facilities that matter. It is the people who are able to have frequent contact with each other that makes research successful."[59] Should Ethiopia decide to prohibit removal of extremely valuable specimens from that country, it would likely attract international funds for building research facilities there, just as occurred in Kenya during the 1970s. According to Pilbeam, "Ethiopia should not see it as denigrating to have international collaborative projects in this area. It is impossible for any single institution to plan such an expedition. What is important is the genuine and productive involvement of local researchers and students."[60]

In conclusion, Lewin states, "It would surely be a tragedy if the unfortunate history of the past few years were to erect an unbreachable barrier between those Ethiopians previously associated with the research mission and those now linked with Berkeley. Such a division in the country's intellectual resources would be of service to no one."[61]

An amendment to the Archeological and Aboriginal Relic Preservation Act, a state law of Victoria, southern Australia, was adopted in 1984. Under this law, the University of Melbourne was directed to relinquish its substantial and important collection of more than eight hundred Aboriginal skeletal remains to the Victoria Museum, where their ultimate fate would be decided by a state-appointed committee of Aboriginal people. The Victoria Museum already had some six hundred skeletal remains. Jim Berg, executive officer of the Victorian Aboriginal Legal Service, declared, "We start from the position that everything must return home, to be reburied."[62] Together, the university and museum collections represent the bulk of Australia's most significant anthropological collections. Berg argues, "The desecration of burial sites and the locking away in museums of our ancestral remains has shown a complete lack of respect for the Aboriginal community. It causes us great anxiety and distress. So you can imagine that arguments by anthropologists that these represent important scientific collections don't go down very well with the Aboriginal community."[63] Alan Thorne, an anthropologist at the Australian National University, Canberra, concedes that the Aborigines' moral case is unassailable. The scientific argument is not likely to be represented in the deliberations because there are no Aboriginal physical anthropologists. And whose fault is that?

About fifty crania from Coobool Creek, all about ten thousand years old, are included in the Melbourne collection. These fossils offer an unparalleled opportunity to help answer one of the most interesting questions of Australian prehistory: Did the indigenous population derive from two separate migrations of Indonesian and South Asian people about forty thousand years ago? Being of intermediate age, the Coobool Creek fossils were considered to be especially important in this respect.

Australian archeologists, who had been active in raising the consciousness of the Aborigines to their cultural heritage, feared that the new law might eventually lead to the reburial of cultural materials in addition to skeletons. This would be a "cruel twist of irony" in that those scientists who had helped the Aborigines the most would also suffer the greatest loss of resources to study. Demands were also being heard that called for removing books that contained sacred Aboriginal pictures from libraries. The ramifications of the new law could be quite sweeping. Lewin recognizes,

> There is a clear parallel between American Indians and Australian Aborigines in terms of the inequities dealt them throughout history, but the contemporary combination of others' guilt and their [Aborigines'] own political clout appears to be handing the Aborigines an opportunity to grasp much more quickly and more completely what they now want: to wrest their heritage from the hands of their colonizers. But whether burying the whole of their heritage is the best way to preserve it, rather than entering into a collaborative scholarly appreciation of it as the American Indians have, is a matter that requires some dispassionate discussion.[64]

The situation in the United States is not quite as devastating to anthropological research as in Australia. Native Americans successfully lobbied politicians to pass a law in 1989 that required release of objects claimed by their tribes. The Smithsonian Institute's collection of American Indian skeletons and artifacts was the largest in the world. The Smithsonian agreed, under the new law, to release an undetermined fraction of its collection for disposal by modern Indian tribes if they file claims. These claims do not have to be based strictly on genealogy; cultural affiliation will do. Not only skeletal remains but also burial artifacts will likewise be released. Museums had been criticized for almost a decade that they were continuing the Indian massacres of the nineteenth century through disrespect for the dead in the twentieth century. Clement Meighan of the University of California at Los Angeles and founder of the American Committee for the Preservation of Archaeological Collections (ACPAC) compared the reburial program to book burning. "Every politician thinks it's their [*sic*] bounden duty to pass legislation controlling archeological research" because each gets credit for favoring minorities without any apparent cost.[65] However, Meighan maintains there is a cost to science. Fossils and artifacts are archeologists' books. Their reburial destroys irreplaceable research materials. He predicts that America's anthropological collections will all disappear unless these kinds of laws are repealed. Universities will have to remove all photos of Native Americans from their library books because some of these people believe the camera captures one's soul. This is undoubtedly an exaggerated claim, indicative of the passionate concerns in both camps concerning the reburial program. Scientists cannot freely investigate any problem they wish, even in a democratic society. Legalities or moral convictions may present restrictions.

PUBLICATIONS

A piece of research does not become a part of the growing body of scientific knowledge until it has been published, preferably in a refereed journal. Journals are the organs of scientific organizations through which their members can communicate their findings to others in the same discipline or to anyone that may be interested. Research reports are often referred to as "papers." Although each journal establishes its own style in which papers must be presented, they all must allow for, and indeed demand, that at least the following subjects be addressed. A statement of the problem must be given, preferably early in the paper, so that the reader may know exactly what the author(s) set out to investigate. After all, an evaluation of the degree of success of a project cannot be made unless one knows the purpose and limits of the work. Did the author do what she set out to do? If so, how well was it done in terms of expedient design, careful data gathering, thoroughness and appropriateness of analysis, logic of deductions, specification of all applicable alternative hypotheses, credit for those on whose work this project was built, and so forth. If any hypothesis guided the research, it too should be presented so that the reader may be apprised of a potential bias in the conduct of the research. One may be tempted to "see" what one wants to see during the conduct of research. This bias has been explored in some detail in chapter 4.

One of the most important parts of a scientific paper is the "materials and methods" section. Here the author presents in detail all of the information needed to repeat the experiment or to make a comparable set of observations and to follow the same type of analytical (perhaps statistical) procedure. It is the absence of this material, more than anything else, that makes magazine articles in *Scientific American* or reviews in *Advances in Genetics* less than truly scientific. Following the presentation of the results, one or more interpretations or inferences must be made. If the results are deemed inconclusive, that too is an interpretation. In this latter case, however, the author should, if at all possible, make suggestions as to how the ambiguity might be resolved. A call for repetition of the work is common, but not very productive. Perhaps a larger sample size or longer term of observation might generate enough new data to allow a tentative judgment to be made either supporting or refuting a particular hypothesis. Maybe choosing a different type of research material, such as a different species of organism, in which to repeat the experiment might produce more definitive results. What if Gregor Mendel had started his research on the hawkweed instead of using peas? Actually, Mendel was advised to work on hawkweed after he had reported his pea research. The hawkweed work led nowhere because, as we know now, that species is *apomictic*, i.e., it sets seed without fertilization. If there is no sex involved, naturally he could not find the same kinds of regularities as he did in sexually reproducing peas. It makes me wonder how many times in the history of science has the unfortunate choice of inappropriate materials been the downfall of research projects. Editorial boards of scientific journals are designed to act as referees that determine whether papers submitted for publication are worthy thereof. The quality of a journal may quickly diminish in the minds of the scientific community if it allows shoddy scholarship to appear in its pages. These reviewers understand the great responsibility that their position wields. They must maintain the integrity of the journal and yet be fair to those whose work might not fit into the mainstream of current scientific thought. Where does one draw the line between "radical thinkers" and just plain crackpots?

The Foundations of Quantitative Genetics

Ronald A. Fisher is acknowledged to be the greatest statistician of his generation, if not of all time.[66] He wrote a paper in 1912, while still a graduate student, that adumbrated the use of maximum likelihood as an estimating method. Despite his obvious potential, he had great difficulty finding a job, but finally was hired to teach physics and mathematics at Rugby and Hailey-bury Schools in England, on a naval training ship, and at Bradfield College. He found teaching frustrating, probably because he could not bring the subject down to the level of his students. There were two schools of thought on heredity in those days. One group was studying "all or none" qualitative traits like those that Mendel had seen in his pea plants (e.g., purple versus white flowers). The other group was concerned with continuously varying quanti-tative traits that had to be measured in some way (such as body height, skin color, and intelligence). Research on quantitative traits had begun in the late 1800s by Francis Galton, a cousin of Charles Darwin. These two schools were at war with each other during the early 1900s, each claiming that they had the key to understanding all forms of heredity. The Mendelian school was represented by William Bateson and R. C. Punnett. The biometrician group was championed by Karl Pearson and W. F. R. Weldon. These two groups were separated by methodological and conceptual gulfs that seemed insur-mountable. Fisher, while still a student at Cambridge, entertained the hypoth-esis that if a large number of Mendelian factors (genes) contributed to a trait it would explain the continuous variation that the biometricians were studying. He began gathering data on the observed correlations between rel-atives to see if they were compatible with his multiple-factor hypothesis. Fisher wrote a paper in 1916 titled "The Correlation between Relatives on the Supposition of Mendelian Inheritance," which he submitted to the Royal Society of London for publication. It so happened that Pearson and Punnett were assigned as reviewers of Fisher's paper. Both reviewers rejected it. It is claimed that this was the only instance in which those two archenemies (Pearson and Punnett) ever agreed on anything. Here was statistical evidence that could unify the two schools into a more comprehensive science of genetics. But Pearson and Punnett apparently had such animosity toward each other and such sensitive egos that they did not want to acknowledge that their differences could ever be reconciled. So they took it upon them-selves to prevent such knowledge from becoming public and recommended against the publication of Fisher's paper. Two years later, with the financial help of Leonard Darwin (Charles Darwin's son), the paper was finally pub-lished in the *Transactions of the Royal Society of Edinburgh*. This 1918 paper, more than any other, established the foundation of quantitative genetics. If papers of this caliber have trouble being accepted for publica-tion, how many worthy papers of lesser import suffer the same fate?

In 1919 Fisher finally got a job he liked at the Rothamsted Experiment Station. It was here that he worked out the statistical procedures for the analysis of variance and covariance, factorial design, field plot arrangements, and other experimental designs that are still used today. His interests later turned to evolution. In 1930 he wrote the monumental book *The Genetical Theory of Natural Selection*. This was one of the most influential works establishing a marriage between Darwinism and modern genetics, which came to be known as the "neodarwinian revolution."

Chaos

Robert May started out as a physicist at Princeton University, but later turned to biology. He was intrigued with designing mathematical equations that mimic the behavior of insect populations. Using an elementary nonlinear equation for his simulation, he made the number of insects in any year dependent upon the amount of food available during the previous year. If the value for the food supply was set below a certain value, the insect populations dwindled as expected. If he raised the food value somewhat higher, the population size would oscillate from year to year around a particular number. This would correspond, in the real world, to an initial overabundance of food that allowed the population to grow. The next year there would be less food for each individual. Many would starve and fewer would survive into the following year. Now food was again abundant, so the cycle would repeat. May then set the food value higher in his equation. The population unexpectedly responded by oscillating around two numbers. If he increased the food value again, the population size oscillated around four numbers and took four times as long to return to the starting point. Eventually he reached a critical food value where the population size became chaotic from year to year and could not be predicted. This bizarre behavior is now known as the "period doubling route to chaos."[67]

Mitchell Feigenbaum was a young physicist at Los Alamos, New Mexico, in 1976. Using May's simple equation, Feigenbaum found that the rate at which a complex system becomes chaotic when a critical variable value (like the food value in May's insect simulations) is raised is always the same. This was dependent on only two specific numbers that appeared to be unrelated to the equation. These values are now known as "Feigenbaum numbers." He concluded that any equation which followed the period doubling route to chaos would behave in the same way at the same rate. This concept was so bizarre to the editors of mathematics journals that for three years Feigenbaum's papers were rejected.

An Italian physicist named Valter Franceschini was trying in 1979 to analyze five messy equations that modeled turbulence in fluids. His simulations revealed period doubling to chaos just as May's insect populations had done. Franceschini was unaware of Feigenbaum's theory when he was told by a friend to look for the Feigenbaum numbers. Franceschini immediately reran his calculations on the computer and out popped the Feigenbaum numbers. Since then, numerous instances of Feigenbaum's theory have been observed. Other routes to chaos have been discovered and chaos theory has become one of the hottest lines of research, not only in mathematics but in all of the sciences. For three years, however, Feigenbaum's theory could not get a fair hearing by the scientific community because of the decision of editorial boards. The advancement of science has enough obstacles to overcome without adding this type of impediment.

Censorship

Let's face it. Newspapers are businesses—they are in operation to make money. Sensational news increases sales. However, newspapers must guard against printing unverifiable statements as facts. In our litigious society, newspapers are fair game for defamation suits that could run into millions of dollars in court costs and damages awarded to plaintiffs. This alone tends to keep them honest. But I like to think that most editors of reputable newspapers take pride in their journalism, and would not want to work for trashy tabloids like we see next to the supermarket checkout stands. However, newspaper editors generally do listen to their readers, but have been known to knuckle under to special-interest pressure groups, just like politicians. The same is true of editors of scientific journals. For example, an article appeared in the April 9, 1982, issue of the *Journal of the American Medical Association* (JAMA) reporting the results of a calcium-blocker type of heart medicine. Another heart drug, called nifedipine, was just about that time being manufactured by Pfizer Laboratories of New York under their trade name Procardia. Pfizer's ads in *JAMA* stopped on May 21, 1982. In an article in the *Los Angeles Times* of January 15, 1984, *JAMA* editor Dr. George D. Lundberg reportedly told associate editor Elizabeth Gonzales "that Pfizer pulled $250,000 in ads and threatened to remove two million dollars more because it thought a *JAMA* article favored a competitor's drug."[68] Gail McBride was a coauthor of the April 9 article and also director of *JAMA*'s medical news department. The *Times* reports that "As the summer wore on without Pfizer's ads, McBride said several options were discussed on how to return Pfizer to the fold, including a letter of apology or a follow-up article explaining that Pfizer was displeased with the original article. The second article was published September 17, and Pfizer ads were resumed on October 8."[69] Procardia was being launched commercially during the interval when its ads disappeared from *JAMA*. Over one hundred pages of Pfizer ads appeared in *JAMA*'s rival publication, the *New England Journal of Medicine*, during that period. McBride claimed that the second article contained "legitimate new information" on calcium blockers. "I was not ashamed of the article. But I was ashamed of the reasons why we did it." Lundberg, however, denied that the second article was published to placate Pfizer. He said that it was meant primarily to update developments with this type of drug and it compensated for a lack of balance in the earlier article. According to the *Times*, "Lundberg insisted that the editorial and advertisements departments are separate and that he would never knuckle under to advertisers." It's too bad that Lundberg didn't remember the adage to "never say never."

Publication Bias

For a long time I have been aware of the fact that studies with positive results are more likely to be published than those producing negative results. There are several factors affecting this publication bias. A major factor is the competition for page space in journals. It costs money to publish journals, and these costs have to fit within the budgets of the sponsoring scientific societies. Editorial boards must choose from a plethora of submitted manuscripts which papers are most meritorious. Those that produce unambiguous positive results are likely to be chosen over inconclusive papers or those with negative results because positive results are viewed as successes that might lead to immediate applications. If the problem is an important one, other investigators will soon repeat the experiment to test its validity. It is well known that experiments can produce negative results for many different reasons, most of which are unknown to the investigator. Thus, there probably is a tendency to view negative results as possibly being due to faulty design or conduct of experiments rather than valid results deserving of publication. Furthermore, whatever hypothesis was being tested would tentatively be rejected when negative results fail to support it. Thus, negative results are viewed as being of little interest or value.

In a perfect world, where costs of publication would not be a limiting factor, I would advocate that the results of scientific investigations should not determine their publishability. If a study has been conducted that gives negative results, I would like to know about it before I embarked on a research program on the same problem. Armed with this information, I may choose to use the same materials and methods if I thought, for example, that my techniques were better than those who tried this approach before. Or I might choose not to follow the same approach, but modify the design more or less. Whatever my decision might be, it would be tempered at the outset by knowledge of what had failed in the past. Without this knowledge, who knows how many other investigators will dream up the same approach to the problem and continually stub their toes on negative results? If these results never get into print, there could be a lot of wasted time, effort, and money poured down the rat hole of sterile approaches to solving scientific problems. What we need is a "journal of negative results." But this, unfortunately, is not a perfect world.

Colin B. Begg of the Dana-Farber Cancer Institute in Boston wanted to assess the efficiency of transplanting bone marrow for the treatment of leukemia by comparing his own statistical analyses of clinical trials at Dana-Farber with other published results. According to Begg, "Many people will do studies and then not publish them. And it's hard to figure out which studies aren't published. The issue of the representativeness of published studies becomes a critical one."[70] Beggs and his associate, Jesse A. Berlin, reported their findings on publication bias in the *Journal of the Royal Statistical Society* near the beginning of 1989. In that paper they state: "Publication bias, the phenomenon in which studies with positive results are more likely to be published than studies with negative [inconclusive] results, is a serious problem in the interpretation of scientific research. It occurs because the decision to publish is often influenced by the results of the study."[71]

As an example, other investigators found that a substantial proportion of studies known to be conducted on the effectiveness of a new therapy (unspecified) remained unpublished. Furthermore, 55 percent of the published trials favored the new therapy, while only 14 percent of the unpublished trials did so. Douglas G. Altman of the Imperial Cancer Research Fund in London, England, stated that: "Publication bias in the medical literature has been widely suspected for many years, but until recently there was little clear evidence. Now, however, there can be no doubt that there is publication bias, and that it is a serious problem."[72] Because the published data is not a random sample of results, Begg argues that the emphasis on tests of "statistical significance" for "proving" theories may be misplaced. Most scientists are aware that publication bias exists, but it is argued that well-informed researchers automatically treat published reports with appropriate caution. Unfortunately, more naive readers of published studies (or of newspaper reports thereof) may not appreciate the bias created by unpublished reports. Scientists and the public alike need access to all of the studies if a fair analysis of the problem is to be made.

Homeopathy

Homeopathy is an alternative medical system developed by Samuel Hahnemann in the late 1700s. Practitioners of homeopathic medicine believe that a substance causing certain unhealthy symptoms when taken in relatively large doses can, if taken in minute doses, be used to cure a person naturally suffering those same symptoms. For example, homeopaths might prescribe a minuscule dose of arsenic to cure a case of arsenic poisoning. This sounds to me like the guy who, after a night of boozing, tries to get rid of his hangover by having a small alcholic beverage ("hair of the dog") the next morning. The practice of homeopathy became popular in the early nineteenth century, but it soon was rejected in the United States. As a matter of fact, it was one of the main reasons for the formation of the American Medical Association in 1847. Its members were specifically forbidden to practice homeopathy.[73] Most Western countries have likewise discounted homeopathy as quackery, but the practice still remains popular today in France. One exceptional example may serve as a model for the acceptance of homeopathy. People who suffer from an allergy may *sometimes* become "desensitized" to the allergen (antigen of allergy) by a series of exposures to it. Initially, the dosage is very small; subsequent dosages are gradually increased until the patient is no longer reactive to amounts of the allegen in the environment.

Jacques Benveniste, a pharmacologist at the University of Paris-Sud, led a team of researchers in an in vitro homeopathic study in 1986. The project involved exposing human basophils (a class of white blood cells involved in allergic reactions) to successively diminishing amounts of immunoglobulin E (the class of antibodies associated with allergies) diluted in distilled water. When basophils are exposed to these antibodies, they can cause the cells to release histamine (a physiological mediator of the immune response). Many allergic people must take antihistamine drugs to combat the watery eyes, runny nose, sneezing, and difficult breathing that commonly occur during allergic attacks. The histamine is sequestered in membrane-bound "granules" (that stain dark blue with Wright's stain) in the cytoplasm of basophils. Basophils have membrane receptors for the tail end of the Y-shaped antibodies. When an allergen (such as a pollen grain) binds to an antibody, a signal is sent into the cell causing its granules to move to and fuse with the cell membrane. The contents of the granules are thus dumped into circulation and the histamine goes off to alter the activities of other host cells, producing the symptoms of allergy. This process of degranulation can be studied under the microscope using stained cells. The punctate appearance of the cytoplasm disappears as the basophil undergoes degranulation.

A serum containing antibodies was diluted in stages by Benveniste, sometimes by factors of ten, other times by factors of one hundred, and the diluted solutions were tested on basophils at each stage. The solution was ultimately diluted to as high as one part in 10^{120}, so that the probability of even one antibody molecule remaining would be vanishingly small. To his surprise, as many as 40 to 60 percent of the basophils continued to degranulate at the highest dilutions.

Please excuse me while I digress for a moment, but I don't see how this experiment, even if it worked as Benveniste found, supports the main goal of homeopathy, namely to provide a cure for an illness by application of tiny amounts of a medicine or drug. Regardless of the antibody dilution at which basophils degranulate, the release of histamine is probably going to do damage rather than help the patient. It seems to me that what Benveniste should have been testing is the effects of small amounts of histamine on the prevention or cure of allergies in patients. Oddly, none of the reviews of his work that I have seen address this point. In all fairness, however, the granules of basophils also contain potent chemicals other than histamine that might be involved in allergies.

The basophil response observed by the Benveniste team was not predictable as the antibody dilutions increased, however. The percentage of basophils responding would decline for several dilutions and then increase for the next few dilutions, and so forth. Benveniste found it necessary to vigorously agitate the solution for at least ten seconds after each dilution in order to obtain a basophil response. Since no antibody molecules should have been present at the highest dilutions, Benveniste proposed that "specific information must have been transmitted during the dilution/shaking process. Water could act as a 'template' for the [antibody] molecule, for example, by an infinite hydrogen-bonded network or electric and magnetic fields."[74] Benveniste submitted a report of his experiments to *Nature*, but it was refused publication for two years, even though the scientific referees could find no flaws in the work.[75] The result was just too bizarre. It flew in the face of conventional wisdom and the explanation for it was pure conjecture that involved processes unknown to chemistry and physics. *Nature* insisted that Benveniste should have his experiments independently repeated before his paper would be considered for publication. In compliance, Benveniste arranged for independent laboratories in Israel, Italy, and Canada to repeat the experiments. All of these labs were able to reproduce Benveniste's results. The names of the researchers from these three laboratories also appeared as coauthors on the final report (a total of thirteen scientists from four countries).

The editor of *Nature*, John Maddox, finally consented to allow the paper to be published. It appeared in the June 30, 1988, issue, and it immediately drew a storm of criticism from the scientific community that was mainly directed against Maddox for allowing such an unbelievable paper to be published in the first place. Maddox replied that he published it because referees could find no flaws with the experimental procedure and the results had been successfully repeated in three independent laboratories. Also the French press had discovered the story and was spreading the details of Benveniste's work, so Maddox might have felt he could be accused of censorship if he refused to publish it. *Nature* had published "unbelievable" results on at least two previous occasions. Both times, the papers were accompanied by remarks from referees. In 1972, *Nature* published a paper claiming that untrained rats could be induced to avoid the dark by injecting their brains with a substance called *scotophobin* that had been extracted from the brains of rats trained to avoid the dark.[76] A vigorous dissent was published in that same issue by the referee Walter Stewart. This incident started Stewart on a crusade for accuracy in scientific publications. The other paper, dealing with alleged paranormal abilities in certain people like Uri Geller, appeared in 1974.[77] It too was accompanied by an editorial explaining the reviewers' reservations. Following these precedents, Maddox wrote an editorial in the same *Nature* issue that carried Benveniste's paper, stating that "there is no physical basis for such an activity." Furthermore, he made it a condition of publication that an investigative team chosen by Maddox would later be allowed to watch Benveniste's group perform the experiments, and the results of that investigation could be published.

The investigative team included Maddox, Stewart (then with the National Institutes of Health), and James "The Amazing" Randi, a magician, professional skeptic, and winner of the MacArthur Award for his work in debunking faith healers and pretenders to paranormal powers.[78] The team went to Benveniste's lab early in July 1988 and reviewed all of the lab's notebooks. They found that the work had been done sloppily and lacked adequate controls. Stewart remarked that, "You're making judgments all the time—how pale something looks, or how red. If you have a concept in mind, you can easily get in trouble."[79] The team also found that the normal bell-shaped curve of random variation one sees in almost any set of observations of this kind was not evident in the lab's experiments. Futhermore, large amounts of data that failed to support Benveniste's conclusions had been omitted from his published paper. The procedure gave variable results from one time of year to the next or with different samples. Some lab workers, such as Elizabeth Davenas, the first author of the *Nature* report, were much better at observing positive results than others. That report failed to mention that one of the coauthors from Israel had backed out at the last minute in a dispute over the results, and failed to acknowledge that a homeopathic drug company was a financial supporter of Benveniste's lab.[80]

The investigative team witnessed seven replicas of the experiment. Both the experimenters and the observers knew the contents of each sample for the first three experiments. Only the observers knew the contents for the fourth experiment. All of these first four experiments seemed to have some effect at very small dilutions. In the last three experiments a coding system known only to Randi was employed, so that neither the experimenters nor the observers knew what the samples contained. In these "double-blind" experiments, the high-dilution effect was not observed. The observers concluded that self-deception rather than fraud probably produced the favorable results of the first three experiments. A report of this investigation was published in the July 28, 1988, issue of *Nature* along with a rebuttal by Benveniste.[81] In this rebuttal he pointed to the results of the fourth experiment, where the contents were unknown to the experimenters as an indication of either a genuine effect of microdilution or of fraud. But no fraud had been detected by the investigating team. He further denounced the behavior and the conclusions of what he called the "almighty antifraud and heterodoxy squad."[82] He pointed out that none of the investigators had intimate knowledge of immunology, and this ignorance caused various mistakes and misunderstandings during the investigation. Benveniste further charged in the French newspaper *Le Monde* that, "This was nothing but a real scientific comedy, a parody of an investigation carried out by a magician and a scientific prosecutor working in the purest style of the witches of Salem or of McCarthyist or Soviet ideology."[83] He asked, "Who, with even the slightest research background, would blot out five years of work and that of five other laboratories" on the basis of five days of investigation? To which Stewart replied, "One good experiment is worth more than five years of poor experiments."[84]

Arnold Relman, editor of the *New England Journal of Medicine*, commented that for such unbelievable results, *Nature* should not have published the Benveniste paper and then undertaken an investigation of its own. "A journal should not be an investigative body," he said. Otherwise, "the editor becomes the judge, the jury, and the plaintiff and—in some sense—the accused."[85] Maddox responded to Relman's criticisms in the letters-to-the-editor section of *Science* September 23, 1988. He pointed out that the editors of *Nature* have traditionally taken an active role in assessing the quality of what they are asked to publish.

Thus my first predecessor, Sir Norman Lockyer (editor, 1869-1919), was deeply engaged in the controversy on whether the solar corona is an attribute of the solar or terrestrial atmosphere (he backed the wrong side) and in 1904 commissioned from R. W. Wood a damning investigation of the spurious phenomenon of N-rays.* . . . I [Maddox] believe that our readers will have been instructed on three important points by the Benveniste paper and its sequel: how easily authentic science may be simulated by the careful selection of data and the judicious use of language, how even "rigorously and fairly" reviewed papers may embody defects recognizable even by people whom Benveniste (rightly, in the context) calls "amateurs," and—more alarming—how likely it is that much second-rate science finds its way into print somewhere.[86]

Two pertinent "letters" appeared in the July 23, 1988, issue of *Science News*. Caroline V. Rider writes,

I am outraged to read that the editors of *Nature* saw fit to characterize the work of six laboratories as "[a]n experiment whose conclusions have no physical basis." Reduced to its essentials, what the editors of *Nature* have said is this: These results are not explainable in the context of our present theories of antigen/membrane interaction. Therefore these results *do not exist.* To their credit, they did at least publish the accounts of the experiments. But it leads one to wonder whether other research is being suppressed by journal editorial boards because the results do not fit in with current theories. How frightening! How can science progress if the censors only allow publication of results that fit in with the currently accepted theories?[87]

The second letter was written by John C. Huffman.

---

*N rays are discussed in chapter 4.

> In puzzling over the dilution mystery, have the researchers forgotten about molecular movement across the liquid-gas interface at the exposed liquid surface of the laboratory containers being used? I submit that molecular transport is occurring between liquid surfaces in containers sharing the same gaseous (air?) ambient environment to an extent that far exceeds the dilution in experiments described. The gas-phase transport will surely exceed by many orders of magnitude dilution attempts of 10 factor only 26 times. This degree of dilution would theoretically leave only one or two molecules of the drug in a 100-milliliter container of solvent. Laboratory equipment capable of avoiding container-to-container molecular contamination would be very expensive, and would resemble facilities in the very best semiconductor-fabrication clean rooms.[88]

I have yet to see a rebuttal to Huffman's argument.

Ugly Science

The lofty goals of science are pursued by scientists who, unfortunately, are as human as anyone else. This means that they can make mistakes, they can fudge a little here or there, they will fight to get and keep their jobs, they will occasionally go beyond acceptable guidelines, and perhaps even perform illegal acts. Of course most of these indiscretions are individually rare, but the fact that they occur at all tends to tarnish the puritanical view of science held in the minds of the public. Philip H. Abelson commented on some of these problems in the editorial column of *Science*:

> A small but growing number of scientists have acted irresponsibly in their zeal to appear in print. Actions have included simultaneous publications of the same material in two or more journals, dual submission of manuscripts, and repeated publication at intervals of material that is different in form but not in substance.[89]

In the competition for research funds and promotions, the number of publications an author has to her credit is often a critical factor. One way to get your name on a lot of papers is to participate, either sequentially or jointly, in large team projects where your contribution can be made relatively quickly. Most of the contributors' names will likely be included in the "et al.'s" when the paper is cited, but when multiple papers are published each participant can take turns at being listed as first author. Some of these "teams" can be very large. For example, an article appeared in the September 1988 issue of *Physical Review Letters* titled "Experimental Mass Limit for a Fourth-Generation Sequential Lepton from e^+e^- Annihilations at s = 56 GeV" by G. N. Kim and 103 coauthors from nineteen universities, most of them in Japan.[90] The list of authors was so long that it took three pages, instead of the normal two, to present the table of contents. Outrageous as this seems, the National Library of Medicine found a 1986 paper in *Kansenshogaku Zasshi* with the title (English translation) "Comparative Study of MK-0791 and Piperacillin in Respiratory Tract Infections" by R. Soejima and 192 others from twenty institutions.

Other ploys that can be used to increase your bibliography are to fragment your research report into two or more papers (a practice that has been dubbed "the least publishable unit") or to submit slightly different versions of the same report to different journals. In rapidly advancing fields of science, it is important to get your results into print as quickly as possible to avoid being scooped by some other research group. No one likes to put all their eggs in one basket, so the temptation exists to submit your paper to more than one journal, with the intent of withdrawing all except one after you learn which journal will publish it first. Naturally, all journals hate this practice. It causes much waste of reviewers' time, slows the processing of meritorious material, and increases the costs of publication that must be borne by others, such as the government, libraries, and members of scientific societies. Until recently, there has been a kind of "gentleman's agreement" that authors would not submit a given paper to more than one journal at a time. But because of the abuse of this agreement, some journals have put this understanding in writing. Formerly, journals have not penalized authors who have violated this rule because, according to Abelson, "editors have enough responsibilities without taking on the task of policeman." Now, however, that practice is going to change. Benjamin Lewin, editor of *Cell*, and Daniel Koshland, editor of the *Proceedings of the National Academy of Sciences*, agreed in 1982 that they will refuse to consider for three years any manuscript by an author who published similar articles in those two journals. Abelson was told by another editor of an author who had submitted the same manuscript to two journals in a particular field. His penalty was severe. All U.S. journals publishing in that field agreed that they would never again consider any manuscript from that author's laboratory.

These problems got so bad that Harvard Medical School finally published guidelines suggesting that a person being considered for promotion to full professor should be judged on no more than ten of his papers, those for associate professor on seven papers, and those for assistant professor on five papers. DeWitt Stetten Jr., former deputy director of the National Institutes of Health (NIH), noted in 1986 that only twelve citations are required in nominations for both the Nobel Prize and membership in the U.S. National Academy of Sciences.[91] In other words, these guidelines were proposing that promotion boards and grant-giving committees should forget the numbers game and start making awards based on merits of the published work. The guidelines further suggest that authors be held responsible for the content of papers that bear their names and that "honorary authorships" should not be given on scientific papers. Only those who have "made a significant intellectual or practical contribution" should be listed as coauthors on these papers. It is important that scientists keep their own house clean because, if they don't, Congress is likely to get into the act to safeguard the integrity of the research community through the passage of regulations (not just guidelines) that would have the force of law.

Two guardians of scientific integrity, Walter W. Stewart (mentioned above in the story of homeopathy) and Ned Feder, both of the NIH, submitted a letter to *Nature* in the fall of 1983 detailing the extent of mistakes and fraud in the scientific literature. Their paper remained unpublished until January 1987.[92] Their study was intended to discover how well coauthors take responsibility for the accuracy of papers bearing their names. Stewart and Feder chose the infamous John Darsee as a test case. As a young medical researcher in 1980, Darsee was discovered to have fabricated much of his laboratory data. He collaborated with forty-seven other medical researchers (twenty-four from Emory University and twenty-three from Harvard) as coauthor on 109 publications (including journal articles, abstracts, and textbook chapters) between 1978 and 1981. The Darsee affair was the subject of three separate investigative reports from Harvard, Emory, and the NIH. These reports found no evidence of "wholesale forgery" on the part of his collaborators, according to Stewart and Feder, but they did find many lesser offenses. As an example, in an Emory paper describing a family with a high incidence of heart disease, a seventeen-year-old man was depicted in a family tree as having children ranging in age between four to eight years, suggesting that he must have been eight or nine years old when he fathered his first child. In the same family tree there was a woman in the preceding generation who apparently had her last child at the age of fifty-two. Actually, the ages given on the pedigrees were meant to be those at which heart disease was diagnosed, but this was not made clear in the original article. Thirty-one of Darsee's forty-seven coauthors were accused of similar "carelessness" by Stewart and Feder. Examples included "failing to check that graphs matched measurements cited in the text, failing to retain identifying information on human subjects and accepting coauthorship on studies to which they did not significantly contribute." Moreover, they charged that "13 of the coauthors were guilty of 'more serious' misconduct, such as publishing statements they knew or should have known were false or misleading, failing to acknowledge outside sources of important research data or publishing the same paper twice—under different titles and with slightly different information—so as to make it seem that there were two distinct studies."[93]

In a rebuttal following the Stewart and Feder paper in *Nature*, Eugene Braunwald, a coauthor on many of Darsee's papers, objects that "the paper repeatedly and unfairly connects Darsee's fraud at Harvard to his coauthors there through a process of guilt by association."[94] Stewart and Feder pointed out that the Darsee investigators at Harvard had found, in a study of heart-tissue recovery in dogs, that data from a previous experiment had been used as a control group, rather than running their own controls in tandem with the test group. Braunwald defended this procedure as a means of avoiding needless sacrifice of dogs. He admitted that it would have been better if the paper had explained the rationale for the use of "historical controls," but it was Darsee who omitted this explanation. It was also noted by Stewart and Feder that the Harvard researchers had twice published the same data on the dog-heart experiments. The first paper described their condition after three day's recovery from heart-tissue damage, and the second paper described their condition after two weeks. Braunwald explained that the early data had such "important clinical implications" that it had to be published quickly before they had time to review the two-weeks' results. Moreover, according to Braunwald, the second paper contained "substantial new information," including data on twenty more dogs. He castigates Stewart and Feder: "The general understanding of scientific fraud is hardly advanced by a discussion which hinges upon typographical errors and similar quibblings."[95] John Maddox, editor of *Nature*, also thought that Stewart and Feder's report had been too hard-nosed. "The recipe implicit in the Stewart and Feder argument, that zealous suspicion of everybody within sight is the way to ensure the integrity of the scientific literature, is of course a recipe for disaster, a road to general mistrust, a license for every would-be whistle-blower and a means by which the literature would be made yet more solemn."[96] He acknowledges that the problem of errors, inconsistencies, and misstatements in the scientific literature is too widespread, probably due to the pressure on scientists to publish in great quantity. Maddox concludes, "It does seem to be that people's promotions have come to depend far too much on what they've published. I think that some decoupling of the two is urgently needed."[97]

The NIH originally had no objection to Stewart and Feder's analysis of the literature along with their laboratory research. But as time went on, their Darsee project came to occupy so much of their time that their laboratory work had essentially come to a halt. William Raub, then newly appointed deputy director of the NIH and head of the office on research fraud, stated: "We agreed that it was a proper scientific inquiry,"[98] adding that no one envisioned that completion of the project would occupy Stewart and Feder full time for two-and-a-half years more. A few of Darsee's coauthors obtained legal counsel to block publication of the original version of the Stewart-Feder paper. The paper submitted to *Nature* in 1983 was withdrawn, and one of NIH's own lawyers had to be assigned to help produce a legally defensible manuscript.[99] By the time the final version was published, less defamatory language had been substituted to head off a legal battle. Stewart and Feder then submitted the revised paper to several other journals including *Science* and the *New England Journal of Medicine*, but it was rejected on editorial grounds. After three years, Maddox finally consented to publish the paper in the January 15, 1987, issue of *Nature* because the notoriety it had acquired gave it the distinction of being "the most famous unpublished paper in science."[100] Stewart and Feder were directed to return to their laboratory research, but the problems they uncovered in the scientific literature remain on the ugly side of science.

FINANCING

In the old days, scientific projects could be financed out of one's own pocket in many cases. The problems were much simpler then. But today many science protocols require sizable sums of money to pay for such things as wages, equipment, materials and supplies, computer time, publication charges, and the like. In addition, institutions take a bite out of the research budget by charging for overhead costs of administration, including secretarial work, accounting, pay roll and services, laboratory space, utilities, janitorial and security services, and other expenses. These are only part of the financial burdens that accompany the advancing sophistication of research. Scientific knowledge is growing exponentially, partly because there are more scientists working today than at any other time in history. Unfortunately, the needs of society have similarly expanded in many other directions such as care of the elderly, the homeless, those who cannot afford health care, prisons, the war on drugs, educational needs, pollution control and cleanup, and so on. The nation's budget simply cannot adequately fund all of these needs. Competition for government-sponsored grants is becoming increasingly severe. Many worthy projects cannot be addressed simply for lack of money. This single factor is probably the most critical "stone in the road" blocking the advancement of science at the rate of which it is capable. It is a problem that is likely to become even more crucial as we move into the twenty-first century.

"Big science" projects cost so much that they can be done only by national or international funding. For example, studying the physics of subatomic particles requires the development of ever more powerful "atom-smashers." The extreme temperatures needed to simulate the fusion reaction of our sun and other stars have required terribly expensive equipment, and we have yet to reach the break-even point where the amount of energy generated by fusion is equivalent to that expended. Modern astronomical tools are also expensive. The Hubble telescope put into Earth orbit in 1990 cost over a billion dollars. Even that expenditure did not assure its success. A flaw in the polishing of the mirror gave a distorted image. The general public and its legislative representatives were outraged, and rightly so in this case. Evidently, a computer error was made in transposing some values that misguided the polishing instrument. Because of this, science will no doubt find it more difficult to obtain money for these kinds of projects in the near future. Biological science embarked on its first really "big science" program in 1989 with the human-genome initiative. This project proposes to sequence the entire human DNA complement of the twenty-two autosomes (nonsex chromosomes) plus the sex chromosomes (X and Y). DNA sequencing involves determining the linear order of about three billion nucleotides (the basic building blocks or monomers) that make up the DNA polymers. The project was estimated to require at least five years, at a cost of about one dollar per nucleotide, or three billion dollars total. Other estimates put the value at several hundred million dollars, depending on the assumptions about hoped-for advances in automated DNA-sequencing technology.[101] The project may be nearing completion by the time this book is published. Oddly enough, the prime mover behind the human-genome project is the Department of Energy. Its rationale for getting involved in this biological problem stems from its interest in following up the effects of radiation on humans from the use of atomic energy. It's not only the human genome that will require funding either, because agriculturists want detailed maps of the genomes of domesticated animals such as cattle and swine or of plants such as corn, wheat, and rice. Immunologists and microbiologists want the DNA sequences of pathogenic bacteria and viruses. Other researchers want maps of their favorite experimental organisms, such as the mouse, the fruit fly, yeast, etc. It seems that there is no end to the demands that science places on the financial resources of the world.

Even in the best of times, there is only so much money available for scientific research. New problems have a way of displacing research funds for older projects. Money designated to solve new problems means less funding for old problems. The pie can be divided only so many ways. For example, a newspaper article[102] reported in 1988 that a San Francisco sickle-cell-anemia foundation was on the verge of closing its doors after twelve years. Its director, Erdine Njie, suspected that competition with AIDS organizations was partly to blame. "AIDS is important," she said, "but there has to be money allocated to respond to other needs as well. I don't see that happening."[103] Sickle cell anemia is primarily a disease of black people. Money was made available for sickle-cell research back in the 1960s and 1970s because, according to Dr. Robert Johnson, "It was a time of social consciousness raising. Prejudice was an issue, to not treat blacks as second class citizens. All that's now passed. . . . There are no advocates for sickle cell anemia. On the national scene the dollars (for sickle cell) are drying off. Money is being siphoned off and going in the AIDS direction."[104] Johnson was the director of the Adult Sickle Cell program at Alta Bates Hospital in Berkeley, California. Dorthy Boswell, executive director of the Los Angeles-based National Sickle-Cell Association, estimates that 60,000 black Americans are born each year with sickle-cell anemia. This is an inherited disease of red blood cells that interferes with their ability to carry oxygen from the lungs to the body tissues, and it can be lethal. The disease is not limited to people of African heritage; it can also afflict those from Mediterranean, Middle Eastern, and Asian countries. Many people mistakenly believe that sickle-cell anemia has been cured. In actuality, there are now more patients with the disease because medical technology has increased the life span of those born with it. Major corporations that had formerly contributed to sickle-cell work now "have kind of taken on AIDS as a major charity," according to leukemia fund-raiser Art Hoffman.[105]

Not only is it difficult to maintain funding for most scientific research, but there are some projects that the government simply refuses to sponsor. As an illustration, in 1986 French scientists developed a drug that could either block implantation of a fertilized human egg or terminate pregnancy.[106] This so-called morning-after pill was designated RU 486. Pills previously developed for use after intercourse contained female sex hormones, or estrogens, that had to be given in such high doses that they induced vomiting and abdominal pain. RU 486, however, is a synthetic steroid with a molecular structure like that of the hormone progesterone that normally maintains pregnancy. This chemical mimicry allows RU 486 to bind to progesterone receptors on uterine cells and thereby block the activity of progesterone. Other hormones called *prostaglandins* and structurally related compounds are used to induce uterine contractions as an aid to normal deliveries. They have also been used to induce abortions, but the large doses required for this purpose produce severe side effects. A 1986 report indicated that when combined with a prostaglandin, RU 486 constitutes a safe, simple, and effective approach to the termination of early pregnancy. It does not require special skills to administer and it is relatively free of side effects. This drug is just one of several contraceptive strategies being investigated by the World Health Organization (WHO) to help control the global population explosion. If RU 486 could be made available at very low cost it might be particularly well suited to developing countries where population growth is rapid, healthcare facilities are limited, and frequent contraceptive dosages are impractical. WHO reports that about 500,000 pregnant women in Third World countries die annually, and 15 to 20 percent of these deaths are the result of illegal abortion attempts. The subject of abortion is a political "hot potato" in the United States. Although RU 486 has been tested in the United States, it has not been approved for use by the Food and Drug Administration. Indeed, the prospect of a drug like RU 486 allowing a woman to abort a pregnancy safely in the privacy of her home alarmed the opponents of abortion. The Center for Biomedical Research of the Population Council in New York was founded in 1952 to explore ways to cope with global population growth. It is funded by a number of foundations and government agencies in the United States and abroad. Its director, Dr. Wayne Bardin, said that he and other researchers had been questioned by members of Congress to insure that no research on drugs like RU 486 was being done with federal funds. Ironically, in the face of the world's most pressing problem, that of overpopulation, the United States government feels that it cannot, given the present political climate, support the kind of research needed to help solve it.

In 1979, California Rural Legal Assistance (CRLA) filed a lawsuit against the University of California on behalf of the California Agrarian Project and seventeen farmworkers. CRLA charged that the university violated state law by spending tax dollars to benefit private interests. The suit also claimed that mechanization research was counter to federal land-grant acts. According to CRLA interpretation, these acts require federally funded research in agriculture to benefit small farmers and laborers, not agribusiness. Mechanization, it claimed, drives small farmers out of business and displaces thousands of farmworkers. The attorney for CRLA, William Hoerger, said, "putting laborers on welfare is not a social plus."[107] The suit sought broad restrictions on agricultural research, and would require that the university weigh the social consequences of mechanization projects, such as loss of jobs, before they are begun. In addition, agribusiness would have to pay for all expenses related to academic research that it now subsidizes in part. The university argued that the lawsuit was a threat to the academic freedom to decide which research projects to pursue. The heart of the dispute, according to the university, is a social issue, not a legal one, and that the legislature is the proper authority, not the courts.

The university tried unsuccessfully for a year to have the case dismissed. California Superior Court Judge Spurgeon Avakian ruled in 1980 that the lawsuit was based on legitimate grounds, stating, "It is a judicial question whether legislative and constitutional mandates are being followed in the allocation of public funds to particular research projects." However, Avakian continues, "[I]t is not a matter for judicial decision in this case whether agricultural mechanization is good or bad for society."[108] Four years later, on March 12, 1984, the case finally went to nonjury trial in the Alameda County Court. On the witness stand, university president James Kendrick Jr. said, "We don't deny that machines take the place of the worker . . . but it is society's role, not the inventor's role to compensate the worker. . . . [It is] society's right to impose a go or no go on a discovery."[109] Charles Hess, dean of the College of Agricultural and Environmental Sciences at the Davis campus of the University of California, argued, "Because public research funds have been rapidly dwindling, [a relationship with industry] is being looked upon with favor by scientists at many public institutions. Can the university protect the public investment in its research programs through patents and receipt of royalty income derived from products of privately sponsored research? . . . There are no simple answers. . . . The challenge is to preserve the beneficial aspects of the relationship . . . and to ensure collaboration through which society will realize the potential benefits."[110] The university officials argued that agricultural research had been unfairly singled out. They claimed that there are few gripes about the electronics industry, whose innovations have replaced thousands of workers.

The trial was interrupted in mid-April 1984 when Judge Avakian was stricken with a respiratory ailment. A new judge, Raymond Marsh, was appointed to the case. A decision was finally handed down by Marsh on November 17, 1987. It required the university to set up a review process to ensure that federally funded agricultural research primarily benefits small farmers. University attorney Christine Helwick, commenting on the court's decision, said that the problem with the judge's order is that he "asks us to predict the downstream effect of research and requires that it impacts the right group. It becomes a political guessing game to predict the impact of research and who is going to get hurt."[111]

Unfortunately, a large segment of our society is constantly bombarded by newspaper and magazine articles charging that our tax dollars are being spent on worthless scientific boondoggles. Much of what the general public knows (or thinks it knows) about science is derived from these unreliable or biased sources. Consider some of the articles taken from the *National Enquirer* with the following titles.

$75,000 TO STUDY HOW SPIDERS EAT— & YOU'RE PAYING FOR IT!

LEAPIN' LIZARDS!
$60,000 of your taxes is wasted studying why they have favorite islands

It stinks! Govt. sinks $35,000 of your taxes into sardine study

$94,000 in taxes wasted to study shrews' sex secrets

HOW DO STARFISH MAKE LOVE?
. . . you're paying $205,000 to find out!

Govt. shells out $275,000 so eggheads can learn why some birds are unfaithful

Govt. spends 34 years and $1.2 million . . . watching pigeons eat

On and on it goes, issue after issue. Why the *Enquirer* chooses to pick on the biological sciences almost exclusively in these diatribes is a puzzle to me. Perhaps there is a feeling that studies in the physical sciences or other scientific disciplines are more likely to bear fruit in the development of new products or technologies. Or it may be that the editorial board of the *Enquirer* is not able to comprehend journal articles in these other fields as easily as those in the biological sciences. It may be difficult to envision how some of these studies could ever be of any practical use to humans aside from satisfying someone's intellectual curiosity. But one never knows when a study in basic science, without any known direct applications to improving human welfare, may lead to very important technological breakthroughs.

One of the greatest success stories growing out of basic research in the biological sciences is found in a book by S. E. Luria titled *A Slot Machine, A Broken Test Tube:An Autobiography*.[112] Salvador Luria was one of the great pioneers of molecular biology. In 1952 he was investigating why some mutant bacterial cells, when infected by a virus (bacteriophage or phage), died but apparently failed to produce progeny phage. In the nonmutant strain of this bacteria, these phage normally would replicate and cause the host cell to lyse or break apart, releasing hundreds of progeny phage. Luria found, however, that phage extracted from the mutant bacteria were able to grow very well in a different species of bacteria. He reasoned, correctly, that his mutant bacteria had not failed to produce phage. Instead, these cells had produced a modified phage that would not complete its life cycle in related bacterial cells, but would do so in bacteria from a different species. Luria had discovered the first instance of the phenomenon of "restriction and modification." Werner Arber and Jean Weigle went on to discover that the phenomenon of restriction was actually an enzymatic attack on the DNA of the infecting phage. The phage DNA was chopped into pieces of varying lengths by these enzymes unless it had been "modified" by the addition of methyl groups in a previous host cell, thus "restricting" the growth of phage only to certain host cells. Cells that exhibit the phenomenon of host-controlled restriction and modification produce two bacterial enzymes: one enzyme modifies the DNA molecules of the host cell (adds methyl groups) and of the infecting phage at specific sites, whereas the other enzyme digests any DNA that is unmethylated at those sites.

So what? Big deal! Who cares? The popular press could have had a field day blasting scientific studies such as this for wasting money on things that seemingly have no practical application. Those who would chastise governmental granting agencies for financially supporting basic research of this kind show an appalling lack of appreciation for science history. The enzymes that cleave unmodified DNA at specific sites are now known as *restriction endonucleases*. These enzymes make it possible to splice together any two unmodified pieces of DNA in the test tube, even from completely unrelated sources such as a bacterium and a human. This capability opened up the entire field of recombinant DNA and biotechnology. We now can use bacterial cells (or other suitable carriers such as yeast cells) as tiny factories for churning out proteins (such as human growth hormone) that are too difficult or expensive to produce commercially for medical or industrial purposes. Arber, together with Hamilton O. Smith and Daniel Nathans, shared the 1978 Nobel Prize in physiology or medicine for the development of techniques utilizing restriction endonucleases to study the organization of genetic systems.

Where would science and technology be today without the pioneering work of people like Luria, who slave away on projects of little or no interest to anyone but themselves? Granted, not all (most likely very little) basic research is going to pay off as handsomely as that initiated by Luria. But no one knows from what source these kinds of major breakthroughs will come. Can our society afford to support the majority of these projects when so few of them prove to be useful in promoting human welfare? What other choices do we have? One alternative is to take the money saved from supporting basic research and spend it on welfare programs to feed, clothe, shelter, and provide medical care for the needy people in our society. This might seem to be the most ethical thing to do. In the short term, that might be true, but over a longer span of time maybe not. Consider all of the basic research currently in full swing on how yeast cells propagate themselves, on how fruit flies develop from eggs, on how worm cells age and die, on how retroviruses grow in monkeys. These may seem like frivolous projects, but scientists are learning many things from these nonhuman organisms that may directly benefit humans someday with cures for cancer, prevention and/or treatment of birth defects, making the aging process less debilitating, and developing a vaccine for AIDS. So in the long run, more human misery and suffering might be avoided by continuing to finance basic science research. One thing is clear. Our technological world is not going to be in any better shape tomorrow than it is today unless we provide the money to support scientific research. For better or for worse, history rarely recalls the times that benevolence prevailed by shunting money from science into public welfare programs. The only things that have made a lasting impression on history are the marvels of engineering (e.g., computers), agriculture (e.g., hybrid corn), and medicine (e.g., polio vaccine). The thing that is usually forgotten is that all of these marvels grew out of public support for basic research.

NOTES

1. *Electrophoresis* is a technique that causes different proteins to migrate through a gel at different rates according to their sizes, net electrical charges, and/or their shapes. Proteins are polymers (i.e., sequences of identical or structurally similar units) called *amino acids*. Each kind of protein has a specific number of amino acids in a specific sequence.

2. Michael A. Gold, *A Conspiracy of Cells* (Albany: State University of New York Press, 1986), 27.

3. Ibid., 33.
4. Ibid., 71–72.
5. Ibid., 105.
6. Ibid., 114.
7. Ibid., 115.
8. Ibid.
9. Ibid., 116.
10. Ibid.
11. Ibid., 117.
12. Ibid.
13. Ibid., 125. Salk's idea "was to try to switch the patient's immune system into combat mode by challenging them with an injection of foreign cells."
14. Ibid., 151.
15. Ibid., 143.
16. I. Tangley, "Mouse Mixup May Alter Research Results," *Science News* 122 (July 24, 1982): 53.
17. Ibid.
18. Peter Carlson, "A Seattle Man Vents His Spleen Against Those Who Would Use It for Profit," *People* 24 (September 23, 1985): 100.
19. Ibid.
20. Ibid.
21. Mark Crawford, "Court Rules Cells Are the Patient's Property," *Science* 241 (August 5, 1988): 653–54.
22. Ibid., 653.
23. Ibid., 654.
24. Barbara J. Culliton, "Mo Cell Case Has Its First Court Hearing," *Science* 226 (November 16, 1984): 814.
25. Crawford, "Court Rules Cells Are the Patient's Property," 654.
26. Peter Huber, "Litigation Thwarts Innovation in the U.S.," *Scientific American* 260 (March 1989): 120. Huber is senior fellow at the Manhattan Institute for Policy Research and author of *Liability: The Legal Revolution and Its Consequences* (Basic Books, 1988).
27. Ibid.
28. Ibid.
29. Ibid.
30. Elizabeth Corcoran, "Patent Medicine: New Patients Challenge Congress and Courts," *Scientific American* 259 (September 1988): 128.
31. Ibid., 121.
32. Ibid.
33. Ibid.
34. Ibid., 120.
35. Janet L. Hopson, "Fins to Feet to Fanclubs: An (Old) Fish Story," *Science News* 109 (January 10, 1976): 28.
36. Ibid.
37. Keith S. Thomson, "Marginalia: The Second Coelacanth," *American Scientist* 77 (November–December 1989): 538.
38. Ibid.

39. Hopson, "Fins to Feet to Fanclubs," 30.

40. Keith S. Thomson, "Marginalia: A Fishy Story," *American Scientist* 74 (March–April 1986): 169.

41. Ibid., 170. Hans Fricke, "Coelacanths, The Fish That Time Forgot," *National Geographic* 173 (June 1988): 825–38.

42. Hopson, "Fins to Feet to Fanclubs," 29.

43. Fricke, "Coelacanths, The Fish That Time Forgot," 830.

44. Thomson, "Marginalia: The Second Coelacanth," 538.

45. Thomson, "Marginalia: A Fishy Story," 171.

46. Ibid.

47. Roger Lewin, "Ethiopia Halts Prehistory Research," *Science* 219 (January 14, 1983): 147.

48. Ibid.

49. Ibid.

50. Ibid., 148.

51. Ibid.

52. Ibid., 149.

53. Donald C. Johanson and Maitland Edey, *Lucy: The Beginnings of Humankind* (New York: Simon and Schuster, 1981).

54. Lewin, "Ethiopia Halts Prehistory Research," 149.

55. Ibid.

56. Ibid.

57. Ibid.

58. Ibid.

59. Ibid.

60. Ibid.

61. Ibid.

62. Roger Lewin, "Extinction Threatens Australian Anthropology," *Science* 225 (July 27, 1984): 393.

63. Ibid.

64. Ibid., 394.

65. Eliot Marshall, "Smithsonian, Indian Leaders Call a Truce," *Science* 245 (September 15, 1989): 1184.

66. James F. Crow, "Eighty Years Ago: The Beginnings of Population Genetics," *Genetics* 119 (July 1988): 473.

67. Physics section, *Discover* 5 (September 1984): 39.

68. "AMA Journal Official Links Story to Ads," *Los Angeles Times*, January 15, 1984, I-18.

69. Ibid.

70. I. Peterson, "Publication Bias: Looking for Missing Data," *Science News* 135 (January 7, 1989): 5.

71. Ibid.

72. Ibid.

73. Andrew C. Revkin, "Dilutions of Grandeur," *Discover* 10 (January 1989): 4.

74. L. Beil, "Dilutions or Delusions?" *Science News* 134 (July 2, 1988): 134.

75. R. Pool, "More Squabbling Over Unbelievable Result," *Science* 241 (August 5, 1988): 658

76. Ibid.

77. Robert Pool, "Unbelievable Results Spark a Controversy," *Science* 241 (July 22, 1988): 407.

78. Revkin, "Dilutions of Grandeur," 74.

79. Ibid., 75.

80. L. Beil, "Nature Douses Dilution Experiment," *Science News* 134 (July 30, 1988): 69.

81. John Maddox, James Randi, and Walter W. Stewart, "High-Dilution Experiments a Delusion," *Nature* 334 (July 28, 1988): 287–91.

82. Pool, "More Squabbling Over Unbelievable Result," 658.

83. Ibid.

84. Revkin, "Dilutions of Grandeur," 75.

85. Pool, "More Squabbling Over Unbelievable Result," 658.

86. John Maddox, "Maddox on the 'Benveniste Affair,' " *Science* 241 (September 23, 1988): 1585.

87. Caroline V. Rider, "Scientific Suppression?—letter #1," *Science News* 134 (July 23, 1988): 51.

88. John C. Huffman, "Scientific Suppression?—letter #2," *Science News* 134 (July 23, 1988): 51.

89. Philip H. Abelson, "Excessive Zeal to Publish," *Science* 218 (December 3, 1982): 953.

90. "Author Proliferation," *Science* 241 (September 16, 1988): 1437.

91. Barbara J. Culliton, "Harvard Tackles the Rush to Publication," *Science* 241 (July 29, 1988): 525.

92. Walter Stewart and Ned Feder, "The Integrity of the Scientific Literature," *Nature* 325 (January 15, 1987): 207–14.

93. Mary Murray, "A Long-Disputed Paper Goes to Press," *Science News* 131 (January 24, 1987): 53.

94. Ibid.

95. Ibid.

96. Ibid.

97. Ibid.

98. Barbara J. Culliton, "Integrity of Research Papers Questioned," *Science* 235 (January 23, 1987): 423.

99. Ibid.

100. Murray, "A Long-Disputed Paper Goes to Press," 52.

101. Roger Lewin, "Genome Projects Ready to Go," *Science* 244 (April 29, 1989): 604.

102. Dexter Waugh, "Sickle Cell Contributions Wane; AIDS Focus Blamed," *San Francisco Examiner*, December 3, 1988.

103. Ibid.

104. Ibid.

105. Ibid.

106. "Safe, Effective 'Morning-After' Pill Developed." *San Luis Obispo Telegram Tribune*, October 14, 1986.

107. Marjorie Sun, "Weighing the Social Costs of Innovation," *Science* 223 (March 30, 1984): 1368.

108. Ibid.

109. Ibid., 1369.

110. Ibid.

111. Marjorie Sun, "UC Told to Review Impact of Research," *Science* 238 (November 27, 1987): 1221.

112. S. E. Luria, *A Slot Machine, A Broken Test Tube: An Autobiography* (New York: Harper & Row, 1984), 98–99.

CHAPTER 8

BRAINSTORMS

The history of science is a series of discoveries and revolutionary ideas. If these events were easy to come by there would be no special glamour or awe associated with being a scientist. Anyone could just put his mind to the task and, with perhaps a little elbow grease, come up with whatever was needed to solve the problem. But that is not the way science works. There is no end to the problems going begging for a solution, and many of these problems are of the highest importance for all of humanity. The best minds of the scientific community have yet to find a way to prevent or effectively treat cancer without any undesirable side effects. We still know very little about normal brain functions or malfunctions. How a set of genetic instructions in a fertilized egg can direct the embryological development of a new individual is essentially unknown. If we knew the answer to this latter problem we might be in a better position to find ways of dealing with birth defects. What causes aging and death? Think of the misery and suffering that might be alleviated or prevented by knowing the answer to this question. Questions abound, but solutions are difficult to find. So how are things discovered, and where do great ideas in science come from? Is there some way that we can teach students how to be creative so we can tip the scale to favor an increased rate of scientific discovery for the future?

THE ORIGIN OF THE POLYMERASE CHAIN REACTION[1]

"Genetic engineering" is the creation of new DNA, usually by the recombination of DNA from different organisms by artificial means using enzymes known as *restriction enzymes*, and the production of many copies of the recombined DNA by a process known as *cloning*. Insulin is a protein hormone manufactured by the pancreas. Patients with diabetes lack sufficient insulin to maintain normal levels of blood sugar. Insulin that is nearly identical to that of humans can be extracted from the pancreas of farm animals, but it is costly and some patients react adversely to nonhuman insulin. By genetic engineering, we can use restriction enzymes to cut the insulin gene from a human chromosome and splice it into the DNA of yeast or bacteria. These host cells multiply rapidly to produce a clone consisting of millions of identical cells, each of which produces thousands of insulin molecules that can be extracted for human therapy. Other proteins, such as human growth hormone, blood-clotting factors, and immunoregulatory proteins, can be obtained in the same way.

It may be desirable for some purposes to clone and extract a large population of a specific DNA segment (such as a gene or some region of interest that does not code for protein) rather than a protein product of a gene. For example, many gene-therapy experiments are currently underway attempting to insert functional copies of human genes into cells that are defective in patients. Determining the sequence of the basic units that are linearly linked together in a DNA segment of interest also requires a large population of identical DNA molecules. Until the invention of the in vitro "polymerase chain reaction" (PCR), the only way to obtain them was by the laborious process of cell cloning.

The PCR can start with even very small amounts of DNA (theoretically, a single DNA molecule) and yet produce millions of replicas of a desired DNA segment. The starting sample doesn't even have to be purified; contaminant DNA from other sources is usually no problem. The process has been streamlined by automation so that millions of identical DNA molecules can be produced in a twenty-four-hour period. Today, PCR technology is used for many purposes. For example, it is used to determine (with very little error) if a sample of DNA from blood, hair, semen, scrapings from under fingernails, or other sources belongs to a specific individual or not. This is often very important in helping solve crimes such as murder or rape. PCR can also be used to resolve cases of disputed parentage. It can be used to determine if a person carries a normal gene or a defective one, and this may aid in early treatment or decisions about having children. The PCR has revolutionized biotechnology and society. So, when, how, and by whom was this process developed and why wasn't it invented at an earlier date? Or was it?

On a Friday night in April 1983, Kary Mullis was driving into northern California for a weekend away from his job. He enjoyed night driving, and it was his custom to get away to his cabin every weekend. Even though he wanted to forget work, his mind kept thinking about a problem he was trying to solve in his job at Cetus Corporation in Emeryville, California. At that time, Cetus was one of the most successful genetic-engineering companies in the United States. Mullis had earned his Ph.D. in biochemistry from the University of California at Berkeley in 1972. His postdoctoral career was episodic, including a stint at waiting tables in a bakery. He was hired by Cetus in 1979 to do what amounted to technician-level work in the preparation of genetic probes.

A molecular *probe* is a relatively short sequence of nucleotides (the building blocks of nucleic acids such as DNA; typically 20–50 nucleotides), artificially constructed by a DNA synthesizer ("gene machine"), that complements a segment of a gene of interest. When these probes are made radioactive (or are linked to other "reporter" molecules such as enzymes), molecular biologists can go fishing for the complementary gene among all of the three billion nucleotide pairs that constitute the genetic endowment of each human cell. As explained in chapter 2, DNA consists of two strings of nucleotides wound around each other in a helical fashion (a double helix). Each nucleotide consists of three parts: (1) a basic, nitrogenous, organic (carbon-containing), ringlike structure; (2) a sugar; and (3) a phosphate (PO_4) group. In each strand of DNA, the phosphate of one nucleotide is bonded to the sugar of the adjacent nucleotide, forming an alternating sugar-phosphate-sugar-phosphate, etc. "backbone." The backbones wind around each other like the handrails of a spiral staircase. A nitrogenous base is covalently bonded (sharing an electron) to each sugar. The steps of this staircase are formed by the pairing of bases from opposite strands. The bases pair with each other in a very specific fashion. There are four kinds of bases (A, T, G, and C). Base A normally pairs with T, and likewise G pairs with C. Thus, if one knows the sequence of the nucleotide bases on one strand, it is possible to infer the sequence on the opposite or complementary strand by applying the rules of base pairing.

Replication of double-stranded DNA molecules requires the synthesis of a relatively short sequence of 9–20 nucleotides (an oligonucleotide) that can pair specifically with a complementary segment on a target strand of DNA. This type of oligonucleotide is called a *primer* because it serves as a starting point for enzymatic polymerization (linking together) of nucleotides that form complementary pairs with nucleotides on the opposite template strand. The essential enzymes for this process are called *DNA polymerases*. The first DNA polymerase was discovered in 1955 by Arthur Kornberg who won a Nobel Prize in physiology or medicine for his in vitro DNA-synthesis experiments four years later. This enzyme cannot polymerize nucleotides unless a primer is bound to the template strand. It can then extend the primer in one specific direction. In nature (in vivo), primers are made of RNA nucleotides (closely related to DNA nucleotides with similar base-pairing properties). For in vitro experiments requiring probes or primers, DNA oligomers could be laboriously synthesized manually. Mullis had been hired by Cetus to synthesize oligonucleotide probes, but the boring repetitiveness of the process of making probes cried out for automation. So Mullis had, within two years on the job, converted the lab to automation. This gave him a lot of time to think about and tinker with new ideas. The first rapid method for determining DNA-nucleotide sequences had been developed by Frederick Sanger and A. R. Coulson in 1975. It then became possible to know, for example, how the gene for sickle-cell anemia differs from the normal sequence. In this case, there is only a single nucleotide difference. Mullis began thinking about how to rapidly determine if the normal or abnormal nucleotide was at any particular position in a gene whose normal sequence was already known. This could be of great importance as a diagnostic medical tool.

It was this problem that occupied his thoughts as he continued his late-night drive. A variation of the Sanger sequencing technique involves separating the two stands of DNA by heating, bonding of a suitable radioactive primer to one of the strands, extending the primer, and finally capping the extended chain so that no further extension can take place. The extension of the primer is accomplished through the action of DNA polymerase, by base pairing and polymerization of nucleotides opposite to the template strand. The "capping" is due to the incorporation of a nucleotide that contains an unusual sugar known as a *dideoxyribose*. Thus a nucleotide containing this unusual sugar and the base T is designated ddT; one containing any unspecified nucleotide base is ddN. In the Sanger dideoxy-sequencing technique, four batches are run simultaneously. In each batch, three of the nucleotides have normal sugars and the fourth has a dideoxy sugar. Mullis was toying with the idea of synthesizing a primer that immediately flanked (i.e., that ended adjacent to) the target nucleotide that one wished to identify. By eliminating all of the normal nucleotides and including all four ddNs (only one of which was radiolabeled) in each of the four batches, only one of these four systems would extend the primer by a single ddN; identifying this ddN would tell him whether the corresponding target base in the template strand was normal or abnormal.

Somewhere around Cloverdale, it came to Mullis that the determination of the target nucleotide would be more definitive if he also ran a control experiment that identified the nucleotide at the same target position on the other strand. To do this, he would construct a different probe that would flank the same target nucleotide on the opposite strand, and subject it to the same capping procedure. By making the two probes of different lengths, he would be able to distinguish them from each other in each of the same four batches.

If the two targets were found to complement each other, then the identification of these bases would be confirmed. Thus, it should be possible to determine if an individual (e.g., one suspected of carrying the sickle-cell gene) has the normal or abnormal nucleotide at any particular position in the hemoglobin gene or any other gene whose normal sequence is already known.

Lots of things could go wrong with such an experiment, and Mullis began to anticipate what they might be. He figured that if loose nucleotides were in the DNA sample, they might be added to the primer before the planned addition of the labeled ddNs. He first considered destroying the polymerizing ability of any loose nucleotides with the enzyme alkaline phosphatase. But that would necessitate removing the phosphatase enzyme before adding any ddNs, otherwise they would also be destroyed. As he began to descend into the Anderson Valley, Mullis "hit on an idea that appealed to [his] sense of aesthetics and economy."[2] He would add DNA polymerase twice; once to remove the loose nucleotides from the sample, and then again to incorporate the labeled ddNs. He reasoned that by pretreating the sample with just primers and polymerase, any loose nucleotides would become incorporated into extension products from the primers. The primed extension fragments could then be separated from the DNA targets by heating. Heating causes DNA strands to separate, but the strands will become paired again as the mixture cools. Although the extended sequences would remain in the sample there would likely be far more unextended primers than extended ones, so the DNA targets would probably hybridize (pair with) unextended primers as the mixture cooled. Unfortunately, heating also destroys enzyme activity. So after the heating phase, Mullis planned on adding more polymerase along with ddNs.

More questions arose in his mind. Would the pretreatment products interfere with the subsequent incorporation of ddNs? Could the pretreatment extension products have gone so far as to create a sequence containing a binding site for another primer molecule? Then it suddenly dawned on him that these problems would not arise because the target strands and the extension strands would have the same base sequences. The pretreatment would have doubled the number of target sequences in the sample. If the system was heated again and primers, polymerase, and free nucleotides were added, the number of targets would double again. (See fig. 8.1.) He began to calculate powers of two in his head, but bogged down at the tenth power when 2^{10} is about 1,000; if true, then 2^{20} is about a million. He stopped the car to get a pencil and paper to check his calculations. After confirming his math, he continued driving and soon realized that after a few rounds of primer extensions, the length of the exponentially accumulating strands would all be identical, with a primer at one end and a sequence complementary to the other primer at the other end. Furthermore, fragments of much greater size could be replicated by designing primers that hybridized

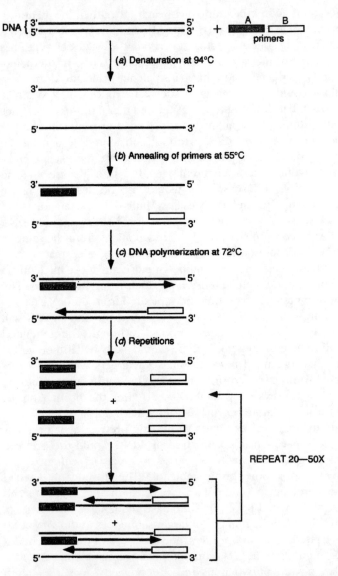

Fig. 8.1. Schematic representation of a typical PCR assay. A PCR assay involves the repeated use of temperature cycling to achieve primer-directed enzymatic amplification of desired segments of DNA. First the target DNA is denatured by heat (a). Upon cooling, the primers, which flank the desired segment of DNA, hybridize with complementary targets (b), and then DNA synthesis takes place (c), starting from the primer. These steps are repeated (d) 20–50 times to obtain large quantities of the desired DNA fragment. From W. D. Stansfield, Jaime S. Colomé, and Raúl J. Cano, *Schaum's Outline of Molecular and Cell Biology*, fig. 9–4. © 1996 McGraw-Hill. Reproduced with permission of The McGraw-Hill Companies.

further apart on the target sequence. At this point, he stopped the car again and began to make drawings of the process. His companion in the car failed to rouse from her sleep by his exclamation, "You're not going to believe this. . . . It's incredible." He then drove on to the cabin without further stops. But that night Mullis confessed that he got little sleep, "with deoxyribonuclear bombs exploding in my brain."[3]

It was difficult for Mullis to believe that the idea of a polymerase chain reaction (PCR) had not been tried before by one or more of the thousands of workers who had been extending single primers with polymerases for years. But if it was known and if it worked, everyone would be using it to amplify or multiply DNA fragments instead of the laborious process of cloning them in bacteria. When he returned to Cetus the following Monday he initiated a literature search and began discussing his idea with anyone who would listen. No one knew of a PCR in the literature and they didn't offer any reason why it should not work. And yet no one was particularly enthusiastic about it either. If it would work, it would revolutionize the biotechnology field. How could they be so blasé? Over the next several months, Mullis planned his first PCR experiment. He had to make a lot of educated guesses about the kinds of buffer solutions, the concentrations of reactants, lengths of reaction times, how much to heat and cool the mixtures, etc. He chose a target bacterial-plasmid DNA sequence of twenty-five base pairs and constructed two primers of eleven and thirteen bases long. When he finally was ready, he ran his favorite kind of experiment, ". . . one involving a single test tube and producing a yes or no answer."[4] To his great surprise and delight, the experiment worked perfectly the first time. Things like this don't happen very often. As he left the lab fairly late that evening, Mullis stopped at the office of Albert Halluin, the patent attorney for Cetus, and explained to him the results of his PCR experiment. Halluin was the first person, in about a hundred to whom Mullis had explained the PCR, that considered it to be significant. Halluin congratulated Mullis on his invention.

In the months that followed, Mullis continued to refine the technique and amplified ever larger sequences. He was even successful in replicating a fragment from a single-copy human gene. The method could work on extremely small amounts of sample DNA, even when it was in an unpurified state. The only remaining problem was that the polymerase enzyme was destroyed in the heating phase, so fresh enzyme had to be added in every cycle. These enzymes are very expensive and it adds one more step to the process. Then in 1987, Cetus extracted the gene for DNA polymerase from a bacterium that lives in the boiling geysers of Yellowstone National Park.[5] The gene was transferred into a common laboratory bacterium for production of the enzyme. This genetically engineered DNA polymerase, called *Taq*, can survive the near boiling phase of the PCR and so it only needs to be added to the PCR once. A machine was then invented to fully automate the PCR.

During the 1984 annual Cetus Scientific Meeting, Mullis presented a poster describing the PCR. Again, nobody seemed very interested. Joshua Lederberg, Nobel laureate, president of the Rockefeller University, and discoverer in 1946 that bacteria can exchange genetic material, was at that meeting. Mullis asked Lederberg what he thought of his discovery. During their discussion, Lederberg mentioned that about twenty years earlier, after Kornberg had discovered DNA polymerase, they jointly had considered how it might be used to make large amounts of DNA. In those days, probes and DNA sequences were not readily available. Mullis ends an account of his discovery in *Scientific American* in the following way: "I think that Josh, after seeing the utter simplicity of the PCR was perhaps the first person to feel what is now an almost universal first response to it among molecular biologists and other DNA workers: 'Why didn't I think of that?' And nobody really knows why; surely I don't. I just ran into it one night."[6]

For 1989, *Science* (the journal of the American Association for the Advancement of Science) chose to honor symbols of scientific progress rather than a personality. The symbols chosen were the DNA polymerase enzyme and the polymerase chain reaction procedure.

Mullis was awarded a Nobel Prize in chemistry in 1993 for developing the PCR.

Postscript

According to Mullis,[7] he both conceptualized and performed his first PCR reaction in 1983 while working at Cetus. After adding refinements to the procedure, Cetus was awarded its first patent on the PCR process in 1987. In 1989, the chemical giant E. I. DuPont de Nemours & Company challenged Cetus's patents in court, claiming that the PCR process had previously been precisely described in several journal papers (1971,[8] 1972, 1974) from the Massachusetts Institute of Technology (MIT) laboratory of H. Gobind Khorana. Together with Robert W. Holley and Marshall W. Nirenberg, Khorana had been awarded the Nobel Prize for physiology or medicine in 1968 "for their interpretation of the genetic code and its function in protein synthesis." In 1990, the U.S. Patent Office proclaimed that the Khorana papers did not invalidate the Cetus patent on the PCR process. Kornberg said that

> anyone with a basic knowledge of biochemistry could have deduced how to perform PCR from [Khorana's] papers. . . . [T]he only reason *no one did* was that the technique was ahead of its time. Neither DNA sequences nor [synthetic] primers were available in the early 1970s, and by the time they were, cloning had been invented and had become the chosen method for copying DNA. PCR may have fallen out of sight, . . . but it had been invented. It was in the public domain. . . . The patent examiner blew it.[9] (Emphasis mine)

Unfortunately, Kornberg erred in stating that "no one did" perform PCR experiments before Mullis.

> The first PCR experiment was performed by Dr. Kjell Kleppe, a postdoctoral fellow in Dr. Khorana's laboratory. This experiment was a reaction involving two cycles of amplification of a 30 nucleotide-long dupex DNA, using primers nine and 10 nucleotides long. Dr. Kleppe described his experiments publicly at the 1969 Gordon Conference on Nucleic Acids. . . .[10]

Mullis[11] claimed to have made a literature search in 1983 for PCR-like reactions, but failed to find any. According to his own account, no one with whom he discussed his PCR concept had heard of such procedures. Why experts in the field of in vitro nucleic-acid synthesis were not well aware of the papers and public announcements about PCR experiments coming out of Khorana's prestigious laboratory as early as 1969 remains a mystery. Certainly Mullis must have been made aware of Khorana's precedence by the time of the DuPont challenge. However, by failing to acknowledge any of Khorana's work in his 1990 historical account of the origin of the PCR process, Mullis gives the impression that he was the first to conceive of and successfully perform a polymerase chain reaction. Is it possible that the Nobel Committee was also unaware of Khorana's contributions when it awarded Mullis the 1993 prize? Most popular accounts of the discovery of the PCR also appear to have ignored the much earlier PCR work of the MIT team. It's never too late to try to set the record straight and give credit where it is properly due.

THE GENE CONCEPT

Almost everyone knows that Gregor Mendel (1822–1884) is considered to be the "father of modern genetics." But even historians of science do not know for sure if or how he came to discover the basic laws of heredity. Many unsuccessful attempts had been made before Mendel's time to solve the riddles of inheritance. However, "Mendel's approach was completely novel and appears to have been uniquely his own."[12]

Johann Mendel was born in Heinzendorf, Moravia, on July 22, 1822. As a boy, Johann obtained much practical botanical experience as he worked with his father on the family farm. Johann wanted to continue his education, but his family was too poor to send him to college. So he decided to join the clergy and have the Church pay for furthering his education. In September 1843, he entered the Augustinian monastery of St. Thomas in the town of Brünn (then called Brno) and took upon himself the name Gregor. At that time, there existed at the monastery a community of scholars and investigators of very considerable stature. In August 1848 he was ordained a priest. About a year later he was appointed a substitute teacher at Znaim High School. He failed an examination for appointment as a regular high school teacher in November of 1850. Mendel wanted to be a teacher rather than a parish priest. It was obvious that he needed both training and a formal teaching license if he was to continue in that profession. Therefore, Mendel was sent to the University of Vienna. He remained there from October 1851 to April 1853. He studied botany under Fenzel, zoology with Kner, and plant physiology and paleontology with Unger.[13] He became familiar with previous work in plant hybridization (crossing different varieties), with recent research on the process of fertilization, with the cell theory of Schleiden, and with papers on cellular reproduction.

Most standard accounts of Mendel's life cite this preparation as important to the later development of his ideas concerning heredity. Floyd Monaghan and Alain Corcos,[14] however, suggest that it was Mendel's training in mathematics and the physical sciences (especially chemistry) that played the most crucial role. Mendel studied physics under Frederick Franz, who said of him, "In my own branch (physics) he is almost the best." Mendel also studied mathematics and experimental physics with Christian Doppler; the same Doppler who discovered the acoustical law now called the Doppler Effect. In addition, Mendel learned how to construct and use physical apparatus, studied higher mathematical physics with E. Ettinghausen, and served as an assistant demonstrator at the Physical Institute of the university. Both Doppler and Ettinghausen taught that the best way to understand natural phenomena was by mathematical analysis. Thus, Mendel had been well schooled in thinking of the world in terms of mathematical descriptions and had undoubtedly learned how to design experiments under these eminent tutors.

After he returned to Brno, he taught physics and natural history at Brno Modern School (Staatsrealschule) until 1868 without ever obtaining a teaching certificate. Despite his obvious prowess in mathematics and physics, he failed to pass the certification test in physics and natural history in May 1856. His examiners said that he lacked the requisite clarity of insight. Later that same year he began experimental crossing of pea varieties, and became a founding member of the Brno Natural History Society in 1862. Mendel gave his first lecture on the results of his pea experiments on February 8, 1865, and a second one on March 8. His report entitled "Experiments on Plant Hybridization" was published in the *Proceedings* of the Brno Natural History Society for 1865, and released in 1866. Neither his lectures nor his paper found a receptive audience. From ten letters, written to Carl von Näegeli, professor of botany at Munich, during the period 1866 to 1873, we learn that Mendel performed hybridization experiments between species or varieties belonging to twenty-six different genera. In some cases, the hybridization rules derived from peas were confirmed; in other cases they were not. The inconsistency of these rules vexed him. In March of 1868, Mendel was elected abbot of the monastery. His second paper was published in the *Proceedings* in 1870. It reported on his failure to find the same kind of regularities in the hawkweeds (*Hieracium*) as he found in peas. These letters prove that Mendel tested the extent to which his "pea laws" could be applied in other plants, and that he did not try to hide his failures by failing to publish negative results. Thereafter Mendel became increasingly involved in administration of the monastery and ceased doing science. He died of Bright's disease (a kidney disease) on January 6, 1884, unaware that what he had discovered would later be considered the founding principles of a new scientific discipline that would be called *genetics*.

Mendel's methodology should have served as a model for other researchers to follow in either confirming his published results and interpretations or in making discoveries about the heredity of organisms other than peas. Unfortunately, for reasons that are still obscure, this did not immediately happen. For thirty-five years after the first public announcement of Mendel's experiments, his work lay essentially unknown and unappreciated. Then, in 1900, three botanists (working independently in three different countries, within about three months of one another) announced their experimental findings in various organisms that confirmed the regularities of heredity that we now refer to as "Mendel's laws." Hugo de Vries of France, Carl Correns of Germany, and Erich von Tschermak-Seysenegg of Austria had essentially completed their experiments and analyses using methods similar to those of Mendel, but without prior knowledge of him or his works. While doing their literature searches prior to submission of their papers for publication, these three men "rediscovered" Mendel's 1866 paper and gave him credit for prior discovery of the laws of heredity that now bear his name. In the lead section of this chapter, it was explained how the discovery of the polymerase chain reaction (PCR), as described by Mullis, was actually a "rediscovery" of the work that had been reported fourteen years earlier by Kjell Kleppe. However, in his historical account (1990) of the origin of the polymerase chain reaction, Mullis gave no credit to Kleppe, Khorana, or other members of the MIT team whose priority of discovery of the PCR is now more widely known.

In his now famous 1866 paper on the hybridization of peas, Mendel noted the "striking regularity with which the same hybrid forms always reappeared whenever fertilizations took place between varieties of the same species. . . ." In order to discover natural laws, one must assume that nature is not capricious, but orderly. Mendel must have been trying to find such a law or laws, because he said that "so far, no generally acceptable law governing the formation and development of hybrids has been successfully formulated. . . ." Contrary to what many people believe, Mendel was not trying to discover basic laws of inheritance, but rather the rules governing the formation and development of intraspecific (within a species) hybrids. During Mendel's time, there were few natural laws known in biology (Darwin's principle of natural selection being one of the most important), but none of them had been conceived of or expressed in mathematical terms. There were, however, numerous examples in physics and chemistry that could have served as models waiting to be adopted by the biological sciences. The success of physics and chemistry in explaining relevant phenomena was achieved by postulating the existence of hypothetical entities that possessed specific properties, and the existence of specific relationships among those entities. Given such a theoretical framework, it is possible to make predictions that can be either confirmed or refuted by subsequent observations or experiments. This is the heart of modern scientific methodology.

The atomic theory, as proposed by John Dalton in 1808 and elaborated by others, was a hotly debated idea at the time that Mendel was studying chemistry at the university. It was generally acknowledged that atoms could not be created, divided, or destroyed in chemical reactions, but the arrangements and associations of the individual atoms could be altered to produce new molecules without any change to the atoms themselves. Furthermore, the indivisible atoms of a given molecule were assumed to combine in small whole number ratios such as 1:1, 2:1, 2:3, 3:1, etc. In most cases, it was still difficult to determine the atomic weights and chemical formulas of molecules. The guiding principle of chemists was the search for the simplest chemical formula that could be derived from their analytic data. Since atoms were assumed to be indivisible units, it was logical that their associations in molecules must be in whole number ratios. So the complex number ratios obtained from analytical data were reduced to their nearest simple whole number ratios. In his last two terms at the university, Mendel had studied chemistry under Redtenbacher. The analogy of Mendel's "gene concept" with the atomic theory and "laws of chemistry" is very striking, and it has been suggested[15] that this is precisely where Mendel got his idea about hereditary "elements" or "factors."

During the period from about 1780 to the 1860s there was a series of replacements of fluid theories by particulate or atomistic theories in chemistry and physics. In chemistry, phlogiston was a hypothetical amorphous substance that was thought to be released as flame in combustion. This idea eventually became replaced by the oxygen theory and the final acceptance that all matter is basically particulate in nature. In physics, the caloric theory of heat was replaced by the particulate kinetic-molecular theory, and fluid theories of electricity were replaced by the electron-flow theory. In biology, the cell theory can be thought of as a particulate view of tissues; all multicellular bodies are made of cells. The cell is the basic unit of life. The seminal fluid theories of reproduction eventually were replaced by the particulate sperm-egg theory. Early use of the sperm-egg theory caused a split among embryologists into either the ovum (egg) or sperm camps. The ovum school thought the egg contained the vital life force that became stimulated to divide and produce an embryo by union with a sperm; the sperm was just an activator of the egg. The sperm school thought that the egg was just a bag of nutrients and the only vital force resided in the sperm itself. Mendel proposed that both male and female gametes contributed equally to the heredity of offspring. His particulate concept of heredity, though not accepted during his lifetime, eventually came to replace the old blending-fluid theories. Thus, Mendel's theory of inheritance should be viewed as one of the most recent in a long series of replacements of fluid theories by particulate theories in science.

R. A. Fisher proposed that Mendel probably knew before he began his experiments what to expect. In other words, his data-gathering stage was probably used to confirm his theory rather than to generate it. In a 1985 paper, Monaghan and Corcos[16] seem to reverse themselves from their 1983 position, and argue that Mendel did not know what to expect from his experiments because he did not have a genetic theory in mind and that his work was not designed to discover the basic laws of inheritance.

Before we evaluate Mendel's contributions, let's first review what is known about the nature of heredity today. We know now that many traits are each governed by a hereditary unit called a *gene*. The classical structural gene is a segment of DNA that codes for the amino acid sequence of a protein. If that protein is an enzyme, then it can catalyze the conversion of one substance (the substrate) into one or more different substances (products). If the gene is present, one character state (presence of the product) can be expressed by the cell, but if that gene is absent, then a different character state (absence of the product) will be exhibited. Each chromosome in the cell nucleus contains a molecule of DNA. Thus, genes reside on chromosomes. Each gene occupies a specific position or locus on a chromosome. Different forms of a gene may be found at the same chromosomal locus on homologous chromosomes (similar in size and genetic content); i.e., any given gene derived from one's mother may not be identical to the gene from one's father. These different gene forms are now called *alleles* (Mendel referred to them as "antagonistic elements"). Thus, a gene that specifies a functional enzyme is an allele of its alternative form that does not code for a functional enzyme. In general, each individual gets its heredity equally from both parents. One set of chromosomes is contributed by the egg and a comparable homologous set by the sperm. The number of chromosomes in a sex cell or *gamete* is called the *haploid* number (n), which is characteristic for each species. For example, humans have twenty-three chromosomes in each egg or sperm. The union of an egg and sperm in fertilization produces a single cell called a *zygote*, and it contains two sets of chromosomes called the *diploid* number (2n = 46 for humans). A zygote divides by the process of *mitosis* to produce all of the nongametic body (somatic) cells of a multicellular organism. The mitotic process functions to make all of the somatic cells of an organism have the same genetic structure (the same number of chromosomes and identical genetic content).

The organs of reproduction (gonads) in animals are the ovaries in females and the testes in males; in flowering plants, the female reproductive part is the pistil (containing the ovary and its ovules), and the male part is the stamen (containing the anther and its pollen grains). When an individual reaches sexual maturity, some of the somatic cells in these organs of reproduction undergo a different kind of division process called *meiosis*. Meiosis is a reductional division that reduces the normal somatic diploid number to the haploid number characteristic of gametes. During meiosis, each paternally derived chromosome pairs with its homologous maternally derived counterpart. Then the homologues normally separate into different gametes. How a member of any one pair of chromosomes gets into a gamete is independent of how a member of any other chromosome pair gets into that same gamete. Thus, each haploid gamete contains a sample half of the parental diploid set. The sexual mechanism contributes to genetic diversity in the offspring or progeny. This diversity provides the raw material from which natural selection functions to preserve the best adapted gene combinations and it is also the source from which artificial selection (by humans) can create new varieties of crop plants or breeds of animals.

Chromosomes were discovered after Mendel died, so he knew nothing about the processes of mitosis and meiosis. After 1900, any observable (empirical) attribute (character or trait) of an organism became known as a *phenotype*. The hereditary units were called *genes* and the branch of biology devoted to the study of heredity became known as *genetics*. The genetic structure of an individual was called its *genotype*. Although Mendel used them in a different context, capital and lowercase letters are now used to symbolize different forms of a gene (alleles). For example, a gene responsible for production of a functional enzyme could be represented by the capital letter *A*, whereas its alternative allele that produces no enzyme could be symbolized by the same letter in lowercase *a*. If a diploid cell contains two identical copies of the same gene or allele (*AA* or *aa*) these genotypes are said to have a *homozygous* genotype. When a cell contains different alleles (*Aa*) it is said to have a *heterozygous* genotype. Suppose that the phenotype of a cell is determined by the presence or absence of the activity of a specific enzyme. A cell of genotype *AA* or *Aa* would possess the enzyme activity whereas a cell with homozygous genotype *aa* would not have the enzyme activity. Borrowing terms from Mendel, we now say that gene (allele) *A* is *dominant* over its *recessive* allele *a*. The phenotype of enzyme activity is a dominant trait, whereas the phenotype lacking enzyme activity is a recessive trait. In a cell of heterozygous genotype (*Aa*), the recessive allele (*a*) cannot show its phenotypic effect (no enzyme activity) because its dominant allele (*A*) allows the cell to make the enzyme.

In the formation of gametes, a heterozygous plant Aa would be expected to form two gametes with equal frequencies, *A* and *a*. Random union of these gametes would be expected to produce *AA, Aa*, and *aa* zygotic genotypes in the ratio 1:2:1, respectively. At the phenotypic level, these offspring would be expected in the proportions of 3/4 with enzyme activity (*A*-type), and 1/4 without enzyme activity (*a*-type). Whenever both male and female parents have *AA* genotypes, all of their gametes would contain *A*. The union of these gametes would create only offspring genetically and phenotypically identical to their parents. We would say that the parents breed true to type. Homozygotes for the same allele are genetically pure and exhibit this pure breeding phenomenon. A similar case is true when individuals of genotype *aa* intercross; all of their offspring should be both genotypically and phenotypically recessive.

Now let us return to the story of Gregor Mendel. Most accounts of Mendel's contributions in contemporary textbooks depict him as a theorist who discovered the basic laws of heredity by postulating hereditary units that we now call genes. They say or imply that Mendel proposed that somatic cells contain two alleles for each trait, and that some mechanism (now known as meiosis) must be responsible for insuring that only one of those two alleles gets into a gamete. These popular accounts were written by the rediscoverers and reinterpreters of Mendel's paper who had understanding of many things that were unknown in Mendel's time. This new knowledge predisposed these authors to attribute to Mendel many things that he did not say and possibly never even considered. The paradigm of heredity in Mendel's day was that the genetic material consisted of a fluid of some sort. Thus, blood in animals and sap in plants were thought to carry genetic information. Remember now, I am using modern terms to explain these historical events. According to the fluid theory, offspring should exhibit a phenotype intermediate between that of its parents, just as when one mixes red and white paint, pink paint is produced. Mendel knew of such cases, but they were not the kind of traits that he chose to investigate.

Contrary to what most textbooks tell us, Mendel was not trying to discover basic laws of heredity, but rather, "to follow the developments of hybrids in their progeny."[17] He acknowledged the fact "that no generally applicable law of the formation of hybrids has yet been successfully formulated."[18] He proposed that the reason why such a law had not been discovered was that "among the numerous experiments [performed to date] not one has been carried out to an extent or in a manner that would make it possible to determine the number of different forms in which hybrid progeny appear, permit classification of these forms in each generation with certainty, and ascertain their numerical interrelationships."[19] The successful completion of such an experimental program should produce a "solution to a question whose significance for the evolutionary history of organic forms must not be underestimated."[20] We now infer that his research was designed to find out if hybrids could breed true, and if so, to show that hybridization could be an important mechanism in the evolution of new species. Nowhere in his paper does Mendel say that he was trying to establish a particulate theory of heredity or that he even had such a theory in mind. Mendel's results, however, obviously could not be accommodated under the existing fluid theory of heredity, and so many historians have concluded that he must have been forced to think in terms of hereditary particles.

Part of Mendel's success in discovering the "laws of hybrids" was in his choice of experimental material. Peas are essentially completely self-fertilizing in nature. That is, the sperms in pollen grains shed by a pea flower will normally fertilize the ovules or egg cells within the pistils of that same flower. We know now that repetitive self-fertilization purifies a genetic lineage so that all members thereof come to have an identical homozygous hereditary endowment (genotype). Mendel collected different varieties of peas that had a long history of breeding true for particular combinations of traits. Thus, he must have felt confident that his starting material was genetically pure (homozygous). One of Mendel's experiments involved the artificial crossing of a pure variety of peas with yellow seeds to a pure variety having green seeds. The hybrid or F_1 (first filial generation) offspring all had yellow seeds. When he allowed these hybrids to fertilize themselves, the next generation (F_2) had some plants with yellow seeds and some plants with green seeds. Mendel wrote in German, and my evaluation of his 1866 paper has been based on the wording from the English translation made by Eva R. Sherwood.[21] Mendel's use of the term "hybrid" would now be called "heterozygous." Today the term "hybrid" usually denotes offspring between different species. Mendel's word "form" has now been replaced by "phenotype." His use of the word "character" was meant to indicate the breeding structure of an individual; i.e., whether on self-fertilization it bred true to form or segregated into dominant and recessive forms; we would use the term "genotype" today. In modern vernacular, the pure breeding (constant, parental) characters would be called homozygous; the hybrid character would be called heterozygous. Mendel said that the yellow-seeded form could exist in either a parental (pure) dominant character state or in a hybrid dominant character state. In general, he represented the pure dominant state by a single-letter symbol (A), the pure recessive state by the single-letter symbol a, and the hybrid state by the two-letter symbol Aa. The recessive green form behaves in breeding as a pure parental character state; i.e., self-fertilization of green-seeded plants of any generation produces only green-seeded progeny. He found that, on average, the F_2 consisted of yellow-

and green-seeded plants in a ratio of approximately 3:1, respectively. On further testing, he also discovered that the F_1 hybrids consisted of pure breeding parental characters and hybrid characters in the ratio of 1:2, respectively. Thus hybrids do not breed true to type, but segregate in the manner just described.

Mendel reported on six other pairs of contrasting phenotypes, each of which individually behaved in hybrids just like the yellow- and green-seed alternatives. When studied jointly, he found that their ratios could be predicted mathematically by a combination series using the expected ratios as determined from crosses where attention was focused only on a single contrasting pair of phenotypes. For example, if a second pair of pure dominant and pure recessive characters were represented by the symbols B and b, respectively, four phenotypic groups would be possible. The F_2 offspring, having the two dominant traits (AB), would be expected $3/4 \times 3/4 = 9/16$; those with one dominant and one recessive trait (of two kinds; Ab, aB) would be $3/4 \times 1/4 = 3/16$ each; those with two recessive traits (ab) would be $1/4 \times 1/4 = 1/16$. This gives rise to the familiar 9:3:3:1 ratio. (Although this ratio does not appear in Mendel's paper, he presents the data from which it can be calculated.) The same principle could thus be applied, in theory, to all seven pairs of contrasting phenotypes. Mendel also concluded that it "appears logical that in the ovaries of the hybrids there are formed as many sorts of egg cells, and in the anthers as many sorts of pollen cells, as there are possible constant combination forms, and that these eggs and pollen cells agree in their internal composition with those of the separate forms."[22] Mendel proposed that the regularities of hybrid formation could be accounted for if we "assume that the various kinds of egg and pollen cells were formed in the hybrids on the average in equal numbers."[23]

It is easy to see how so many people could be tempted to interpret Mendel's archaic language in terms of modern genetic knowledge. Since crossing pure varieties of contrasting phenotype produced hybrids showing the dominant traits, not intermediate traits, and selfing of hybrids produced some offspring with dominant traits and others with recessive traits, how could Mendel escape the conclusion that the hereditary material must be particulate rather than fluid? Fluids such as red and white paints in pure parents would blend into uniform pink hybrids, from which the pure parental colors could not be recovered in any subsequent generation. As logical as this deduction would appear to be, we cannot know for sure that Mendel used this line of reasoning. We simply do not know what was going on in his mind regarding the nature of the hereditary material.

He was aware that both parents contributed in some hereditary way to their offspring and that it didn't make any difference, as Karl Fredrich von Gäertner had found previously, whether the dominant trait in hybrids was derived from the maternal or the paternal parent. (It was discovered in 1984 that the expression of some genes depend upon the gender of the parent from which they were derived; the phenomenon is known as *genomic imprinting*.) Mendel was behaving strictly as an empiricist[24] by reporting only what he could observe, namely the phenotypes and the breeding structure of individuals (i.e., whether they would, on selfing, breed true or segregate). There is no rationalism in his paper. He does not propose any hypothetical entities that cannot be seen, such as genes. This is why some modern critics of Mendel cannot, in good faith, attribute to him the title of "father of genetics." He of course was unaware of the diploid nature of somatic cells and the haploid nature of gametes. His symbolic use of *Aa* for the hybrid character (i.e., a breeding nature in which segregation occurs) does not necessarily indicate that he held concepts compatible with our present theories, namely that each somatic cell contains two alleles at the same genetic locus on a pair of homologous chromosomes or that gametes contain only one of these alleles. The fact that he represented a pure parental character by a single-letter symbol has been excused in the past by others as simply Mendel's short way of expressing the diploid homozygous genotype. But that is putting words into Mendel's mouth. He also used the same single letter to represent the potential character state of a parental gamete. Apparently he did not entertain any theory regarding the number of genes that were present for a trait in either somatic cells or gametes. Even if Mendel had a genetic theory, his single-letter symbol *A* may simply have implied that there was no other allelic form in the somatic cells or in the gametes. There may have been one, two, three, or any number of *A* genes in these cells. Representing the hybrid condition by a pair of symbols *Aa* likewise does not necessarily imply that only two alleles existed in body cells. He might have thought that hybrids could be genetically *AAAaaa* or *Aaa* or who knows what other combinations? Only in one context does Mendel use two identical symbols for homozygotes. "The result of the fertilization may be made clear by putting the signs of the conjoined egg and pollen cells in the form of fractions, those for the pollen cells above and those for the egg cells below the line. We then have

$$\frac{A}{A} + \frac{A}{a} + \frac{a}{A} + \frac{a}{a}.\text{"}^{25}$$

After 1900, Mendel's report on the role of hybridization in peas was misinterpreted, in the light of new discoveries, as presenting the fundamental rules of particulate heredity. Genes were labeled dominant or recessive. After the chromosome theory of heredity was proposed, it became easy to understand how different alleles of a genetic locus could be caused by meiosis to separate into different gametes. This separation process was called "allelic segregation." Mendel was falsely identified as the formulator of "the law of the segregation of alleles." How a member of one pair of chromosomes segregates into a gamete is independent of how a member of any other pair of chromosomes gets into that same gamete. Thus, genes occupying loci on different chromosomes exhibit this same property. This genetic rule was also falsely attributed to Mendel. It became known as "the law of independent assortment."

The penultimate section of Mendel's 1866 paper is entitled "Experiments on Hybrids of Other Plant Species." This is the only part of the paper where Mendel uses the term "element." What these elements are and their function(s) is not made clear, but from the following passages, one can hardly escape the inference of their particulate nature. "This development (of a fertilized egg) proceeds in accord with a constant law based on the material composition and arrangement of the elements that attained a viable union in the cell."[26] And later on he states, "we must further conclude that the differing elements succeed in escaping from the enforced association only at the stage at which the reproductive cells develop. In the formation of these cells all elements present participate in completely free and uniform fashion, and only those that differ separate from each other."[27] From this last statement, I infer that either there is only one such element in constant forms or that two or more identical elements do not separate from each other in the formation of reproductive cells. If he thought of his "elements" as hereditary determiners, then he apparently had no idea about how many elements existed in either somatic cells or in gametes, and did not envision a reductional process such as meiosis in which identical as well as nonidentical alleles segregate. We now know that even identical alleles separate during meiosis as homologous chromosomes move into different reproductive cells. Since Mendel did not make a clear symbolic distinction between the "material composition" (genotype?) of somatic cells and gametes, he probably was confused. Unless you read this last segment of the paper, this confusion is not as readily apparent. I therefore was surprised when I found out that Peters,[28] in reprinting an English translation of Mendel's paper, decided to omit this last section because "these paragraphs have little bearing on the principles Mendel proposed in this paper, and I have found from experience with my students that these pages serve primarily to confuse than to clarify." Amen! That is precisely why students should be exposed to that material, so that they can feel as confused about heredity as Mendel appears to have been. In studying science history, let's examine all the data before we jump to analyses and interpretations. It is always dangerous for anyone to "sanitize" the data that is presented in a paper, regardless of good intentions, lest information be removed that others may consider vital for an accurate evaluation of the problem. Of course it goes without saying that the best job of analyzing any piece of scientific literature can be done only by going back to the primary source (the original

266

paper or manuscript, if possible) as written in the author's language. Errors are often made in translations that might thereby be avoided. Unfortunately, in Mendel's case, many of my readers are not fluent in German, so they (as I) will have to depend on the English translations that are available.

If we are charitable and assume that Mendel did have a particulate genetic theory in mind, either before or after he performed his pea experiments, suggestions have already been offered in this essay to explain where he (or anyone else) might have obtained such a theory of heredity in the 1860s. Whether he had or used such a theory is equivocal, so the origin of such an idea in his case is a moot question. Mendel's uniqueness and creativity are so well established that there is no need to attribute to him discoveries and ideas that were not his. If Mendel were alive today, I don't think he would appreciate the distortion of his contributions that have been made by the rediscoverers and reinterpreters of his work. As historians, we should use the Mendel case as an object lesson of what happens when we misinterpret the record and fail to present as accurate a picture of what happened as possible, or let what we now know influence our interpretation of historical documents.

Today, a biological species is defined as a collection of freely interbreeding individuals that are reproductively isolated from other such groups. In Mendel's day, however, species were defined quite differently. According to Mendel, "If we adopt the strictest definition of a species, according to which only those individuals belong to a species which under precisely the same circumstances display precisely similar characters, no two of these (pea) varieties could be referred to one species."[29] While most of his peas belonged to the species *Pisum sativum*, others had been classified as belonging to other species (*P. quadratum, P. saccharatum, P. umbellatum*). Mendel did not think it important, for the purpose of his experiments, to which of these species his varieties belonged. "It has so far been found to be just as impossible to draw a sharp line between the hybrids of species and varieties as between species and varieties themselves."[30] His results demonstrated that pea hybrids could not breed true, but that true breeding forms could be recovered in the progeny of hybrids. According to Mendel, Gärtner had found some species hybrids "that remain constant in their progeny and propagate like pure strains. . . . This feature is of particular importance to the evolutionary history of plants, because constant hybrids attain the status of *new species*."[31] By this definition, Mendel's pea hybrids did not qualify as new species, and thus were (I infer) of no evolutionary importance. I can't imagine Mendel being very happy about this, because one of his main purposes in doing his experiments was to shed light on the importance of hybridization in the evolution of organic forms.

POSTSCRIPT

There is no question about the sorry current state of science education in the United States. Student literacy in science and mathematics is lagging far behind many other less affluent countries. The blame for this intolerable situation has been attributed to everything from poor teacher performance to students' watching too much television. According to the many studies made on our educational crisis, a major goal of science education should be to teach students how to think critically and how to solve problems. It is true that science involves a certain way of logical thinking, but it may also require insight and the ability to put things together in patterns unique to each problem. My genetics course is built around a problem-solving approach. I try to keep the amount of factual material that the student must remember to a minimum. I try to emphasize the utilization of a few basic principles to solve a wide range of problems. All too often, I fear, scientific literacy is evaluated by asking about established scientific facts. Science does depend on facts. But without theories to organize and explain the facts, there is no science. Where do scientific theories come from?

In the sections of this chapter, I have tried to give the reader a sense of the difficulty one often has in trying to identify exactly how scientists make new discoveries and come up with new theories. One thing is clear. Revolutionary science requires a break with conformity. Unconventional thinking is a rare commodity. If our society wants to enhance the pace of scientific discoveries in the future, our educational systems should be teaching students not only how to think logically, but also how to think unconventionally. Is this something that can be taught? I have trouble teaching many of my students to think conventionally in logical ways. I have no idea how to teach them to think in unconventional ways. I am not alone in this respect. I have consulted many of my colleagues in the teaching profession about this problem, and they all concur that insight and unconventional thinking can be illustrated, but cannot be taught and learned by reading books or attending lectures.

For how many human generations have people watched the Sun rise in the east, arch across the sky, and set in the west without questioning the notion that it was the Sun that moved and not Earth? Finally in 1543 an unconventional thinker by the name of Nicolaus Copernicus dared to suggest a radically new way of interpreting the age-old observations. He proposed that Earth rotated on its axis as it moved around a stationary sun. Can you imagine the response he got to that theory? There was no end to the objections. Common sense tells us that if Earth rotated on its axis it would cause violent winds to constantly sweep the landscape. The centrifugal force of a spinning Earth would cause people, the waters of Earth, and everything else that wasn't nailed down to fly off into space. Even the Bible indicates that it is the Sun that moves and not Earth: "So the sun stood still in the midst of heaven, and hasted not to go down about a whole day" (Josh. 10:13).

Cherished beliefs die hard. Radical new ideas upset the orderliness of our former worldviews. We can't or don't want to think in new ways because of what we have been taught, what we think we know, and what we want to believe.

Each time I teach the genetics course, I see the difficulty that students have in letting go of what they already know in approaching a new problem. For example, after explaining basic Mendelian principles, I use coat colors in guinea pigs to illustrate them. They are told that black coat is determined by a dominant gene and brown coat by its recessive allele. I explain that six genetically different kinds of matings can be made in a population having two alleles at the coat-color locus. I then ask them to forget anything that they have been told about the inheritance of guinea pig coat colors. They are to imagine that they were the ones who discovered guinea pigs. About half of the population is black and the other half is brown; males and females of both colors are almost equally frequent. They are told that hundreds of matings have been made, all of which had both parents black, and anytime two black guinea pigs are crossed, all of their offspring are black. What is the simplest genetic hypothesis consistent with these data? Often the response is that black pigs are homozygous. I ask them, "Homozygous for what kind of gene?" They often answer "Homozygous for the dominant black gene." Then I ask, "Are there no heterozygous black pigs in the population?" After acknowledging that because of random mating there should be lots of heterozygous black pigs in the population, I ask, "What happens when heterozygous black pigs mate?" They answer correctly that about one-fourth of the offspring are expected to be brown. "Shouldn't some of the hundreds of matings between black pigs have involved heterozygous parents? And since no browns have ever been found in their progeny, do not the results conflict with your proposed theory?" Then they get their thinking caps on and propose that black coat color is due to a recessive gene. Thus, any black pig would have to be homozygous for the recessive gene. When recessive blacks are intercrossed, all of their progeny would be expected to be black. That theory is consistent with the observations. Unfortunately, some students leave class with a quizzical look on their face, like they heard the explanation but somehow they can't believe that it could be correct. It seems as though once they had been told how the system works in the real world, they could not let go of that knowledge, as asked to do, and approach a new hypothetical problem with an open mind ready to evaluate new data and formulate new explanations if necessary. In this pedantic approach, I try to get students to think in unusual ways (e.g., forgetting what they already "know") and to think about the consequences that should logically follow from the hypotheses they are proposing. This should be the meat of any science course. It is my fondest hope that long after my students may have forgotten the details of genetic lore, their appreciation of the scientific method and the discipline of trying to think in new ways will last a lifetime.

NOTES

1. Kary B. Mullis, "The Unusual Origin of the Polymerase Chain Reaction," *Scientific American* 262 (April 1990): 56–65.
2. Ibid., 61.
3. Ibid.
4. Ibid., 69. On September 8, 1983, Mullis attempted his first PCR experiment using the gene for human nerve growth factor as his target. It failed, as did several other attempts over the next two months. In December, he switched targets to a segment of the bacterial plasmid pBR322 and obtained his first success. Paul Rabinow, *Making PCR: A Story of Biotechnology* (Chicago: University of Chicago Press, 1996), 100–101.
5. Yvonne Baskin, "DNA Unlimited," *Discover* 11 (July 1990): 78.
6. Mullis, "The Unusual Origin of the Polymerase Chain Reaction," 65.
7. Ibid., 56.
8. K. Kleppe et al., "Studies on Polynucleotides. XCVI. Repair Replication of Short Synthetic DNA's as Catalyzed by DNA Polymerases," *Journal of Molecular Biology* 56 (1971): 341.
9. Marcia Barinaga, "Biotech Nightmare: Does Cetus Own PCR?" *Science* 251 (1991): 739.
10. Nancy Smyth Templeton, "The Polymerase Chain Reaction: History, Methods, and Applications," *Diagnostic Molecular Pathology* 1, no. 1 (1992): 58.
11. Mullis, "The Unusual Origin of the Polymerase Chain Reaction," 61.
12. Floyd V. Monaghan and Alan F. Corcos, "Possible Influences of Some Nineteenth-Century Chemical Concepts on Mendel's Ideas About Heredity," *Journal of Heredity* 74 (July–August 1983): 297.
13. Ibid.
14. Ibid.
15. Ibid., 297–98.
16. F. V. Monaghan and A. F. Corcos, "Mendel, the Empiricist," *Journal of Heredity* 76 (January–February 1985): 49–54.
17. Curt Stem and E. R. Sherwood, eds., *The Origin of Genetics; A Mendel Source Book* (San Francisco: W. H. Freeman and Company, 1966), 1.
18. Ibid.
19. Ibid., 2.
20. Ibid.
21. Ibid., 1–48.
22. James A. Peters, *Classic Papers in Genetics* (Englewood Cliffs, N.J.: Prentice-Hall, 1959), 15.
23. Ibid.
24. Monaghan and Corcos, "Mendel, the Empiricist," 49–54.

25. Stem and Sherwood, *The Origin of Genetics*, 30.
26. Ibid., 42.
27. Ibid., 43.
28. Peters, *Classic Papers in Genetics*, 2–20.
29. Ibid., 3.
30. Ibid., 4.
31. Stem and Sherwood, *The Origin of Genetics*, 40.

CHAPTER 9

NO MAN
IS AN ISLAND

If I have seen farther it is by standing on the shoulders of giants.
—Sir Isaac Newton (1642-1727)[1]

T o many people, history is just a list of names and dates to be associated with great events: Galileo discovered the moons of Jupiter, Newton invented (independently of Leibnitz, according to the latest consensus of historians) the calculus, Mendel formulated the laws of heredity, Watson and Crick deciphered the structure of DNA, and so forth. Prizes are usually awarded to individuals, not to groups of people. Deceased contributors to the advancement of scientific knowledge normally do not get prizes (e.g., Nobel Prizes are not awarded posthumously). No wonder that we are accustomed to thinking that each advance in science should be associated with a certain person or small team; e.g., a Nobel Prize in one of the sciences (chemistry, physics, or physiology or medicine) is awarded to no more than three people. It's too bad that we have grown conditioned to this way of thinking because, as Newton reminds us, no scientific advance can be completely attributed to a single individual. We are all products of our separate cultures. Our education in those cultures cannot help but influence how we think and what we can accomplish by building on what we have been taught and what we have experienced. What debt did Galileo owe to those who had perfected lens grinding to an art before he built his telescope? Could Newton have invented the calculus if he had not gained a thorough knowledge of mathematical principles whose origins can be found in the ancient Egyptian and Greek civilizations? Would Mendel have carried out his pea hybridizations if he had not been acquainted with the plant-breeding work of Gärtner and Kolreuter? Could Watson and Crick have built their model of DNA without having access to crystallographic data that they themselves did not generate? Every scientist owes a great debt to innumerable other folks who usually go nameless unless they were intimately connected with a published piece of work. How far removed from a science project does one have to be before he may be ignored by the author(s) of a paper in citing credits? For example, should technicians who only do assigned laboratory experiments or fieldwork receive credit for a scientific discovery when they neither defined the problem, nor planned the experiments, nor analyzed and interpreted the results, nor wrote the report, nor were responsible for obtaining the grant that supported the research? This can be a very tricky problem that in some cases has led not only to bruised egos, but to federal investigations and international agreements.

THE HUMAN IMMUNODEFICIENCY VIRUS

In the May 4, 1984, issue of *Science*, Robert C. Gallo et al. published four papers[2] that established the viral nature of acquired immune deficiency syndrome (AIDS) and outlined a process for developing a blood test for detecting one's exposure to the virus. It was on this basis that Gallo has been acclaimed (by some sources) as the man who discovered the cause of AIDS. Ever since AIDS was recognized as a distinctive clinical disease in 1981, two teams of workers had dominated the race to find the causative agent. Gallo and his team were working at the National Cancer Institute (NCI) in the United States, while Luc Montagnier and his colleagues were working at the Pasteur Institute in Paris, France. In the May 20, 1983, issue of *Science*, there appeared several papers on AIDS, including two from Gallo[3] and one from Montagnier.[4] The Montagnier paper reported on a virus he called LAV (lymphadenotropic virus), but acknowledged that "the role of the virus in the etiology of AIDS remains to be determined." In July and September of 1983, Montagnier had sent Gallo (at his request) samples of the French virus LAV. Less than a year later, the Gallo team published their bombshell reports in the May 4, 1984, issue of *Science* that presented the proof that Montagnier's group was seeking. Gallo had several virus candidates that he named (for the patients from whom they were isolated) CC, MoV, RF, MN, and SN. Conspicuously absent from this list of viruses was Montagnier's LAV. In April 1984, the *New York Times* reported in a front-page article that the head of the Centers for Disease Control (CDC) in Atlanta had claimed that the CDC researchers had found evidence that the French virus LAV was the cause of AIDS.[5] The next day, Margaret Heckler, former Health and Human Services Secretary, announced at a press conference that the American team under Gallo had found the cause of AIDS and developed a blood test to detect if a person had been infected with the virus. Apparently this was done to announce to the world that America had triumphed in the race to find the cause of AIDS. The Pasteur Institute challenged the "official" American position in 1985 by claiming that Gallo had "misappropriated" LAV in the development of the blood test.[6] The allegation was hotly contested over the next two years. Involving as it did cooperation between different nations, the political ramifications of the squabble had to be solved by a negotiated agreement, not only between Gallo and Montagnier, but also by then U.S. president Ronald Reagan and French premier Jacques Chirac. The international agreement declared that Gallo and Montagnier be recognized as "codiscoverers" of the AIDS virus.[7] This was the end of round one.

Round two began when an article by journalist John Crewdson appeared in the November 19, 1989, issue of the *Chicago Tribune*. Crewdson had been dogging this case for more than three years. In this particular article, he suggested that the only AIDS virus that was yielding productive experimental data in Gallo's lab was the French virus LAV, implying that Gallo had used it to make the blood test. The article caught the attention of Representative John Dingell (D-Mich.), who demanded that the National Institutes of Health (NIH) initiate an inquiry. An extensive article by Ellis Rubinstein[8] was published in the June 22, 1990, issue of *Science* that reviewed what at that time had been obtained under the Freedom of Information Act (FOIA) and from the "Washington leak." The NIH's inquiry panel, which had been conducting a "preliminary fact-finding" mission since December 1989, had its first meeting with Gallo on April 8, 1990. The panel was under the administration of the NIH Office of Scientific Integrity (OSI), and was composed of scholars (unassociated with the Gallo affair) nominated by the National Academy of Sciences and the Institute of Medicine. At the time Rubinstein published his article, the committee had been in secret deliberations for six months.

Before continuing, it might be advantageous to review a little information about the nature of the AIDS test. The blood test for AIDS detects antibodies against the virus in the blood of a patient. It does not detect the virus itself. These antibodies are produced by a person only after being infected with the virus. The evidence that the AIDS virus has infected an individual does not necessarily mean that the individual has developed the clinical signs of AIDS, such as a decrease in the number of white blood cells (leukocytes) designated T4. Nor does it necessarily mean that the infected person will develop these clinical signs later. There is much that we have yet to learn about the disease called AIDS. One of the current theories postulates that a cofactor is necessary to activate the virus, perhaps a tiny bacterium known as a *mycoplasma*. The virus itself can be produced in large amounts by growth in laboratory cultures of cells. Viruses are obligate intracellular parasites, i.e., they cannot be grown outside of cells. Each kind of virus normally infects and reproduces ("grows") only within certain cells of specific species. For example, the AIDS virus can infect both humans and chimpanzees, but it does not cause the clinical symptoms of AIDS in the ape. Viruses are constructed fully formed within the cell. They do not "grow" (in the conventional sense of increasing size or mass) once they have been constructed. Only certain cells grow well continuously in glassware (in vitro) and allow productive infections resulting in viral replication. The virus recovered from such cells can be processed to extract some of its protein components, which are then used as the "antigens" to which antibodies of the patient will bind. This is the basis of the "AIDS test" that is now used to screen all blood for transfusion in the United States. In order to detect the virus itself within a patient, rather than its footprint antibody, the AIDS virus can be purposefully injected into a mouse. After the mouse has manufactured antibody-producing cells (mature B lymphocytes or plasma cells) against the virus, some of these cells are made to fuse with cancer cells called *lymphomas*, thereby creating immortalized cells known as *hybridomas*. These hybridoma cells grow well and indefinitely in the laboratory, and from them a pure population of antibodies can be extracted. This so-called *monoclonal antibody* (MoAb) can then be used to determine if a cell sample contains the AIDS virus. The AIDS virus is able to mutate or change some of its antigens quite easily, so that some MoAbs developed against one strain of the virus might react poorly or not at all with a different strain of the same virus.

Montagnier had isolated his LAV strain of the AIDS virus in December 1982. It is a matter of record that Gallo "had three new AIDS samples in culture" in May 1982, a fourth in August, and five more in October (a total of nine cultures). None of them "grew" very well, however; i.e., not many virus progeny were produced. No one knew at the time how the virus reproduced, but Gallo predicted that it was a retrovirus, probably a variant of the first known human retrovirus, *human T-cell lymphotropic virus* (HTLV-I), that his team had discovered. This discovery had been made in 1979 by Gallo researchers Bernard Poiesz and Frank Ruscetti. A *retrovirus* uses RNA as its genetic material rather than DNA. Each retrovirus particle has packaged into it an enzyme called *reverse transcriptase*. This enzyme, found in cells infected with a retrovirus, can make a DNA copy from the viral RNA. This DNA copy inserts into the host's chromosome and becomes, in essence, part of the host-cell's genetic endowment (genotype). In this integrated state the virus DNA is called a *provirus*. It may stay in the proviral state for many years without causing any problems. Then something is thought to trigger the provirus to begin replicating and making new viruses (viral RNA and its associated proteins). Gallo's prediction was confirmed in December 1982 when his crew detected reverse transcriptase in two AIDS patients. However, these samples tested negative for two proteins (p19 and p24) that characterize HTLV-I. (The numbers following the "p" represent the molecular weights of the proteins in thousands of daltons. A dalton is equal to the mass of a hydrogen atom.) By February 1983, two additional samples were found with the same characteristics, i.e., positive for reverse transcriptase and negative for the HTLV-I proteins. Some of these viral samples later proved to be the cause of AIDS, and came to be called, by international agreement, *human immunodeficiency virus* (HIV). The sample designated CC contained viruses that did not appear in electron micrographs to be HTLV-I, but nonetheless tested positive for the HTLV-I proteins. The CC line of cells was growing poorly, and by May 16, 1983, it was dying. This was very strange because the human retroviruses known at the time (HTLV-I and HTLV-II) didn't kill cells, they immortalized them, making them able to grow continuously in cell cultures. Later it was discovered "that CC was in fact doubly infected with both HTLV-I and HIV."[9]

Montagnier detected reverse transcriptase in his LAV in February of 1983 and asked Gallo to send him the MoAb reagents that would allow him to determine if LAV could be distinguished from the HTLV viruses. Gallo complied, and Montagnier found that LAV was distinct from the HTLVs. Gallo received from Montagnier the first sample of LAV in July 1983, but he couldn't get it to grow in cell culture. He received a second sample in September of the same year. Although the two labs were exchanging materials, Gallo stated to a review panel that Montagnier did not wish to establish a collaboration. At this time neither of these labs was having much success in growing viruses in continuous mass culture. Such mass cultures were needed, as explained previously, to produce the reagents required to identify or "type" the various viral isolates, proving that they were all identical, and thereby establishing the cause of AIDS. By November 1983 Gallo's lab had a viral isolate named MoV (from patient Mo) being cultured in cells designated HUT78. MoV appeared to be a variant of HTLV-II. But by February 1984 evidence was available that MoV was neither HTLV-I nor HTLV-II. Gallo now wasn't sure whether MoV was a new virus or a contaminant. Contamination between cell lines is a constant problem in even the best cell-culture laboratories (see "The HeLa Debacle" in chapter 7). Crewdson reported that MoV may, in fact, be a LAV contaminant. On November 15, 1983, Gallo's team had put into culture eleven samples from patients with AIDS. One of these, designated RF, was growing vigorously by April 1984. In January 1984 Gallo decided to put his LAV cultures in the freezer and continue working with his own isolates.

While Betsy Reed was working with RF in one of Gallo's labs, Mikulas Popovic, working in another, decided to do an unusual experiment. Trying to make a blood test, he pooled as many as ten samples from different patients (LAV was reportedly not knowingly added to the pool). "Popovic reasoned that the failure (of the virus) to grow might be tied to the fact that none of the viruses individually was producing high enough concentrations of reverse transcriptase. Maybe if he had dumped ten viruses into the same pot, the reverse transcriptase level would be enough to jolt one of them into action."[10] Probably most of his colleagues thought that the logic behind it was crazy, but one virus, designated IIIB, came out of that pool and "grew like a charm." This was the virus that ultimately was used for the blood test. Later it was shown that IIIB and LAV were genetically very similar. Most of the viruses isolated from different AIDS patients were similar at, say, the level of cousins, but IIIB and LAV were more like siblings. Popovic, however, claims that "biologically they were not the same in the way they behaved."

Many questions have been raised concerning this research. Did Gallo "steal" Montagnier's virus to make the blood test? Is IIIB a LAV variant? Did LAV get into the pool by accident or was it added deliberately, but covertly? Gallo responded indignantly to the idea that he consciously used LAV for his blood test, but admitted that it is possible that LAV could "have inadvertently contaminated our cultures and suddenly dominated the culture by rapid growth."[11] He steadfastly maintained, "Our work never depended on a single isolate."[12] For example, he had the MoV and RF strains. When asked why he decided to use a virus (IIIB) derived from a pool for the blood test instead of one of his other strains, Gallo replied that he was aware that MoV might have been contaminated and RF didn't grow quite as vigorously as IIIB. Also IIIB had been growing a couple of weeks longer than RF. He also assumed that IIIB was an "American" virus derived from an American patient, whereas RF was known to be Haitian. His intuition told him that there might be differences in the viruses related to geographic origin that could be important in a blood test. In 1985, it was found that the genetic sequences of IIIB and LAV differed only by about one hundred nucleotides (1 percent of the genome), whereas most other isolates differed by 5 percent or more. This confirmed the suspicions of many people that the two viruses were essentially the same. In March 1990, however, it was discovered that even viruses which come from the same cohort or population can be much more alike than anyone had realized, and therefore it might be possible that IIIB and LAV are actually different isolates.

The credit for discovering HIV was, by international agreement in 1987, given jointly to Gallo and Montagnier. Both the French and American teams agreed to share credit and royalties from the discovery that led to a blood test for HIV antibodies.[13] The discovery of HTLV-I in 1979 was made in a cell line that had been established in John Minna's lab in the Clinical Oncology Branch of the then Veteran's Administration by researchers Adi Gazdar and Paul Bunn. All three (Minna, Gazdar, and Bunn) were listed as coauthors on Gallo's 1979 paper on HTLV-I. In the May 4, 1984, issue of *Science*, the Gallo team announced the isolation and continuous production of a cytopathic retrovirus designated HTLV-III from patients with AIDS. Mikulas Popovic had grown it in large quantities in cells he named H9. These cells were not killed by the virus. The origin of the H9 cells was not explained in that paper. Years later, H9 cells were shown to have been cloned from a line designated HUT78 that had been established by Gazdar and Bunn in Minna's lab. They had given HUT78 to Gallo, just as they had freely donated the cell line containing HTLV-I earlier.

Several vials of a sample, later to be designated as HUT78L (for Litton Bionetics, an NIH subcontractor), were left in the freezer of Gallo's lab when Bernie Poiesz transferred to Syracuse in 1980. By 1981, at least two additional samples marked HUT78 had been deposited there (by whom is not clear). All three samples were assayed (in research unrelated to AIDS) by Dean Mann for their HLA type. HLA stands for *human leukocyte antigen.* These are molecular markers on cells that are distinctive for each individual. All of the cells from a given source should be of identical HLA type. Mann was working in the Laboratory of Human Carcinogenesis and was concerned with interferon production by HTLV-I-infected cell lines. *Interferon* is a protein made by virus-infected cells that inhibits viral replication. Mann found that, in two of the three samples, it appeared that more than one person's cells had contaminated them. A year later (1982), Mann tested the three samples again plus a fourth sample designated HUT78B ("B" for Paul Bunn who, at Gallo's request, had directed a Gazdar technician to send a sample to Gallo's lab). Some of these samples had markers for B lymphocytes on them (indicating they were bone-marrow-derived antibody-producing cells) rather than being T (thymus-derived lymphocytes) cell lines. This mix-up was of no concern to Mann because he was only concerned with lymphokine production. (*Lymphokines* are proteins, such as interferons, produced by lymphocytes, such as T cells, that help mediate the immune response.)

In 1983 Mika Popovic had been assigned to find a cell line in which the AIDS virus would grow. He was advised by Mann to look for a "CD-4+ cell line that had MHC class 2 (markers) on it." T cells with these characteristics were the ones that tended to disappear in AIDS patients, suggesting that they were the targets of the AIDS virus. It was not known at that time that some cell lines with these properties would allow the virus to grow without killing the host cells. Popovic knew that HUT78 was a cell line with the proper cell markers, but he didn't know which of the samples labeled HUT78 in Gallo's lab freezers he should use, because Mann had shown that they were all different. Popovic chose one, for reasons undisclosed to *Science,* and *cloned* (i.e., made identical copies of) individual cells from it so that each isolate would be a pure sample. The cloning, he explained, was necessary to find a population of cells that was uncontaminated with any retrovirus. Popovic was able to isolate several cell lines that were resistant to the killing power of HTLV-III (later to be called HIV) and yet allowed the virus to grow without exposure to the extremely expensive growth factor known as interleukin-2. One of Popovic's lines was designated HT, and from it several clones were made, including those named H4 and H9. The Gallo team's famous 1984 paper reported in *Science*, "We have used clones H4 and H9 for the long-term propagation of HTLV-III from patients with AIDS and pre-AIDS."[14] The origin of cell line HT was said to have been derived from an adult with lymphoid leukemia. The two clinician researchers who worked with patient HT would not have diagnosed HT as being leukemic from the Gallo team's description. The use of the designator HT for the key cell line obscured the fact that it derived from HUT78. Gallo's critics want to know why he changed the name of HUT78 to HT. Why did he give credit neither to Adi Gazdar nor Paul Bunn in his 1984 paper when his work was done in derivatives from their cell lines? Both Gallo and Popovic claim that they thought HT was HUT78, but they could not be sure, based upon Mann's HLA-typing report. Needless to say, Gallo's critics don't accept this explanation.

Popovic claimed that he had never heard of Gazdar, even though Gazdar was the lead author on the paper describing the origin of the HUT78 cell line. This admission puts Popovic in a bad light because he should have read that paper sometime during all those years when he was trying to figure out what HUT78 actually was. Not only that, but when he contacted Paul Bunn in 1984 at Gallo's suggestion, he did not explain his problem to Bunn, who may have been able to help him solve it. Both Popovic and Gallo cited the rush to produce the AIDS test as the reason for publishing their results before the origin of H9 could be resolved. People were dying, and the sooner the AIDS test could be produced the fewer people would be infected by contaminated blood transfusions. Gallo stated, "Maybe I'm insensitive to somebody's feelings about this. It never dawned on me that this was so tremendously important."[15] Gallo was quoted as saying that Gazdar's plea for credit is a "pathetic joke" and "I don't consider it so brilliant. In my mind, there is no credit for a cell line. If it happens by accident that you have a cell line, so freaking what? We didn't patent the cell line; we patented the process."[16] In response to which Gazdar said, "Well, then the next question is: Does it matter who discovered the virus? That does matter, doesn't it?" Gazdar laughed at Gallo's claim that cell lines are discovered by chance and said, "I'd say the same [about discovering a virus]."[17]

In the June 22, 1990, issue of *Science*, Ellis Rubinstein asks the following poignant questions: "What constitutes a scientific achievement deserving of credit? Were reasonable efforts made to publish H9's paternity? And most important, what lessons are to be learned from the way this small group of outstanding scientists did their science . . . and from the way they treated their scientific colleagues?"[18]

This complex story doesn't end here. In the May 10, 1991, issue of *Sci-*

ence[19] it was reported that Montagnier's LAV was thought to have been isolated from an AIDS patient with the code name BRU early in 1983. It was later shown that LAV and Gallo's HTLV-IIIB were genetically identical. This is what led to the controversy over whether Gallo had inadvertently or deliberately developed his AIDS antibody test from LAV-BRU. During February 1991, Gallo's team reported in the British journal *Nature* that three of the five samples that the French had sent him in 1983 had DNA sequences differing from those previously published. Thus, BRU could not be the source of either LAV or HTLV-IIIB and it raised the possibility that Gallo's virus may have contaminated the French lab. Montagnier's group subsequently did their own tests and confirmed that Gallo was correct. The early samples of BRU, taken from deep-freeze storage, were missing a specific amino-acid sequence (designated the V-3 loop) in its envelope protein. Some viruses, including those in Gallo's study, acquire a portion of the host cell membrane as they leave the cell. This outer covering of the virus is called the viral "envelope." There was, however, another viral isolate growing in the French lab in 1983 from a patient code-named LAI that did have the characteristic V3 loop. It was concluded that LAI had contaminated the BRU culture, and because it grew faster than BRU it soon displaced BRU. The samples of LAV that Gallo received in September 1983 were actually doubly infected with both BRU and LAI. Thus, contamination must have occurred in both labs. Gallo conceded that the original source of his HTLV-IIIB was probably LAV-LAI. But ironically, LAV-LAI may not have been a "French virus" because Frenchman "LAI" was attending college on a scholarship in the United States when he reportedly had his first homosexual experience. So it may have been an "American virus" that was eventually isolated by his French compatriots. The question now arises, is this kind of slipshod science worthy of a Nobel Prize, and if so who deserves the credit?

YELLOW JACK

Yellow fever (or "yellow jack" in the vernacular) is an acute, often fatal disease characterized by fever, jaundice, hemorrhage, and vomiting of blood ("black vomit"). Jaundice is a yellowish staining of the skin, the white of the eye, and elsewhere with bile pigments that are derived from products of red blood cell destruction. The earliest record of yellow fever goes back to 1506 in Central America. It appeared among the Indians of New England in 1618. On the Caribbean island of St. Lucia in 1664, out of a population of fifteen hundred soldiers, it killed 1,411. On that same island, it killed two hundred of the five hundred sailors there during the next year. In 1668 it appeared in New York, in Boston in 1601, and in Philadelphia in 1695. The worst epidemic of yellow jack in the United States occurred in Philadelphia in 1793,

DEATH OF A RAT

causing complete dissolution of society. All roads leading out of the city were jammed with refugees fleeing the plagued area. At least one hundred thousand lives have been lost to yellow jack since that 1793 epidemic, involving such major cities as New Orleans, Philadelphia, Memphis, Charleston, Norfolk, Galveston, New York, and Baltimore. No one fears yellow fever in this country anymore, thanks to the work of a team headed by Major Walter Reed in 1900. Dr. James Carroll, Dr. Aristides Agramonte, and Dr. Jesse W. Lazear were also appointed as members of a board of medical officers assigned to investigate acute infectious diseases, and especially yellow fever, on the island of Cuba. During the Spanish-American War, soldiers were dropping like flies from diseases such as typhoid fever and yellow fever. In the spring of 1900, after yellow fever wrought havoc again, Surgeon General George Sternberg dispatched Reed and Carroll to Cuba. They arrived there in June 1900. Lazear and Agramonte were already in Cuba working as Army contract surgeons.

At the tender age of sixteen, Walter Reed (1851–1902) was the youngest student to ever graduate from the medical school at Charlottesville, Virginia. After graduation, he went to Bellevue Hospital Medical College in New York, and received his medical degree a year later. In 1874 Reed entered the U.S. Army as a surgeon in hopes that he would have more time to carry on scientific research. Also, he wanted to have an assured income so that he could marry his sweetheart, Emilie Lawrence. He passed the examinations for the Army Medical Corps in February 1875 and received his commission the following June. He was married on April 25, 1876. His first assignment was eighteen years of garrison duty in Arizona. Finally in 1890, he was assigned to duty in Baltimore. He did bacteriological studies at Johns Hopkins University. Three years later he was appointed professor of bacteriology at the Army Medical School in Washington. Between 1893 and 1900, Reed had helped identify the bacterial cause of typhoid fever and determined that it could be carried by the common housefly. Reed became deeply involved with yellow-fever research in 1897 when he set about trying to confirm the claim by an Italian scientist, Giuseppe Sanarelli, that he had isolated a bacterium from yellow-fever patients. Reed soon proved that Sanarelli's germ was a hog-cholera bacillus, not the cause of yellow fever.

As soon as Reed's team had assembled in Cuba, they began to search for an organism that was found only in victims of yellow fever. Improved sanitary conditions had failed to diminish the incidence of yellow fever. Almost everyone was convinced that it was a contagious disease that could be spread from person to person by intimate contact with clothing or bedding used by victims and therefore contaminated with their secretions and/or excretions (*fomites*). This was known as the "fomite theory." Few gave credence to an alternative theory that yellow fever was transmitted by mosquitoes. This idea was first proposed by Nott in 1848, later by Beauperthuy in 1854 and Carlos Finlay in 1881.[20] Finlay was a Cuban physician who deduced not only that the disease was transmitted by mosquitoes, but also that the culprit was the common household mosquito now known as *Aedes aegypti* (formerly called *Culex fasciatus* and later *Stegomyia fasciata*). Finlay came to this conclusion because this species is the only one that cohabits with humans in urban environments. They must breed in pure, still water such as in rain barrels or open drinking vessels. Finlay maintained his convictions in spite of his failures on many attempts to transmit the disease with *Aedes* that had bitten yellow-fever patients. Where he got his subjects for these experiments is not known. There were at least two research reports that gave some support for Finlay's theory. One was by the British physician Ronald Ross in 1897 showing that the malarial protozoan parasite is transmitted by the *Anopheles* mosquito after a period of incubation in its stomach.

Jesse Lazear, who was working at the Johns Hopkins University at the time, was one of the first American scientists to confirm Ross's work. The second line of support came from an epidemiological study by Henry Rose Carter on yellow-fever outbreaks in two small Mississippi towns. Carter reported that an outbreak would usually begin with a single case involving a person who had just returned from an infected city like New Orleans. There would be a gap (which Carter called the "extrinsic incubation" period) of about two or three weeks before new cases appeared. Reed later confessed in a letter to Carter that "your own work in Mississippi did more to impress me with the importance of an intermediate host than everything else put together."[21] After Reed's team had tried to find the microbe of yellow fever and failed, they decided to devote some of their work to the investigation of Finlay's theory. Lazear was placed in charge of the mosquito work because of his experience with the malarial parasite. Carroll and Agramonte, unaware of the viral cause of yellow fever, continued to search for a bacterium. (A year later, the viral etiology of yellow fever would be proven, largely due to the work of Carroll.) Reed's team visited Finlay at his home in Havana and received from him some eggs of female *Aedes aegypti* mosquitoes.

At that time it was assumed that yellow fever was strictly a human disease.[22] So the only way to test Finlay's theory was to risk human life. Reed's group concurred that it was necessary to risk the lives of a few volunteers in order to potentially save thousands from the ravages of yellow fever in the future. The team members planned to participate as volunteers also. Since Agramonte was born in Cuba, he was assumed to be immune to yellow fever, and therefore he was not considered to be a suitable candidate for experimentation. Before the experiments began, Reed was called back to Washington in early August 1900 by Sternberg to finish a report on typhoid fever that was begun before he went to Cuba. In the meantime, Lazear hatched Finlay's eggs and kept the mosquitoes in glass tubes. The vial could be unstoppered and placed over the skin of a yellow-fever patient, allowing the mosquitoes to feed on the blood that was thought to be contaminated with

DEATH OF A RAT

the germ that caused yellow fever. Then after a few days the infected mosquitoes were allowed to bite healthy volunteers. None of these early experiments successfully transmitted yellow fever (to the relief of the volunteers, but to the disappointment of the researchers). Nevertheless, Lazear continued to experiment. Finally on August 27, 1900, he used a mosquito that had fed on the blood of a yellow-fever patient twelve days earlier, and exposed it to the skin of Dr. James Carroll. Within a few days, Carroll became very ill of yellow fever and almost died. Now Lazear had a protocol that seemed to work. The secret was to have the mosquito feed on a yellow-fever patient early in the illness and then allow at least twelve days before its feeding on an uninfected volunteer. Later it was shown that this time is necessary for the virus that causes yellow fever to move from the mosquito's gut to its salivary glands where it can be transmitted to the next victim.

The success with Carroll supported Finlay's theory, but the evidence wasn't airtight. It could not be ruled out that Carroll might have been infected by some other route because he had been in yellow-fever wards and even in the autopsy room just a few days before the experiment. So the next step was to repeat the experiment on someone who was known to have been nowhere near yellow fever for a long time. A volunteer soldier, Private William Dean, was such a person. A few days after being bitten with Lazear's mosquitoes, Dean came down with yellow fever, but he survived. This success was short-lived, however, because on September 13, 1900, either on purpose or accidentally (the reports are conflicting), Lazear was bitten by an infected mosquito. He became ill on September 18 and died a week later after going through the tortures of yellow fever. Albert Truby was attached to the military hospital in Cuba at the time and wrote a biography of Reed forty years later. Truby thinks that Lazear purposefully exposed himself to infected mosquitoes, but his colleagues kept the self-experimentation quiet because his death might be ruled a suicide and his wife would then be denied his insurance benefits. When Reed learned of Lazear's death, he hurried back to Cuba, arriving on October 3, 1900. Lazear's notebook contained important clues that led to his success; namely, a female *Aedes* must feed on a yellow-fever patient in the first three days of his illness and then be kept for at least twelve days before it was allowed to bite an uninfected person. Reed quickly wrote a report which he delivered at a meeting of the American Public Health Association in Indianapolis on October 23, 1990.

The report of the yellow-fever commission was immediately challenged because their findings were based on such a small sample and the experiments were not controlled. Within a month, Reed was back in Cuba designing a set of controlled experiments that could provide definitive proof of the mosquito theory. Reed had a camp built about six miles from Havana for these experiments. He named it Camp Lazear, after his dead comrade. American soldier volunteers who had never had yellow fever were quartered at the camp along with Spanish immigrants who offered their services for pay. If a person is to develop yellow fever, he does so within six days after exposure. Therefore, anyone kept in quarantine at the camp for two weeks without developing the disease would be proven to be unexposed before they entered camp. Two American soldiers, John R. Kissinger and John J. Moran, were the first to volunteer for the experiments. Major Reed informed them of the importance of the experiments and the risks that they would be taking. They refused to take any money for their services. Reed then touched his hat and said, "Gentlemen, I salute you."[23] Kissinger said that he volunteered "solely in the interest of humanity and the cause of science."[24] Reed replied, "In my opinion this exhibition of moral courage has never been surpassed in the annals of the Army of the United States."[25] Kissinger was infected on December 5, 1900. Four days later he had developed the clinical signs of yellow fever, from which he eventually recovered. Thirteen men at Camp Lazear were infected in this way; ten of them came down with yellow fever, but none of them died. None of the other thirty or forty men at the camp that were not purposefully infected became ill.

To address the fomite theory, Reed then had a small house built containing one room, fourteen by twenty feet. It was tightly constructed to be insect-proof. The doors and windows were placed so as to admit as little sunlight and air as possible. A coal-oil stove was used to keep the room temperature at ninety degrees during the day and the air was humidified so as to resemble the hold of a ship in the tropics—warm, dark, and moist. On November 30, 1900, Dr. R. P. Cooke, acting Assistant Surgeon, United States Army, and two privates from the Hospital Corps, all uninfected young Americans, entered the building. They put on the soiled and unwashed bedclothes that had been worn by yellow-fever patients. The unwashed sheets, pillow slips, and blankets likewise were from yellow-fever patients. They shook these articles each day so that the air would be filled with the germs of yellow fever if they were present in these fomites. They ate all of their meals and drank all of their water from common containers. They used the same privy and washed out of the same bucket. After twenty days and nights of continuous existence under these conditions, not one of the men developed yellow fever. This experiment was repeated two more times using different volunteers, with the same results. The fomite theory was dead as far as yellow fever was concerned.

A second house was constructed to answer the question, How does a house become infected? This house was also insect-proof, but was well ventilated. The house was divided in two by a wire netting from floor to ceiling, allowing air (but not mosquitoes) to pass from one side to the other. Thus, if there were any germs, putrid airs, or "miasmas" floating in the air that could cause yellow fever, they would be equally found on both sides of the screen. Two men stayed on each side of the netting for two weeks, remaining perfectly well. Reed then took out the two men from one side of the room and set free fifteen mosquitoes that had fed on yellow-fever patients. John J. Moran then entered the mosquito-infested space for a short time on three successive days. Four days after his first visit, he developed yellow fever, and eventually recovered. The two men on the other side of the netting remained in the house for eighteen days and nights, including the whole time that Moran was coming and going. They remained healthy. By this experiment Reed proved conclusively that only the presence of the infected mosquitoes was associated with yellow fever.

To show how the house could be disinfected and made safe from yellow fever, Reed had all of the mosquitoes captured and removed from the infected side of the house. Now, according to Reed, the house was decontaminated. To prove that it was safe, the two men who had lived in the infected side returned there and continued to live in the house together with the two men on the other side of the screen as controls. They all remained healthy. Reed had convincingly demonstrated that *Aedes* was a vector of yellow fever. These results were reported in the February 16, 1901, issue of the *Journal of the American Medical Association*. Reed's team did not, however, discover the causative agent of yellow fever. The virus of yellow fever was not isolated until 1927, long after Reed had died.

Major William C. Gorgas, Medical Corps, United States Army, was Chief Sanitary Officer of Havana at the time of Reed's discovery. He took charge of the military campaign to eradicate *Aedes* from Havana in February 1901. Every house and yard was inspected and all tin cans, empty bottles, and similar vessels that could hold rainwater and serve to hatch mosquito eggs were removed from the city. The tops of cisterns were covered with mosquito netting. Covers were placed over rain barrels, and a wooden spigot was placed near the bottom of each barrel so that water could be drawn off without lifting the cover. As each case of yellow fever was reported, workers from the Health Department would go to the house and install screening so that no mosquitoes could get in or out. They then fumigated the house to kill any mosquitoes that might be inside. By September 1901, Havana was declared free of yellow fever, the first time such a statement could be made in 150 years. With that victory under his belt, Gorgas proceeded to the Panama Canal Zone and spearheaded the drive to eradicate *Aedes* there. Early attempts to build the canal met with failure partly because of yellow fever. Gorgas had challenged Reed's team that if they found the cause of yellow fever he would give them the Panama Canal. And so it was to be. The Panama Canal might never have been constructed if old "yellow jack" had not been conquered. When one thinks of the lives that have been saved, the suffering that has been prevented, and the strategic importance of the canal to the maritime commerce of many nations, one can begin to properly appreciate how meritorious the solution to this biological problem is.

The subject of human experimentation was the focus of a book titled *Guinea Pig Doctors*, by Jon Franklin and John Southerland.[26] These authors claim that Reed, while in Washington, took credit for the key discoveries made in Cuba by Lazear. Reed returned to Cuba shortly after Lazear died. Reed was entrusted with Lazear's notebook. There were two clues to Lazear's successful transfer of yellow fever in his notebook. One was the fact that the mosquitoes that had bitten Carroll and Dean had fed on yellow-fever patients at least twelve days earlier, thereby confirming Carter's theory of a period of extrinsic incubation. The second clue, that Reed failed to mention in the preliminary report, was that those same mosquitoes had bitten yellow-fever patients in the early stages of their disease. It is now known that successful transfers of the virus can occur only if the mosquitoes feed during the first three days of infection and then at least twelve days must elapse before the insect is able to pass it on to another person. Franklin and Sutherland maintain that without Lazear's notes Reed would not have been able to carry out his controlled experiments. Reed died of appendicitis on November 23, 1902. Lazear's notebook could not be found among Reed's belongings. If they could be found, it is likely that we would learn if Lazear had died by the purposeful or accidental bite of a mosquito. Franklin and Sutherland "smell a rat" in the mysterious disappearance of Lazear's notebook, as if Reed had tried to hide the fact that Lazear had made the crucial discoveries on which Reed's glory was built. They claim that Lazear has not been given the credit due him.

This contention was challenged in an article by Colin Norman in 1984. According to Norman,

> Although Franklin and Sutherland accuse Reed of stealing all the credit, the Preliminary Report (published in the *Philadelphia Medical Journal* in 1900) was presented as a joint report of the yellow fever board, with all four members listed as coauthors. Carter and Finlay were also given full credit for steering the board toward the mosquito theory. Franklin and Sutherland imply that Reed gave Finlay no credit at all. . . . Lazear clearly played a key role in a discovery for which Reed is popularly remembered. But Reed's contribution was far from negligible, contrary to what Franklin and Sutherland imply.[27]

Norman cites Reed's biographer, William Bean, as saying, "There is plenty of credit to go around for everybody, without taking anything away from Reed." Norman also criticizes Franklin and Sutherland for failing to provide sources for their information. "It is thus sometimes difficult to tell where established fact ends and imaginative reconstruction begins."[28]

Two letters appeared in the June 8, 1984, issue of *Science* in response to Norman's article, plus Norman's reply. Franklin and Sutherland state,

Earlier experiments with mosquitoes failed because researchers, not wanting to clutter their data with misdiagnosed patients, always used late-state cases. By that time the patients definitely had yellow fever but, unknown to the scientists, could no longer pass it on. . . . In any event, William Bean suggests that Reed should get the credit for brilliant confirming experiments, and that after all there was plenty of credit to go around. By that argument we could quit making so much of mere discovery and start awarding Nobel Prizes to folks who do the follow-up.[29]

As to Norman's complaint about the scholarship of *The Guinea Pig Doctors*, they said,

Granted, the lack of footnotes means that Norman must call us and ask us questions. But it emphatically does not mean that we skimped on our homework. The hard reality is that, in a world of great ignorance about science, scholarly tomes are notorious for going unread. Our purpose in writing *Guinea Pig Doctors* was to share some of the human drama in science with the lay public—to excite them with our vision of research as a deeply human process.[30]

In writing this book, I find myself in the same boat, trying not to get too technical, giving sufficient documentation without overloading, and yet sticking to the script of historical accounts of what actually happened.

The second letter was written by J. A. del Regato:

Norman's article intimates a dispute as to whether Reed or Lazear discovered the spread of yellow fever by the mosquito; the fact is, neither one did! Reed planned and carried out a brilliant, controlled human experiment that definitely proved that the mosquito was the vector of yellow fever, but he did not discover the fact. Lazear did discover a method to produce experimental cases of the disease, without which Reed would not have been able to stage his experiments.[31]

I was unable in this letter to find out who del Regato thinks did discover how yellow fever was spread. Del Regato tells how William Gorgas publicly cited Finlay's contributions: "His inspiration was laughed at, his theory was mistaken for the fanciful illusion of a tropical imagination. . . . I cannot pronounce his name with due reverence." Reed's response to Gorgas was, "I know you did not intend to say it, but somehow, I suppose being back in Havana, you feel it is your duty to honey-fuggle the simpering old idiot."[32] If this is the way Reed felt about Finlay's contributions, it makes me wonder how he regarded the contributions of any of the other key players in this story.

Norman's letter, however, indicates Gorgas's admiration for both Finlay and Reed:

Finlay published his mosquito theory nearly two decades before the Reed board began its work; by "a piece of brilliant deduction" he singled out the right mosquito; and he provided the board with ova for its experiments. Del Regato maintains Reed did not give Finlay sufficient homage in the Preliminary Report. The Report actually stated that the theory was "first advanced and ingeniously discussed by Dr. Carlos J. Finlay, of Havana, in 1881." It expresses "sincere thanks" to Finlay for discussing his work with the board and providing ova "of the variety of mosquito with which he had made his several inoculations."
... In his response to Reed, Gorgas affirmed his respect for Finlay, as del Regato states, but went on to say: "You [Reed] are the great man in the matter. His [Finlay's] theory would have remained an idle dream except for your work."[33]

Elsewhere in that same letter Norman states, "Lazear is certainly not the first junior researcher who did key work for which the head of the research team is remembered."[34]

THE ANTISCURVY FACTOR

Who hasn't heard of Captain James Cook (1728-1779) and his early sailing adventures in the South Seas? Previous attempts by others had been made to explore the South Seas, but they had all failed. Relatively few people know why Cook's voyages were successful. These previous attempts failed because of that dreaded disease of long sailing voyages—scurvy. Those who know that Cook prevented his crew from getting scurvy by (among other innovations) providing foods that contained vitamin C probably are unaware of how he came to such knowledge. The real hero of this story is a man whose contributions in this regard have been sadly neglected in the history books—James Lind.

Scurvy has been known from antiquity, but it wasn't until after 1500, when long navigational voyages far from land were attempted, that it became a common disease among sailors. Prior to that time, ships sailed close to shore and could put in to port to replenish food and water supplies as needed. These fresh stores could include fruits and vegetables that contain the antiscurvy factor now known as vitamin C or ascorbic acid (not to be confused with citric acid). Of course, in those early days, the cause of scurvy was unknown and the somewhat random sample of fresh foods taken on ships usually contained enough vitamin C to prevent scurvy by chance rather than by design. For long voyages, however, fresh food was soon depleted and the menu shifted to salted or dried meats and fish, dried peas, hardtack (a hard biscuit or bread made only with flour and water), a few dried fruits such as figs and dates, some cheese, pickled foods (such as olives), oatmeal gruel, and the like. These foods contained little if any vitamin C. The best of the lot went to the officers; the rest of the crew subsisted on the least appetizing portions. English sea captain Alan Villiers,[35] author of "That Extraordinary Sea Genius, Captain James Cook," tells of the hardships and depravations of these early sailors in the British Navy in graphic terms.

> Ships' cooks, often known without affection as "Slushy," had the time-honored right to save such rancid fat as might be recoverable from the evil-smelling meat they boiled. This was slush, indeed. As one contemporary writer, Ned Ward, wrote in *The Wooden World Dissected* (1752), a sea cook was "an excellent mess mate for a bear, being the only other two-legged brute that lives by his own grease." His common practice was to share some of this fat out to favored shipmates to melt into their jawbreaking "bread" to give the stuff some semblance of taste, if none of nutriment. A worse scurvy-maker for men bereft of fresh food would be hard to find.[36]

In addition to poor food, the crew slept in damp quarters on low, crowded working-decks, often on the deck itself between the guns. One might wonder why anyone in their right mind would want to be a sailor in those days. Villers explains that the "ordinary seaman in the Navy was usually 'recruited' by the men of the Press Gang, who slugged them or drugged them in waterfront taverns and dragged them aboard when needed." They were paid a small sum if they survived the bloody battles, but little prize money ever filtered down to the crew. They were seldom allowed to go ashore when in port, for fear they would "jump ship" (desert). Complaints were answered with a whip called a "cat o' nine tails." Britain "was a flogging navy in a flogging age."

At the beginning of the sixteenth century, the expansion of colonial empires, competition for trade in far-off lands, and voyages of exploration and navigation spurred many European countries to build fleets of sailing ships that would be sent on journeys of months or even years. The success of such missions depended upon maintaining the health of each ship's crew. It is estimated that between the years 1500 and 1800, at least a million lives were lost to scurvy. This disease caused more deaths than all other diseases, naval engagements, marine disasters, shipwrecks, and accidents combined. The romantic stories of sailing off to faraway places have been popularized in books and films. Few people realize the hardships and privations that had to be endured on such trips—the "hell that no poet's imagination ever was able to invent. . . . Materials and stores, including food, were subject to mold, to putrefaction, and to infestation with vermin. Rats infested ships; and men had to live in a terrible odor of bilge water, stale cooking, dry rot, dead rats, and unwashed humanity."[37]

James Lind was born into such a world on October 4, 1716, in Edinburgh, Scotland. He was apprenticed to a well-known Edinburgh physician, George Langlands, at the age of fifteen. By this time, the University of Edinburgh had become a famous medical center. Lind joined the Royal Navy as a Surgeon's Mate in 1739. During ten years of service, Lind saw many cases of scurvy exhibiting symptoms which he described as follows:

> lassitude, weakness, swelling of legs and arms, softening and hemorrhage of gums, and hemorrhages under the skin, producing purple or black patches. In patients in advanced stages, teeth became loose and sometimes fell out; the breath became foul; strength failed to the point where the victim was unable to stand or to be up and about; and the slightest touch or movement was excruciatingly painful. If no relief could be provided, death resulted from exhaustion, from heart failure, or from acute infection such as pneumonia.[38]

In 1498, Vasco de Gama led the first four European ships around the Cape of Good Hope. After ten weeks at sea many of his men came down with scurvy, "their feet and hands swelling and their gums growing over their teeth so they could not eat."[39] After oranges became available in the East African ports, scurvy disappeared. On the return trip, scurvy struck again. As soon as they could obtain oranges, the symptoms quickly abated. Almost half his crew died at sea, but by chance a cure for scurvy had been discovered. Unfortunately, this bit of medical lore did not become well known. The value of citrus juice, sauerkraut, and fresh fruits and vegetables for treating scurvy had been reported by various sources more than a century before Lind's time. Those that had heard of such treatments either did not believe them, did not want to try them, or tried them but mixed them with other treatments so that their effectiveness could not be judged in isolation. This is where Lind's genius was shown. On May 20, 1747, Lind designed and began what "may have been the first of all controlled clinical experiments"[40] to test the effectiveness of various dietary supplements in treating scurvy. He describes his experiments as follows.

> I took twelve patients in the scurvy, on board the man-of-war HMS *Salisbury* at sea. Their cases were as similar as I could have them. . . . They lay together in one place in the forehold; and had one common diet, watergruel sweetened with sugar in the morning; fresh mutton-broth often times for dinner; and for supper, barley and raisins, rice and currants, sago and wine, or the like. Two of these were ordered each a quart of cider a-day. Two others took twenty-five gutts (drops) of *elixir vitriol* (sulfuric acid) three times a-day, upon an empty stomach. . . . Two others took two spoonfuls of vinegar three times a-day, upon an empty stomach. . . . Two of the worst patients. . . . were put under a course of sea water. Of this they drank half a pint every day. . . . Two others had each two oranges and one lemon given them every day. These they ate with greediness, at different times, upon an empty stomach. They continued but six days under this course, having consumed the quantity that could be spared. The two remaining patients took the bigness of a nutmeg three times a-day, of an

electuary recommended by a hospital surgeon, made of garlic, mustard-seed, horse-radish, balsam of Peru, and gum myrrh. . . . The consequence was, that the most sudden and visible good effects were perceived from the use of the oranges and lemons; one of those who had taken them, being at the end of six days fit for duty. . . . The other was the best recovered of any in his condition; and being now deemed pretty well, was appointed nurse to the rest of the sick.[41]

Cider was next most effective, but the other treatments were worthless. Lind's experiments demonstrated the effectiveness of a treatment involving a combination of two citrus fruits, but was not designed to prove that either one alone could also be effective.

Lind retired from the Royal Navy in 1748. He obtained his doctorate in medicine from the University of Edinburgh and his license to practice from the Royal College of Physicians in that city. He was elected a fellow of the college in 1750, and became its treasurer in 1757. His classic book *A Treatise on the Scurvy* was published in 1753, followed by three more editions in the next twenty years. His second book, *An Essay on the Most Effectual Means of Preserving the Health of Seamen in the Royal Navy*, was published in 1757; it went through two later editions. In 1758 Dr. Lind was appointed as chief physician at the new Royal Naval hospital, Haslar Hospital, near Portsmouth. At that time Hasler Hospital was the largest hospital in Europe, capable of servicing 2,200 patients. Lind continued in this capacity for twenty-five years. He saw not only patients with scurvy but those with a variety of diseases from all over the world. These experiences led him in 1768 to publish his third book, *An Essay on Diseases Incidental to Europeans in Hot Climates*. This set of three books addressed most of the problems of nautical medicine.

The accomplishments of James Lind go far beyond his proof of the effectiveness of citrus fruits in the treatment of scurvy. He also recommended that new recruits first be brought to receiving ships to be quarantined, bathed, and issued clean clothing. These measures did more to eliminate typhus fever than any others. He recommended that fighting ships at sea be regularly replenished with fresh provisions of fruits and green vegetables. The adoption of these recommendations is thought to have been a decisive factor in Britain's wars with Napoleon where ships had to remain on blockade duty for long periods. He outlined a practical method of distilling fresh water from sea water, adapting utensils normally found on board the ship. He recommended that physical examinations be given naval recruits, and that health records thereof should be maintained. He recommended that naval uniforms be issued to seamen as a means to improve both health and morale of the men. He insisted upon physical exercise and cold baths as "toughening-up" processes. He prescribed the use of bark from the chinchona tree, as a source of quinine, for the prevention of malaria. He recommended that in the tropics ships should be anchored well off shore, and that crews should avoid work details or liberty ashore at night. This greatly helped reduce the incidence of malaria. Lind died on July 18, 1794. Even though many of his recommendations were put into effect during his lifetime, he died before his recommendation on the use of lemon juice to prevent scurvy was officially adopted by the British Admiralty. For all his contributions, James Lind was never elected to the Royal Society or to the Royal College of Physicians of either London or Edinburgh. However, he had a cousin by the same name who was a fellow of the Royal Society.

The Royal Society of London notified the king of England in November of 1767 that the planet Venus would pass over the disc of the Sun on June 3, 1769, and that valuable navigational information could be obtained from that event at a location near Tahiti in the South Pacific. The British government was reluctant to send an expedition to the South Seas because so many other long voyages had failed in the past. But the venture was considered by the king to be worth the gamble. The Royal Society was given four thousand pounds and directed to work jointly with the Admiralty in preparing the expedition. A three-year-old vessel weighing 308 tons was purchased and renamed *The Endeavour Bark*. A naval warrant officer named James Cook was commissioned a lieutenant and appointed commander of the expedition. Cook and an astronomer were then appointed by the Royal Society to make the required observations of the transit of Venus. The Royal Society granted a request from one of its young wealthy fellows, Joseph Banks (1743–1820), to go at his own expense on the trip. He was allowed to take ten additional people with him as naturalists and technicians to help him with his biological studies. Banks put up ten thousand pounds of his own money to supply the ship properly for his needs, making the *Endeavour* the best-equipped ship ever to set out for the purpose of natural-history studies. Three years after Banks had completed his studies at Oxford, he became a ship's naturalist on a seven-month assignment to Labrador and Newfoundland. Banks was a botanist and made many collections there and elsewhere in the western part of England. Banks was later to become president of the Royal Society.

The *Endeavour* set sail from Plymouth, England, in August of 1768. Its voyage was to last three years. It returned home in July 1771 after sailing more than thirty thousand miles. It went around Cape Horn and on to New Zealand, Australia, New Guinea, and Tahiti. James Cook gave his name to a part of the Great Barrier Reef off the northeastern coast of Australia. He sailed the waters between Australia and New Guinea, naming it Endeavour Strait after his ship, and proving that the two landmasses were separate. Botany Bay near Sydney was so named because of the wealth of plant specimens collected by Banks's team. There is a Joseph Banks Island to the south of Australia. Cook gave the name New South Wales to the lands of south-eastern Australia. The expedition was a great success. One of the reasons for this success was that Cook insisted on sanitation and proper nutrition. He provided sauerkraut to prevent scurvy when fresh fruits and vegetables were not available. Cook had another advantage in being the first to use the position of stars to accurately determine longitude. Before his voyage, ships could determine latitude, but not longitude, so east-west distances could only be estimated. The tables from which longitude could be calculated were worked out by astronomers and have been some of the most practical results to ever come from the science of astronomy. This allowed Cook to make the most accurate maps then available of the coastlines he visited. I can also imagine that morale would be better on a ship that knew exactly where it was at all times. In any event, the *Endeavour* returned to England without the loss of a single life due to scurvy, something that had never happened on any extensive cruise before. Cook's name became legendary. He had found a way to prevent the dreaded disease.

But where did he get the idea of a nutritional approach to the problem? Cook's journals make "no mention of Lind, and it is thought that he [Cook] had never seen his [Lind's] *Treatise*."[42] Cook had received instructions from the Admiralty to test and report on the value of the same variety of antiscorbutic remedies that had been reported by earlier explorers (specifically S. Wallis and P. Carteret). These included sauerkraut, portable soup, saloup, malt, and rob of oranges and lemons. "Portable soup" was prepared "from the offals of cattle, flavored with salt and vegetables, and then evaporated down to form hard, gluelike cakes."[43] "Saloup" is a powdered form of dried orchid roots used as a thickening agent in broths. Lind himself had proposed a method for concentrating citrus juice to the consistency of a syrup (called "rob") by heating. Unfortunately, neither he nor anyone else at the time was aware that heating destroys vitamin C. Those whose dietary regimens contained rob as the only source of vitamin C usually fell victim to scurvy for this reason. Vitamin C can also be destroyed by storage in or passage through copper containers. It also loses potency on prolonged contact with oxygen. Hence, after six months or so when scurvy often broke out, treatment with stored citrus juice was less effective or perhaps not at all, depending on how long it had aged. Lind's attempts to repeat his treatment for scurvy at Hasler Hospital were inconclusive because he did not always use fresh citrus fruit or freshly squeezed juice. Long before Lind's report, the Admiralty had received so many conflicting reports regarding the effectiveness of citrus juice that it was reluctant to issue regulations concerning its use. The few recommendations that had been issued were not strictly followed. For example, the Admiralty had agreed to allow issuance of a daily ration of three-quarters of an ounce of lemon juice per man, but it was still up to the individual admirals to request their allotment. Into the twentieth century, some voyages continued to have success while others failed to prevent scurvy.

Cook made two more voyages. He was promoted to the rank of commander. His second voyage left in 1772 to determine if a continent extended from the South Pacific to the South Pole, an idea that was popular at the time. On this exploration there were two ships: Cook was on the ship *Resolution* and his appointed consort, Tobias Furneaux, commanded the sloop *Adventure*. Furneaux had been given strict orders by Cook to use "good discipline, ventilation, cleanliness, common sense, and careful attention to diet."[44] But "he did not always carry them out—certainly not when he was sailing out of company with his senior."[45] Furneaux lost contact with the *Resolution* during bad weather in a sub-Antarctic fog. According to plan, he then sailed to an appointed rendezvous in New Zealand. He was slack about insisting that greens and "sellery" [sic] and other antiscorbutics be eaten by the crew. So when Cook finally met up with the *Adventure*, many of its men were down with scurvy. Only one man on the *Resolution* became ill with scurvy because Cook was a strict disciplinarian. The surgeon and surgeon's mate from the *Resolution* came over to the *Adventure* to help get its crew back on its feet again. On this trip, Cook was the first to sail across the Antarctic Circle. He discovered the large island of South Georgia. His speculation that a frozen continent lay further south in the Antarctic was confirmed by later explorers. Cook returned from his second voyage in 1776, having again lost no man to scurvy.

After his return from this second voyage, Cook prepared a paper presenting his protocol for preventing scurvy, and read it at a meeting of the Royal Society of London. Cook did not view scurvy as strictly a nutritional defect. He gave just as much attention to sanitation, keeping his men clean and dry, as he did to a varied diet. In his report he said, "I am convinced, that with plenty of fresh water, and a close attention to cleanliness, a ship's company will seldom be much afflicted with scurvy, though they should not be provided with any of the antiscorbutics before mentioned [e.g., sauerkraut, sugar]."[46] Thus, the fact that Cook succeeded is now attributed to the fact that his program involved such a wide range of measures that neither he nor anyone else could be sure which one of them or combination(s) of them, were really effective in preventing or treating scurvy. Cook had already been elected a fellow of the Royal Society. Now it voted to bestow on him its Copley Gold Medal for his contribution to the health of seamen. This medal is the highest honor that Britain awards for intellectual achievement. This is the only time that the medal has been awarded for an achievement in the discipline of nutrition.

Before Cook received the Copley Medal, he was promoted to captain and began his third voyage in 1776, in search of a direct water route from the Pacific Ocean to Hudson Bay (the so-called Northwest Passage). Master of the *Resolution* on this trip was William Bligh, who would later become infamous as the captain of the *Bounty*. From England they sailed around the Cape of

Good Hope, then to New Zealand, and continuing northeastward they accidentally bumped into the Sandwich Islands (named after the Earl of Sandwich), later to be called the Hawaiian Islands. Cook landed on the west coast of North America in 1778, and then went on north to the Bering Strait looking for a northwest passage from the Pacific Ocean to the Atlantic Ocean through the Arctic Archipelago of Canada and north of Alaska. He was unable to find any passage because, as we now know, it is choked with ice during most of the year. He then returned to Hawaii for the winter. He was killed by the natives at Kealakakua Bay on the big island of Hawaii in 1779 in, of all things, an argument over a rowboat. A few years ago I caught a motorized boat called *Captain Cook VII* out of Kona, and headed south to Kealakakua Bay to do some snorkeling and to visit the monument to Captain Cook on its shore. The water teamed with a wide variety of fish species. The water was so clear there was no restriction to my vision. The fish were so tame they would eat out of my hand. This is only because it is now a protected marine reserve. I couldn't help but reflect on the tragedy that took place over two-hundred years earlier very near where I was standing. The inscription on the monument states: "Near this spot Capt. James Cook was killed February 14, 1779."

Now the primary question is, Why weren't the recommendations of James Lind taken seriously? No one really knows the answer to this question, but Villiers offers some possibilities. (Remember that Lind was just a Surgeon's Mate, not a medical doctor.)

> It certainly didn't reach the ears of the ships' ranks. They'd never believe a ship's surgeon anyway. In their opinion, 'Sawbones' (as he was called and with good reason, too) could never discover anything but new ways to inflict pain. Medical doctors afloat in those days did not even rate as commissioned officers. As for the fruit Dr. Lind wrote about, it was regarded as a luxury that most sailors suspected as likely to upset even a seaman's cast-iron stomach. Seamen who had seen service in the West Indies station and survived it declared that lime juice, which they confused with lemon juice, was weakening stuff, bad for a man's potency.[47]

Villiers, also a seaman, in relating some of his own remembrances, says:

> Our older twentieth-century seamen, though compelled by the British Merchant Shipping Act to accept the watered lime juice that was dished out daily at noon, often could not be made to drink much of it. The younger men sipped a little and used the rest to clean their teeth. To some, the mere fact that the ship owner, through his servants, was giving them lime juice was enough to condemn it.[48]

It seems odd now that seamen could have been so stubborn and unreasoning. However,

Seamen were at the mercy of the ship, which meant her purser and such stores as he could get by with. Traditionally, what he could make on the food was a source of illicit but necessary revenue to him and the almost always deplorable cook. Pursers were such appalling skinflints and stores fiddlers that they gave their whole profession an ill name that lasted into the twentieth century, in British ships at any rate. . . . James Cook may have set the fight against scurvy back a little by trying so many "cures," not all of them worth much by any means. Cook tried anything and everything and was determined that, between the lot—cleanliness and fresh air and adequate exercise and all that—his men would not be victims of this unnecessary evil.[49]

Dr. J. Elliotson (lecturer at St. Thomas's Hospital in London) published a paper in 1831 in which he concluded that scurvy was a dietary deficiency disease. Ten years later, Dr. George Budd (at another of London's teaching hospitals) came to the same conclusion, and was the first to explain why Lind had observed scurvy early on voyages sailing from England in the spring as being due to the depletion of the sailors' reserves of the antiscorbutic element during the preceding winter. Lind had attributed those cases to cold, wet weather that had interfered with the patients' perspiration. Thus, while Lind was able to treat scurvy with appropriate foods, he apparently did not conceptualize that the cause of scurvy was solely a nutritional deficiency. Budd was also prescient in viewing not only scurvy, but also rickets (now known to be a vitamin D deficiency) and a visual impairment (due to what we would now call vitamin A deficiency) in the same light. Nevertheless, the idea of nutritional deficiency diseases in general (particularly as it applied to scurvy) almost disappeared from the literature over the next fifty years.

Carpenter[50] offers some possible factors that impeded progress on the prevention of scurvy in the nineteenth century. One factor was the development of methods for analyzing the chemical composition of foods for carbon and nitrogen. The great chemist Baron Justus von Liebig (1803–1873) had divided foods into proteins (the nitrogenous fraction of which he considered "truly nutritive") and other organic compounds which were merely a source of calories. Even in 1902 the U.S. Department of Agriculture had classified green vegetables and fruits as "expensive sources of both protein and energy, with special value only in providing laxative bulk."[51] A second factor might have been the discovery of alkaloids by French chemists and the making of an analogy between morbid causes and poisons. A third factor was the observation that dogs did not develop scurvy when fed a diet that would be scorbutic in humans. Of course they didn't know that, other than guinea pigs, primates, and humans, few other mammals require vitamin C in their diets. An additional very likely factor was the influence of Pasteur's germ theory of disease. That theory was so successful in explaining many contagious diseases that it may have served as a model for explaining all diseases. Other historians think that the germ theory may not have been an obstacle, but actually enhanced the discovery of dietary-deficiency diseases. For example, the work on beriberi was initiated as a hunt for a causative microbe, but it eventually led to the discovery of the true cause of the disease; viz., lack of thiamine (vitamin B_1). Still another factor that may have weakened confidence in the "deficiency theory" was the failure to refute earlier claims that the antiscorbutic factor was citric acid or potassium. Then there was the "diminished alkalinity theory," but like the "potassium theory" it did not explain why foods lost antiscorbutic properties as a result of drying or processing. Carpenter believes that it was due to the repeated failure of lime juice after 1860 to unfailingly cure patients with scurvy. Lemon juice was thought to be a specific cure for scurvy. Lemons and limes were often confused, and it was unknown at the time that their antiscorbutic content could be lost during storage or processing. Thus, citrus juice fell into disfavor and the search remained open for other cures.

Scurvy research took off in 1907 after Axel Holst and Theodor Frölich found that the disease could be induced in guinea pigs. At last an animal model had been found that could now be used to isolate the antiscurvy factor. Albert Szent-Györgyi purified ascorbic acid by crystallization in 1928. Its chemical structure was soon determined by a team under the leadership of Sir Norman Haworth. Tadeus Reichstein first synthesized vitamin C on a commercial scale. Both Haworth and Szent-Györgyi received Nobel Prizes in 1937. These are only a few of the multitude of people who have worked on various aspects of scurvy and vitamin C. Most of their names have been lost in abbreviated histories of vitamin C. The familiar account of how James Cook discovered that he could prevent scurvy by daily rations of fresh lime juice is certainly incorrect, and tends to perpetuate a myth that denies Lind of his deserved recognition in the conquest of scurvy. The glory heaped upon the few who are usually remembered partly belongs to a multitude of investigators whom history has forgotten, because, "No man is an island."

NOTES

1. Norman J. W. Thrower, ed., *Standing on the Shoulders of Giants* (Berkeley: University of California Press, 1990), ix.

2. M. Papovic, M. G. Sarngadharan, and Elizabeth Reed, "Detection, Isolation, and Continuous Production of Cytopathic Retroviruses from Patients with AIDS and Pre-AIDS," *Science* 224 (May 4, 1984): 497–500; R. C. Gallo et al., "Frequent Detection and Isolation of Cytopathic Retroviruses (HTLV-III) from Patients with AIDS and at Risk for AIDS," *Science* 224 (May 4, 1984): 500–503; Jörg Schüpbach et al., "Serological Analysis of a Subgroup of Human T-Lymphotropic Retroviruses (HTLV-III) Asso-

ciated with AIDS," *Science* 224 (May 4, 1984): 503–505; M. G. Sarngadharan et al., "Antibodies Reactive with Human T-Lymphotropic Retroviruses (HTLV-III) in the Serum of Patients with AIDS," *Science* 224 (May 4, 1984): 506–508.

3. Edward P. Gelmann et al., "Proviral DNA of a Retrovirus, Human T-Cell Leukemia Virus, in Two Patients with AIDS," *Science* 220 (May 20, 1993): 862–65; Robert C. Gallo et al., "Isolation of Human T-Cell Leukemia Virus in Acquired Immune Deficiency Syndrome (AIDS)," *Science* 220 (May 20, 1993): 865–67.

4. F. Barré Sinoussi et al., "Isolation of a T-Lymphotropic Retrovirus from a Patient at Risk for Acquired Immune Deficiency Syndrome (AIDS)," *Science* 220 (May 20, 1993): 868–71.

5. Barbara J. Culliton, "Inside the Gallo Probe," *Science* 248 (June 22, 1990): 1495.

6. Ibid.

7. Ibid.

8. Ellis Rubinstein, "The Untold Story of HUT-78," *Science* 248 (June 22, 1990): 1499–1507.

9. Barbara J. Culliton, "Inside the Gallo Probe," *Science* 248 (June 22, 1990): 1495.

10. Ibid., 1496–97.

11. Ibid., 1497.

12. Ibid.

13. Joseph Palca, "The True Source of HIV," *Science* 252 (May 10, 1991): 771.

14. Ellis Rubinstein, "The Untold Story of HUT-78," 1502.

15. Ibid., 1504.

16. Ibid., 1500.

17. Ibid.

18. Ibid.

19. Joseph Palca, "The True Source of HIV," 771.

20. T. P. Monath, "Glad Tidings from Yellow Fever Research," *Science* 229 (August 23, 1985): 734.

21. Colin Norman, "The Unsung Hero of Yellow Fever," *Science* 223 (March 30, 1984): 1371.

22. Ibid., 1372. In the 1930s it was discovered that South American and African monkeys harbor the yellow-fever virus and it can be transmitted among them by several species of mosquitoes. Occasionally the virus can be transmitted from monkey to humans. Smallpox, a viral disease that only affected humans, was officially declared to have been eradicated from the earth in 1979. A yellow-fever vaccine was developed in the 1930s, but yellow fever can never be eliminated by a vaccination program because there exists an ineradicable reservoir of the virus in wild monkeys and ineradicable vectors in several species of mosquitoes. Yellow fever has never appeared in Asia even though *Aedes aegypti* is common there.

23. Grace T. Hallock and C. E. Turner, *Health Heroes: Walter Reed* (New York: Metropolitan Life Insurance Company, n.d.), 16.

24. Ibid., 17.

25. Ibid.

26. Jon Franklin and John Sutherland, *Guinea Pig Doctors: The Drama of Medical Research Through Self-Experimentation* (New York: William Morrow and Company, 1984), 225.

27. Norman, "The Unsung Hero of Yellow Fever," 1372.

28. Ibid., 1370.

29. Jon Franklin and John Sutherland, "Letters: Yellow Fever Research," *Science* 224 (June 8, 1984): 1125.

30. Ibid., 1126.

31. J. A. del Regato, "Letters: Yellow Fever Research," *Science* 224 (June 8, 1984): 1126.

32. Ibid.

33. Colin Norman, "Letters: Yellow Fever Research," *Science* 224 (June 8, 1984): 1126.

34. Ibid., 1128.

35. Alan Villiers, "That Extraordinary Sea Genius, Captain James Cook," *Nutrition Today* 4, no. 3 (Autumn 1969): 8-16.

36. Ibid., 11.

37. Park, Davis & Company, "James Lind: Conquest of Scurvy," *A History of Medicine in Pictures* (1959; n. p.).

38. Ibid.

39. Review of *The History of Scurvy and Vitamin C*, by Kenneth J. Carpenter, *Scientific American* 260 (March 1989): 118.

40. Ibid.

41. Park, Davis & Company, "James Lind: Conquest of Scurvy."

42. Kenneth J. Carpenter, *The History of Scurvy and Vitamin C* (Cambridge, Mass.: Harvard University Press, 1986), 77.

43. Ibid., 66.

44. Villiers, "That Extraordinary Sea Genius, Captain James Cook," 14.

45. Ibid.

46. Carpenter, *The History of Scurvy and Vitamin C*, 82.

47. Villiers, "That Extraordinary Sea Genius, Captain James Cook," 11.

48. Ibid.

49. Ibid.

50. Carpenter, *The History of Scurvy and Vitamin C*, 249-51.

51. Ibid., 249-50.

CHAPTER 10

DÉJÀ VU

George Santayana said that "Those who cannot remember the past are condemned to repeat it." This admonition is often cited as one of the foremost reasons for studying history. Certain events that have occurred in the past sometimes may appear to be replayed with new characters, perhaps in different locations, and under more or less altered conditions. The French call it déjà vu ("already seen"). For example, some people who believe in reincarnation might feel, on visiting a castle, that they have been there in a previous life. Perhaps you have had the feeling, on meeting someone for the first time, that you have met that same person before, though not necessarily in a previous life. If this were really true, then history could repeat itself. Of course, in the strict sense, history does not repeat itself exactly because time's arrow cannot be reversed to replay a scene from the past in all of its details. Nonetheless, the concept persists that a series of events can have so many correlations with another set of events from the past that it stretches the bounds of credulity to explain them all as mere coincidences.

The best example I have come across appeared in an Ann Landers column in my local newspaper. The comparison involved the assassinations of presidents Abraham Lincoln and John F. Kennedy.

Both Lincoln and Kennedy were concerned with civil rights. Lincoln was elected president in 1860; Kennedy in 1960. Both were slain on a Friday and in the presence of their wives. Both were shot from behind and in the head. Their successors, both named Johnson, were Southern Democrats and both were in the Senate. Andrew Johnson was born in 1808 and

Lyndon Johnson was born in 1908. John Wilkes Booth was born in 1839 and Lee Harvey Oswald was born in 1939. Booth and Oswald were Southerners who favored unpopular ideas. Both presidents' wives lost children through death while in the White House. Lincoln's secretary, whose name was Kennedy, advised him not to go to the theater. Kennedy's secretary, whose name was Lincoln, advised him not to go to Dallas. John Wilkes Booth shot Lincoln in a theater and ran into a warehouse. Lee Harvey Oswald shot Kennedy from a warehouse and ran to a theater. The names Lincoln and Kennedy each contain seven letters. The names Andrew Johnson and Lyndon Johnson each contain 13 letters. The names John Wilkes Booth and Lee Harvey Oswald each contain 15 letters. Both assassins were killed before being brought to trial. Both Johnsons were opposed for re-election by men whose names started with G.[1]

Are there any examples from science history that have this "spooky" (as Ann Landers called it) quality of déjà vu? I will let you decide for yourself after the following stories have been told.

THE MIDWIFE TOAD[2]

Our tale begins in Vienna, Austria, on August 17, 1880, with the birth of a boy named Paul Kammerer. Paul originally studied music at the Vienna Academy. He then read zoology at the University of Vienna and was awarded his doctorate in 1904. Two years later he became a lecturer at the university. In 1902, at the age of twenty-two, Kammerer joined the Biologische Versuchsanstalt (the Institute for Experimental Biology) and began a collaboration with Hanz Przibram (pronounced "Prischi-brahm"), the founder of the institute. It was one of the first research institutes to have controlled temperature and humidity in its vivarium (a place for rearing animals, usually small ones such as frogs or mice).

In 1903 Kammerer began to raise and experiment with yellow-spotted salamanders (*Salamandra maculosa* forma *typica*). One of his most notable experiments involved the rearing of two groups of salamanders, one on black soil and the other on yellow soil. Kammerer believed at that time in Mendelism (i.e., that biological traits are governed by individual hereditary units now known as *genes*) and in Weismannism (i.e., the concept that environmental modifications of body characteristics or phenotypes cannot change the heredity of the gametes (sex cells; eggs and sperm). Evidently Kammerer was trying to achieve a reconciliation of Lamarckism with Mendelism; i.e., attempting to discover to what extent the environment could modify the expression of Mendelian genes within the lifetime of an individual. His results showed that the group reared on black soil gradually reduced the size of their yellow spots until at full maturity (about six years of age) the spots were quite small. The group reared on yellow soil gradually expanded their yellow spots until they merged into large stripes.[3] However, when Kammerer observed that the progeny of black-adapted parents tended to have less yellow than did their parents at a similar age, and progeny of yellow-adapted parents tended to have more yellow than did their parents at a similar age, he began to think that there might be some truth to Lamarck's theory of evolution (i.e., that both somatic and gametic heredity is somewhat plastic, capable of being directionally changed by the environment).

Kammerer also experimented with the midwife toad[4] (*Alytes obstetricans*), attempting to force it to breed in the water (like most amphibians) instead of on land as it normally does. In 1909, he accidentally discovered (and subsequently published) in the third generation of water-mating *Alytes* that the normally smooth nuptial pads on the thumb and first finger of males had developed rough skin equipped with spines (also called rugosities, asperities, or Brunftschwielen). These rough pads are used by several species of amphibians other than Alytes, that normally breed in water, to hold the slippery female while in the mating embrace known as *amplexus*. The reappearance of an ancient ancestral trait (such as the Brunftschwielen) within a species whose more recent ancestors lacked that trait is referred to as *atavism*. Occasionally rudiments of rugosities will develop even on *Alytes* that are not forced to breed in the water, but they never reach the extent seen in the later generations of the water-bred toads. Kammerer observed that the rugosities persisted through three subsequent generations,[5] suggesting to him that Lamarck was probably right about the inheritance of acquired characteristics. If an animal has a need imposed upon it by its environment necessitating the use of certain body parts (e.g., being forced to breed in water), a directional change occurs in its heredity to produce an adaptation (such as the Brunftschwielen) that can be transmitted to subsequent generations.

Kammerer also exposed the blind newt *Proteus anguinius* to red light and claimed to have caused it to produce large functional eyes.[6] Many inhabitants of caves and other dark abodes tend to lose light-sensory structures. Such examples of degeneration were part of the Lamarckian concept of "use and disuse of parts"; if an organism uses part of its body it should become more perfected for its particular function in subsequent generations; if a part is not used it would be expected to degenerate in subsequent generations. The reappearance of functional eyes in *Proteus* is another example of atavism, but Kammerer did not claim that this experiment had anything to do with the inheritance of acquired characteristics.[7]

The English experimental zoologist William Bateson (1861–1926) was the champion of the Mendelian school of heredity in England during the early part of the twentieth century. In 1910, Bateson at Cambridge University asked Kammerer to loan him a specimen of the midwife toad that had Brunftswielen. He received the following promise from Kammerer: "I will send to you any objects you may need . . . ,"[8] but that promise was not kept. That same year, Bateson visited the institute, but was not shown a specimen of the toad with Brunftswielen. This aroused Bateson's suspicion. Kammerer offered an explanation that the Brunftswielen only develop during the breeding season and tend to disappear during the nonbreeding season. Kammerer did not want to kill and preserve any breeding toads that had the rugosities because the colony was always in danger of failing to reproduce itself.

In 1913 Kammerer published the results of his *Salamandra* experiments. Until 1919 he had only relatively short publications containing diagrams and drawings of the Alytes Brunftswielen. In that year he finally published a detailed report on the toad in the prestigious journal *Archiv fur Entwicklungs Mechanik* (the *Journal of Developmental Mechanics*). E. W. MacBride, professor of zoology at Imperial College London, was a devoted Kammererian. He reviewed Kammerer's experiments in a letter to *Nature* published May 22, 1919. (*Nature* is the most widely read general journal of science in Europe.) Bateson then responded to McBride in the July 3, 1919, issue of *Nature*, implying that Kammerer was involved in fraud concerning his experiments that claimed to transform black salamanders into the spotted variety, and suggested that they may simply have been purchased from commercial sources. Bateson also criticized the photographs of the toe pads as being of poor quality and doubted that they were in the correct location to be functionally effective.

Przibram took full responsibility for all of Kammerer's experiments and claimed to be present when tissue sections of the pads were taken for histological (tissue) study. Przibram sent some of the tissue-section slides to Bateson, but Bateson thought they might have been taken from some amphibian other than *Aletes*. Evidently, Bateson did not trust Przibram any more than he did Kammerer.

On April 30, 1923, Kammerer gave a lecture to the Cambridge Society and brought with him specimens of *Proteus, Salamandra*, and a fifth-generation male toad preserved in alcohol, and showing the nuptial pads on the right forelimb (the left side having been used for histological preparations). This was the only preserved whole specimen remaining in Kammerer's collection, and most of his experimental animals had died. Bateson refused to attend Kammerer's lecture. Kammerer thought that Bateson did not want to examine the specimen for fear of being convinced that fully developed Brunftswielen do exist on the midwife toad. However, Bateson's excuse for not attending Kammerer's lecture was given in a letter to *Nature* (June 2, 1923): "Knowing that Dr. Kammerer had abstained from appearing at the Congress of Geneticists which met in Vienna in September last, I inferred that he had no new evidence to produce. . . ."[9] Many reputable scientists at that meeting had an opportunity to examine the toad specimen out of its jar under a binocular microscope, and none of them thought it was a fake. The press made so much ado over the Cambridge meeting that the Linnean Society of London invited Kammerer to repeat his lecture on May 10, 1923. This time Bateson did attend. However, he reportedly did not examine the specimen thoroughly; he did not even ask to see it out of the jar. Then, in September 1923, he wrote a letter in *Nature* requesting that the specimen be sent to him and offered to pay for special handling. Since the specimens were the property of the institute, Kammerer communicated Bateson's proposal to the directorate. Przibram, naturally wishing to safeguard the only remaining preserved specimen of *Alytes*, declined, but offered to show it to Bateson if he would come to Vienna. Bateson published Przibram's reply and his own in *Nature* (December 22, 1923), and there the controversy was to rest for the next three years.

After World War I, inflation hit Kammerer hard, causing him to resign his post at the institute and to attempt to support his family by writing and lecturing. He made two lecture tours in the United States during 1923 and 1924. In 1926 he published a book based on three shipboard expeditions during 1909, 1911, and 1914 to the Dalmatian Isles (near the east coast of the Adriatic Sea, an area rich in lizards and virtually uninhabited by people). In subsequent experiments with these lizards, Paul claimed that they changed their colors according to their background environment just as the salamanders had done. He also claimed that these changes were hereditary. Following the exploration approach of Darwin and Wallace, Bateson had also looked for evidence of Lamarckism in isolated Asian lakes early in his career, but failed to find any such support. Bateson endured great suffering during his ventures abroad, and it has been suggested[10] that he may have been jealous of Kammerer's success.

Early in 1926 Dr. Gladwyn Kingsley Noble (curator of reptiles at the American Museum of Natural History) visited the institute in Vienna and obtained permission from Przibram to study the preserved toad. No nuptial pads were observed. In the general area where pads should have been was a black color that later proved to be an injection of India ink.[11] Noble told the scientific world of his findings in the August 7, 1926, *Nature*. A second preserved specimen[12] was later discovered by Przibram and his assistant Paul Weiss; Weiss said it too, "showed an equally aberrant 'pigment patch' under the skin. . . ."[13] Noble suggested that Kammerer's histological sections of the Brunftschwielen fell within the range of variability shown by other toad species in the genus *Bombinatur* (*Bombina*).[14]

Arthur Koestler, an English journalist born in Budapest, authored a book titled *The Case of the Midwife Toad*.[15] Koestler believes that a conspiracy of at least three people would be implicated if this was a hoax, including the histologist who made the tissue sections, Przibram, and Kammerer himself. Furthermore, Koestler believes that the adulteration of the specimen was done after its return from Cambridge in a dilapidated state. The two motives he suggests for whoever injected the India ink into the toad's fingers was either (1) to preserve the critical features of the near-vanished nuptial pads before the remaining natural pigment disappeared completely, or (2) to discredit Kammerer. However, McBride later attempted to repeat Kammerer's *Alytes* experiment, but apparently had no success.

Kammerer left the institute at the end of 1922. He was granted one year's leave of absence in December of 1922. In October 1923 an agreement was reached on his final retirement. Most of his last three preretirement years were spent on lecture tours or in the Soviet Union where he had been invited to head up a new experimental institute in Moscow. He was preparing to move to Moscow in September 1926 and had hoped that his lady friend, Grete Wiesenthal, would go with him. When she refused to do so, he wrote four farewell letters on September 23, 1926, walked out into the countryside, put a gun to his head, and committed suicide (or so the reports went). The gun was found in his right hand, but the bullet entered his skull from the left side above the ear.[16] "It was a difficult feat to achieve, and it earned the risk of botching the job."[17]

Most people took Kammerer's suicide as evidence of his guilt over the fakery in his preserved toad. But Koestler sees too many inconsistencies to accept this explanation. In his book he challenges contemporary scientists to repeat Kammerer's work rather than condemn him without a trial. Koestler offers at least four factors which he believes contributed to Kammerer's growing despair: (1) financial penury, (2) the relentless harassment and veiled accusations of fraud by Bateson and his allies, (3) the wrong kind of notoriety, which indirectly played into the hands of his detractors, and (4) his problems with women.

Kammerer was a romantic, if not a dandy. He was married twice, with both marriages ending in divorce. At one time he was infatuated with his assistant Alma Mahler Werfel and (according to her) had threatened to shoot himself on the grave of her husband, Gustav Mahler, unless she married him.[18] Then he fell in love with the painter Anna Walt. His first wife, Felicitas, gave him an amicable divorce. His subsequent marriage to Anna Walt was tumultuous and lasted only a few months. After one of their intense marital disputes, Paul took an overdose of sleeping pills, but saved his own life by vomiting them out.[19] Then there was a series of falling in and out of love successively with the five famous Wiesenthal sisters. The eldest, Grete, was ballet master of the Vienna Opera. It was Grete's refusal to go to Moscow with Kammerer that may have precipitated the last of his periodic depressions.

Kammerer believed that the crucial experiments supplying proof of the inheritance of acquired characteristics were not the salamanders or the midwife toad, but rather those he performed with the invertebrate sea squirt *Ciona intestinalis*.[20] Yet he never did publish his "best evidence" in any primary source (a refereed journal).[21] These little creatures have separate tubes by which they bring water into their bodies for extraction of suspended food particles and then for expelling it. When these incurrent and excurrent siphons were severed, he reportedly observed them to regenerate longer siphons. The siphons were then cut again, and they grew back longer as before. Furthermore, he claimed that the progeny of such surgically treated parents had longer siphons naturally. This experiment is reminiscent of experiments performed by the German biologist August Weismann just prior to the beginning of the twentieth century. In an attempt to demonstrate that modifications of body cells (somatic cells) are not heritable (i.e., not transmissible through the gametes, eggs, or sperm), Weismann cut the tails off mice in each generation for about twenty generations. No diminution of tail length was seen in any generation of the treated mice. He reasoned that if such drastic modifications of the body cells did not become heritable, then the more subtle modifications that natural environmental effects might induce would not be heritable either.

Koestler's challenge to the scientific community to repeat Kammerer's experiments was answered much later by marine biologist J. R. Whittaker of the Wistar Institute of Philadelphia and the Marine Biological Laboratory at Woods Hole, Massachusetts.[22] Following the siphon-amputation experiments, Kammerer removed the lower portion of the animal containing the ovary. From this stub he claimed to observe regeneration of a complete individual with long siphons. Whittaker amputated siphons on two groups of fifty animals. After about a month, the siphons were regenerated, but none of them were elongated. Therefore no attempt was made to reamputate the new siphons. When Whittaker attempted to regrow sea squirts from the lower part of the animal (containing the ovary), all of the animals died after a few days. Kammerer was famous for his skills in rearing delicate species, so Whittaker was asked, "Have you considered the possibility that Kammerer was a more skillful aquarian then yourself, and that this might explain your negative results?"[23] In reply, Whittaker noted that the equipment at Woods Hole was much more sophisticated than that available to Kammerer; furthermore, Whittaker had fifteen years of well-documented expertise in handling *Ciona* in the laboratory compared with Kammerer's two or three years. According to Whittaker, Ciona is a very delicate animal that rarely survives more than six months in captivity, even under optimal conditions. In Whittaker's hands, excision of the lower portion of any *Ciona* invariably resulted in its death. He is now convinced that Kammerer either did not do the *Ciona* experiments or faked the results. Kammerer did publish some photographs of *Ciona* with long siphons. However, Whittaker knew of a local subpopulation of *Ciona* in the Mediterranean Sea near Naples with naturally long siphons. "When something real like that already exists in nature, one becomes suspicious of photographic evidence."[24]

Whittaker was not the first to try to repeat Kammerer's work on *Ciona*. In 1923 H. Monro Fox of the Biological Institute in Brittany amputated the oral siphon from 102 specimens and observed no superregeneration. Later that same year E. W. MacBride explained in a letter to *Nature* that Kammerer had also failed to observe siphon superregeneration when only the oral tube was amputated.[25] In Kammerer's book *The Inheritance of Acquired Characteristics,* published in 1924, he hypothesized that perhaps only southern populations of *Ciona*, similar to those used in his experiments, had "superregenerative" potential. Jose J. Valdes, in a letter to *Science News* June 21, 1975, speculated that Kammerer may have "emphasized and practiced removal of both siphons in his successful 'experiments' (contrary to the techniques used by Mingazzini, his predecessor along these experimental lines) because he had used variety *'macrosiphonica' Ciona*—thus the need for removing both siphons to avoid detection of the naturally long tubes." Valdes claimed to have amputated both siphons or only oral siphons on about twenty specimens of *Ciona* varying in size from six to twelve millimeters without observing any abnormal lengthening of regrown tubes.

Koestler could not accept the fact that Kammerer faked his results with the midwife toad because he had openly displayed a specimen for scientific scrutiny at his April 1923 lecture before the Cambridge Society. It is indeed difficult to imagine he would do such a thing if the specimen did not possess the nuptial pads as claimed. Just prior to his death he was looking forward with great anticipation to his new duties at the Moscow Institute, which was built especially for him. Then there are the very unusual circumstances surrounding the manner of his death. Koestler smelled foul play; if not in Kammerer's death, then certainly in a conspiracy to discredit his work.

THE PATCHWORK MOUSE[26]

Our story now shifts to the year 1966 when a Ph.D. named Marvin A. Karasek at Stanford University in California perfected the technique of skin-tissue culture. An M.D. named William Talley Summerlin was also then doing medical research at Stanford. He was able to apply Karasek's new technique to human clinical medicine because he was a physician and Karasek was not. After growing a bit of human skin in tissue culture, it was grafted back onto the same patient from which it was derived (autografting) with success; i.e., the graft healed and was retained permanently.[27] Live *autografts* present no immunological problems and their success should be complete if the surgery is done correctly and infection is prevented. In 1969, Summerlin took his project to the Veterans' Administration Hospital out of the reach of his senior research collaborator (Karasek) and began to experiment with *allografts* grown in tissue culture. An *allograft* is a tissue (or organ) derived from a genetically different individual of the same species. Normally allografts are rejected sooner or later by the host's immune system. But Bill Summerlin claimed that four of his human allografts grown in tissue culture somehow lost much of their ability to stimulate an immune response in the recipients so that the grafts remained viable on the hosts for abnormally long periods of time (more than a year) without the aid of immunosuppresive drugs.[28] The popular press gave him wide publicity, but many transplantation researchers remained skeptical.

Summerlin's fame soon attracted the attention of immunologist Dr. Robert Alan Good at the University of Minnesota, who invited Summerlin to join his research team at Minneapolis. Bill said he would be delighted to join the group and asked Good if he had a research grant to support his work. This "handout attitude" surprised Good because any researcher worth his salt is expected to obtain his own source of financial support. Nonetheless, Good offered to share some of his own grant money with Summerlin because he thought Bill was onto something potentially important in transplantation immunology, and he wanted any breakthroughs to be made by his team. Summerlin moved to the University of Minnesota in August 1971. Good arranged for Summerlin to visit several of the more experienced transplantation research centers both in the United States and in Europe. He then set Summerlin up with his own staff of postdoctoral research fellows and his own highly pedigreed mice from the Minneapolis medical school.[29] Mice within a highly inbred line are essentially genetically uniform (homozygous for all genetic loci; *loci* are positions on the chromosomes where genes reside). This genetic uniformity is a great advantage in doing biological research because it rules out one of the greatest sources of experimental variation (viz., genetic variation) and thereby increases the sensitivity of any experimental test. These special strains of mice afforded an ideal opportunity to demonstrate the validity of his earlier work with the requisite scientific rigor to make a convincing argument for his theory.

During the fall of 1972 Good became director of the Memorial Sloan-Kettering Cancer Center in New York, one of the world's most prestigious centers for the study of cancer, immunology, and related phenomena. Good brought Summerlin and a staff of about fifty researchers with him to New York. He appointed Summerlin, age thirty-four, to the rank of full professor, a decision that did not set well with some of the more senior members of the institute. Summerlin thereby became the youngest full member at the Sloan-Kettering Institute (SKI). His job was to run the dermatology clinic and to supervise the research of a half-dozen postdoctoral fellows who were working on his transplantation projects.

Summerlin claimed in 1973 that tissue-culture transplants behave as universal donors even across species lines.[30] This would be truly remarkable because it was well known that as the genetic dissimilarity between donor and recipient widened within a species, the chances for long-term survival became more remote; grafts between different species (*xenografts*) were invariably and rapidly rejected. He also later claimed that human corneas (eye membranes) grown in tissue culture had remained viable on rabbits for six months, as compared to only six weeks for fresh transplants. Good's name, as director of the Sloan-Kettering Cancer Center, appeared as coauthor on this report. Bob Good became an instant public celebrity after his picture appeared on the cover of *Time* magazine March 19, 1973. The lead article was titled, "Toward Control of Cancer." In that article, the cancer work of Good's research teams at SKI was reported along with the possibility of conquering the problem of skin-graft rejection through the use of the Summerlin technique.

Attempts were soon made by others outside SKI to try to duplicate Summerlin's results. The Nobel Prize–winning immunologist Sir Peter Medawar was among this group, and he reported his lack of success to Bob Good at a meeting of the SKI consultants in October 1973. Good stated that he had been assured by Summerlin that the rabbit corneal grafts done under his supervision were limbus-to-limbus (complete edge-to-edge). Such grafts would truly be remarkable because they no longer would be privileged sites due to the absence of a blood supply. A *privileged site* such as the cornea of the eye is an anatomical location into which transplants from the same or different species may be made with success as long as care is taken to avoid contact with vascular tissue. However, Dr. Michel Prunieras of the Rothschild Foundation in Paris made it known at the 1973 SKI consultants meeting that he had obtained 60- to 75-day graft retentions using Summerlin's techniques, and he published these observations in a letter to *Science* November 2, 1973. Summerland even displayed a skin graft from a white mouse that had regrown white hair on an otherwise black recipient mouse. Medawar remained skeptical, however, because loss of melanin pigment commonly occurs during healing of a transplant.

Summerlin explained that in his experimental design (protocol) one eye on each rabbit had been grafted with fresh corneal tissue from a human cadaver; the other eye had been grafted with a cornea from the same cadaver only after first being maintained in tissue culture for several weeks. In the two rabbits displayed by Summerlin at that meeting, one eye was clear and the other was cloudy (opaque). Summerlin claimed that the clear eye had the cultured graft and the cloudiness of the other eye was due to rejection of the fresh cornea. Dr. John L. Ninnemann, who had joined the SKI staff in July 1973, thought it strange that the fresh tissue would not sensitize the rabbit (i.e., stimulate its immune system) and thereby cause rejection of both corneas. The corneal transplants had been performed by two Cornell-New York Hospital ophthalmologists, Drs. Peter Laine and Bartley Mondino.

Medawar was so disturbed by the reported results that he phoned Mondino after the meeting and was told that only one eye on each rabbit had been grafted.[31] Medawar then confronted Summerlin with this, and they both went to the animal room to see who was right. Suture lines were visible on only one eye of each rabbit. Summerlin was chagrined and told Medawar he would check with his staff and give him an answer as to what happened. Medawar claimed that Summerlin never did respond. Meanwhile, Summerlin continued to speak and write about those unoperated (control) eyes as though they had been successfully grafted.

Earlier in 1973 Bob Good had been made aware of Summerlin's administrative deficiencies and the lack of success that other researchers were having in using Summerlin's techniques. One of those who complained was Leslie Brent of London University's Department of Zoology who said, "Bob, I just can't communicate with your man Summerlin. He will not give me details; he will not tell me how he does things."[32] Good phoned Summerlin and essentially told him "to get on the ball" and "take care of business." Summerlin then began to correspond with Brent, but it must have galled Good to have to prod one of his department heads like that.

Summerlin's clinical research nurse, Joyce Solomon, was also unhappy with him, complaining that "We could never seem to get an organized protocol out of Bill. His methods of doing the transplants would change and he would never come out and explain to us why he was changing. . . . He was supposed to be the leader and he did not lead. . . . We had a group of patients who were being told that laboratory studies were being done when, in essence, they were not."[33] Good weathered all of this flak without dumping Summerlin because he continued to believe in his potential: "a negative doesn't disprove a positive,"[34] and "if [the work] was repeated it would be of the greatest importance."[35]

A National Institute of Health (NIH) "site committee" visited Summerlin on January 7, 1974, to evaluate his request for more than $600,000. The committee informed Good of their doubts concerning Summerlin's ability to identify the ownership of cells after grafting. Later in January, the committee rejected Summerlin's grant proposal.

Dr. Laino informed Summerlin in writing that about twenty rabbits had been grafted according to the Summerlin technique with no evidence of graft prolongation. Ninnemann was having no success either and he told Good so, asking him if it wasn't time to publish these negative results. As chief of Ninnemann's section, Summerlin's name would have to appear as coauthor of the report that refuted his own research. Summerlin knew he was under pressure to produce concrete evidence to support his claims.[36]

Early on the morning of March 26, 1974, Summerlin had been called to meet with Good to review his work and that of the research fellows under his supervision. Summerlin went to the animal room and obtained from the animal technician, James Martin, mice bearing skin grafts. Two of the mice he had chosen to display to Good were albinos with black skin grafts. Dark cells, when grown in tissue culture, tend to lose pigmentation, so the black on white patches appeared a dirty gray. Impulsively, and in private, Summerlin took a black felt-tipped pen and "touched up" the black grafts so that they would appear more impressive.[37] He then carried the mice into the meeting. Good barely glanced at the mice and then began interrogating Summerlin concerning the details of his work. The meeting lasted about forty-five minutes. At the end of the meeting, Summerlin returned the mice to the animal room.

A short time later, when Martin returned the mice to their storage pens, he noticed some unusually dark patches on the white mice, and found that the dark color could be removed with an alcohol swab. He was astonished and told one of the research technicians, William Walter. The story spread like wildfire from Walter to fellow Dr. Geoffrey O'Neill; from O'Neill to Dr. John Raff; and from Raff to Good, who was in another meeting at the time. Raff insisted on seeing Good immediately. When Good was informed of the "forgery" he immediately called an audience with Summerlin and told him to bring his mice back. Summerlin met with Good and Dr. Lloyd J. Old about noon and admitted having added black ink to the two grafts. In the past, Summerlin had shown Good many mice grafts of different colors and even mice bearing guinea pig skin grafts. But now Good began to doubt Summerlin's integrity and to question all that he had been shown previously. Summerlin was suspended for two weeks while a review committee was established to investigate all of his research. Good initiated a check on a mouse known as the "Old Man," the only remaining long-term survivor still bearing grafted skin from Summerlin's Minnesota work. On April 3, 1974, Good received a report from Dr. Edward Boyce of the Sloan-Kettering Institute that the "Old Man" (an agouti mouse bearing a white graft) was actually a hybrid between two highly inbred strains.[38] It is well known that hybrids between genetically pure highly inbred lines can accept transplants from either parent. The review committee found it difficult to believe that Summerlin was unaware of this possibility and failed to have it checked, because it was the only concrete evidence of success that he had to show.

When the review committee asked Summerlin what had happened to all of the other mice he claimed had also been successfully skin-grafted, he replied that they had been sacrificed at intervals to produce serum to be stored for H-2 (histocompatibility) antibody tests at a later date.[39] It all sounded suspicious, however, because one need not kill mice to obtain a sample of serum. Good wanted to keep a lid on the scandal and handle the matter internally because he was already in trouble with many of the old guard at SKI whom he bypassed in appointments, and also because he had to constantly fight for a piece of the ever-dwindling federal research grants.

If this news were to leak out, it might increase the suspicion with which bureaucrats and the general public view science and scientists, thereby making it more difficult to obtain the requisite funding. But this kind of news could not be kept from the press. Barbara Yuncker, medical reporter for the *New York Post*, somehow heard about the affair on Wednesday, April 10, 1974, and the story broke on Monday afternoon, April 15. The story also appeared in the *New York Times* on April 17. The *Times* article listed some criticisms against Good in the Summerlin affair:

- Good had touted too highly Summerlin's early results before they had been adequately substantiated.
- Good had failed to provide adequate supervision over Summerlin.
- Good had staffed his institution with unproven scientists and placed them under great pressure to produce results.
- Good was shooting for spectacular results in order to attract national fame and money for the Institute.[40]

"The Sloan-Kettering Affair: A Story Without A Hero" was the title of a lead article by Barbara Culliton in the May 10, 1974, issue of *Science*. This is the most widely read scientific journal in the United States and is the weekly organ of the American Association for the Advancement of Science (AAAS). Culliton's article suggests, "If the present crisis generated by the Summerlin case is any indication, it appears that a high-pressure environment that drives individuals to exaggeration and fosters hostility is not ideal for the kind of achievements in research that Good, like everyone else, would like to see. . . ."[41]

Summerlin had the opportunity to explain his side of the story to his review committee on May 10, 1974. But on Friday, May 24, his committee informed the press that the trustees of the center had decided to terminate Summerlin's relationship with the hospital and the institute. Dr. Lewis Thomas, president of SKI, proposed that the most rational explanation for Summerlin's performance was an emotional disturbance that rendered him not fully responsible for his actions. In the terms of his dismissal, the center was to provide Summerlin with up to one year's medical leave with full salary ($40,000) so that he could obtain such rest and professional care as might be required.

When the press asked Good if he had examined any of Summerlin's human grafts from his Stanford days, he claimed to have seen slides from two successful grafts.[42] One was a black male recipient bearing a skin graft from a white person. This was not very impressive because healing skin can easily lose pigment. However, the second one was a male recipient bearing a female skin graft showing Barr bodies four years after surgery. That had really impressed Good. A *Barr body* (or *sex chromatin*) is a darkly-staining, inactivated sex (X) chromosome. Female cells have two X chromosomes (XX) whereas males (XY) have only one X chromosome and the male-determining Y chromosome. Normal male cells should not have Barr bodies. Traits governed by genes on the X chromosome (sex-linked genes) are not more intensely expressed in females (with a double dose of sex-linked genes) than in males (with only a single dose of sex-linked genes) because one of the X chromosomes is randomly inactivated in each female cell as a dosage compensation mechanism. Therefore, finding transplanted cells with Barr bodies on a male host four years after grafting (as Summerlin had claimed) is highly suggestive evidence. However, mitotic (cell divisional) accidents can produce XXY cells in some male tissues; the extra X chromosome would then appear as a Barr body. Stanford's Karasek had supported Summerlin's claims at the May 1973 meeting of the American Society for Clinical Investigation. Thus there were some grounds for Good's support of the Summerlin method, but it was far from conclusive.

On Tuesday, May 28, 1974, Summerlin held a three-hour press conference at his Connecticut home. He was under psychiatric care at the time of this interview and reportedly was feeling better. He said that he made no effort to conceal from Good the inking of the mice. He bitterly disputed the statements of Raff and Ninneman, who claimed that he knew about the single-eye operations as early as October 8 or 9, 1973. He said that two protocols were initially planned. In single-eye transplants, only the right eye would be used; in double-eye transplants, the cultured cornea would be placed on the right eye and the fresh cornea on the left. When he saw the rabbits with the clear right eyes and opaque left eyes, he assumed that the second protocol had been followed.[43] The ophthalmologists who performed these experiments admitted that they did not follow the original protocols and claimed to have informed Summerlin that no bilateral transplants had been performed. Nevertheless, Summerlin's approach to the science of transplantation is revealed by the blithe assumption that the results would appear as he had anticipated. It is difficult to understand how he could be so busy that he could not make a brief phone call to the ophthalmologists to confirm his assumptions. Summerlin thought that Good had saddled him with too many responsibilities and had placed excessive pressure on him to produce positive results.

Early in 1973 the National Cancer Advisory Board had been made aware of the dissatisfactions voiced by virologists at the seeming exclusivity of the National Cancer Institute's program in their field. A committee was established, with the distinguished microbial geneticist Norton Zinder as chairman, to investigate the problem. One of the most significant conclusions made by the Zinder committee even before the Summerlin scandal broke was that "Success in science is an irregular and unpredictable phenomenon. When and where important discoveries are and will be made is almost impossible to determine in advance. The goal should be to maximize opportunities."[44]

ANALYSIS

There are many obvious differences between the Kammerer and Summerlin stories. Kammerer may have committed biological suicide, whereas Summerlin committed professional suicide. Each was experimenting with different organisms in different biological disciplines. Kammerer was not under pressure to produce results of potential value to human welfare the way that Summerlin was. Summerlin admitted adding black pigment to his mice hair, but Kammerer never admitted injecting India ink into the nuptial pads of his toads. Many other differences could be cited. But what about the similarities?

Kammerer had only a single preserved male midwife toad and no live specimens to support his claims; Summerlin had a single live male mouse (the "Old Man") as evidence, but it turned out to be a hybrid. Kammerer boasted of success in the environmental induction of heritable variations (i.e., producing acquired characteristics) in toads (*Alytes*), salamanders (*Salamandra*), and sea squirts (*Ciona*); Summerlin claimed that subjecting pieces of skin to the unusual environment of tissue culture could change their antigenic structure; neither claim could be substantiated by independent investigation with sufficient rigor to convince the scientific establishment. Bateson and others were skeptical of nuptial pads on the toads because they do occur readily in nature in related genera; he was also suspicious of the experimental induction of long siphons in *Ciona* because such variations exist in nature. Likewise, Medawar and others were skeptical of white hairs on a black mouse because this occurs naturally and may be due to many different causes. Kammerer had promised Bateson that he would supply him with whatever materials he wanted, but never did so. Summerlin told Medawar that he would check on the rabbit data and get back to him, but he didn't follow through. Many who viewed the toad specimen were unqualified to judge on the authenticity of the nuptial pads; those who viewed the rabbit specimens were not ophthalmologists and therefore unqualified to make judgments on the authenticity of the corneal transplants being limbus-to-limbus. Kammerer offered histological sections of the male nuptial pads possessing Brunftschwielen in support of his *Alytes* experiments, whereas Summerlin presented slides of male human skin possessing Barr bodies as evidence of the success of his allografting technique. Media "hype" preceded Kammerer's U.S. tours and made his colleagues suspicious of his motives; too much hype in the media was also promulgated before attempts could be made to duplicate Summerlin's experiments.

Koestler suggested that Kammerer's emotional reaction to being rejected by his lady friend was at least partly responsible for his "suicide." Dr. Thomas (head of SKI) suggested that Summerlin was emotionally disturbed due to the pressures of his work and was therefore not fully responsible for his actions. Kammerer claimed his breeding colony of *Alytes* was so weak he could not afford to sacrifice any animals to preserve the Brunftschwielen; Summerlin said that all of his "successfully grafted" mice (other than the "Old Man") had been sacrificed to provide serum for serological histocompatibility tests. Kammerer failed to indicate how to cut the siphons in *Ciona* and (possibly) how to care for the eggs of *Alytes*. Likewise, Summerlin failed to give explicit directions for repeating his experiments to people like Leslie Brent. Kammerer felt that his work was made to appear as a "fraud" by a person or persons unknown; whereas Summerlin felt that he was made to look fraudulent by Raff and Ninneman, who failed to follow their original protocol. Kammerer was granted one year's leave of absence before he was finally retired; the Sloan-Kettering Institute gave Summerlin up to one year's medical leave on full salary so that he could obtain needed professional care.

Does history repeat itself? You be the judge.

NOTES

1. Ann Landers, "Lincoln, Kennedy Oddly Alike," *San Luis Obispo Telegram Tribune*, January 14, 1995, E2.

2. Arthur Koestler, *The Case of the Midwife Toad* (New York: Random House, 1971). Koestler was a prolific writer. Perhaps his most famous book was *Darkness at Noon*, a novel of the 1930s Stalin purges in the old USSR. The Associated Press reported on March 3, 1983, that both Koestler (age seventy-seven) and his wife, Cynthia (age fifty-six), had died from a drug overdose in an apparent suicide pact. He had been suffering from leukemia and Parkinson's disease. No information was given on his wife's health. He was a member of Britain's Voluntary Euthanasia Society, formerly called EXIT. This group published a booklet called *Guide to Self-Deliverance* on how to commit suicide. In its preface Koestler wrote: "The prospect of falling peacefully, blissfully asleep is not only soothing but can make it positively desirable to quit this pain-racked mortal frame and become unborn again."

3. Ibid., 42.

4. Ibid., 43. The female lays many eggs attached to long strings of jelly. After fertilizing these eggs, the male winds these strings around his hind legs and carries them around with him until they hatch—hence the name "midwife toad."

5. Ibid., 45.

6. Ibid., 47.

7. Ibid., 47, 67.

8. Ibid., 61.

9. Ibid., 82.

10. Ibid., 64-65.

11. Ibid., 99, 108.

12. Ibid., 109.

13. Ibid.

14. Ibid., 102.

15. See note 2.

16. Ibid., 120.

17. Ibid.

18. Ibid., 21.

19. Ibid., 94.

20. Ibid., 45.

21. "The Case of the Suspicious Siphons; 'Ciona' Revisited," *Science News* 107 (May 31, 1975): 349.

22. Ibid., 348-49.

23. Ibid., 348.

24. Ibid., 349.

25. Jose J. Valdes, "The Kammerer Case," *Science News* 107 (June 21, 1975): 395.

26. Joseph R. Hixson, *The Patchwork Mouse* (New York: Anchor Press, 1976).

27. Ibid., 23.

28. Ibid., 32. Immunosuppressive drugs inhibit the host's immune system from attacking the foreign graft and causing its rejection.

29. Ibid., 25.

30. Ibid., 32.

31. Ibid., 42.

32. Ibid., 105.

33. Ibid., 108.

34. Ibid., 105.

35. Ibid., 96.

36. Ibid., 48.

37. Ibid., 5.

38. Ibid., 95. The agouti pattern is black hair with yellow tips, typical of wild mice.

39. Ibid. Histology is the study of tissues. If the H-2 antigens of graft and host are not identical, the graft will usually be rejected; these tissues are said to be immunologically incompatible.

40. Ibid., 87.

41. Ibid., 88.

42. Ibid., 97.

43. Ibid., 101.

44. Ibid., 133.

CHAPTER 11

HONESTY IS THE BEST POLICY

Scientific theories may or may not stand the test of time. People who understand what science is all about can live with that. Scientists can get a lot of mileage out of a theory by accepting it tentatively as a working hypothesis and investigating its logical consequences. However, no one but a fool would waste his time investigating predictions made by a theory that was known to be incorrect. A scientific investigation might have been designed better, or perhaps should have had larger numbers in each sample, or the analysis could have profited from a different statistical procedure, or a different interpretation might possibly have been drawn from the same data, and so forth. But one thing that everyone expects of any piece of science is that all of the data or facts have been correctly reported. In other words, scientists are assumed to be honest in their presentation of the observations on which their theses stand. And yet we know from history that this assumption has been violated to various degrees in a small percentage of cases. Is it possible that the actual amount of undiscovered fraudulent science is much larger? If any data crunching, correction, conversion, doctoring, or bashing was done in good faith, it should be clearly explained in the report. However, if only some of the data has been reported, if the data has been consciously tampered with and not reported, or if the data has been fabricated, it is not just poor science, it is not science at all. There is a community of trust among scientists that despite our limitations, biases, and ineptitudes, at least we honestly report what we did and what we observed. When that spirit of trust is violated, it not only harms the individual who did so, but also reflects on the entire scientific establishment. Our society expects science to be truthful even though its theories may be faulty. When the general public loses confidence in the "truthfulness" of scientists, science and society as a whole may suffer irreparable damage. To illustrate what I mean, let's have a look at some historical examples of scientific fraud and try to analyze their effects on science and society.

TOO GOOD TO BE TRUE

In chapter 8, "Brainstorms," it was argued that historians may have misinterpreted the work of Gregor Mendel. His "laws of hybrid formation" may or may not have represented "laws of heredity" in his own mind. The essentials of Mendelian heredity were presented in that same chapter and need not be repeated here. Suffice it to say that, for the purposes of the present chapter, the results of Mendel's hybridization experiments were thought to approximate idealistic whole-number ratios such as 1:1 and 3:1.

Mendel did his experiments on hybridization between various varieties of peas before chromosomes had been discovered and before it was known how reproductive cells are produced. Of course today we accept the theory that the gene is the basic unit of heredity, that genes are on the chromosomes in the nucleus of the cell, that most cells of the body (somatic cells) contain two sets of homologous chromosomes (diploid condition; one set derived from each parent), and that the process of meiosis reduces the diploid number to the haploid (one set) number in the formation of gametes (sex cells; eggs and sperm). We also know that each chromosome contains many genes, and that each gene has a specific locus or position on a given chromosome. If different DNA sequences reside at a given genetic locus, these alternative forms of a gene are called *alleles*. For example, one allele may determine that a pea plant will produce yellow seeds, whereas an alternative allele may produce green seeds. Genes occupying different loci on the same chromosome are said to be *linked*. Linked genes tend to be transmitted to offspring in the same linkage arrangements as they existed in the parents. Occasionally linked genes can be recombined in the formation of gametes by a process called *crossing over* during meiosis. Pieces of homologous chromosomes break and rejoin with one another to shuffle the genes into new linkage combinations. The frequency with which linked genes recombine in this way is a function of the physical distance between them. Loci that are far apart tend to be recombined more often than loci that are close together. However, if genes are so far apart that a crossover always occurs between them in every cell undergoing meiosis, they behave as though they were unlinked; i.e., the genes *assort independently*, as if they were on different chromosomes. Some critics have claimed that if Mendel had investigated one more trait than those reported in his paper, he would have discovered the phenomenon of linkage. Let's see if that claim will stand up in the light of critical analysis.

The garden pea (*Pisum sativum*) has a diploid number (2n) of fourteen. This means that each body cell contains two haploid (n) pairs of seven chromosomes; one set being derived from each parent via their gametes. Thus, peas have seven groups of linked genes. Mendel reported on seven traits: (1) length of vine, (2) color of seed endosperm, (3) color of seed coat, (4) seed shape, (5) pod shape, (6) pod color, and (7) position of flower on the stem. Among the several varieties of peas used in Mendel's experiments, some were found to differ with respect to two of these traits; e.g., one variety might have only yellow, wrinkled seeds, while another variety had only green, smooth seeds. Mendel found that the genes for all seven traits assorted independently of one another in the formation of gametes. We now know that this is characteristic for genes that are on different chromosomes. How members of one pair of homologous chromosomes segregate into different gametes is independent of how a member of another pair of chromosomes gets into that same gamete. Thus, some contend, Mendel could not have investigated an eighth trait without running into the analytical problems created by linkage. Therefore Mendel must have had extraordinary good luck in choosing just seven traits to investigate, and then finding that all of them assort independently of one another.

Others suggest that this probability is so low that Mendel must have investigated eight or more traits and found that some of the data could not be accommodated under his hypotheses, so he only bothered to report those cases that conformed to his theory. Now if that is what really happened, Mendel's genius is still evident, but his integrity as a scientist would be seriously compromised. In doing science, one does not report only those observations that happen to conform to one's preconceived ideas of how things should turn out. That would be intellectually dishonest and a death sentence to one's scientific career if it became known. When you write up your report, you present every bit of the data, "warts and all." Some scientists may wish to transform the data by conversion to logs or some other mathematical manipulation. That is perfectly all right as long as you report all of the data and explain to your audience exactly what you have done to the original data. Some very eminent biologists have come under fire recently for failure to follow these commonsense rules of scientific conduct.

What these critics of Mendel forget (or possibly, what they do not understand) is that loosely linked genes can mimic independent assortment if they are far enough apart on the chromosome. Mendel's experiments have been independently replicated many times, and their interpretations have led to essentially the same conclusions. So his "laws of hybrids" have been well confirmed. Modern analyses have revealed that three of the gene loci are on chromosome 4, a pair of loci are linked on chromosome 1, and one locus each is on chromosomes 5 and 7. The linked loci that Mendel studied jointly are so far apart that they assort independently, as if they were on different chromosomes. So it is not inevitable that Mendel would have discovered the phenomenon of linkage if he had investigated an eighth trait. As a matter of fact, Mendel reported that he actually did investigate some other traits. Pea flower color, for example, is either violet or white. He recognized, in this case, that white flower color is always correlated with white seed coats, whereas violet flower color is associated with darker seed coat colors (gray, gray-brown, or leather brown). Today we know that a given gene may contribute to more than one character, a phenomenon called *pleiotropism*.

In his first two years of preliminary work, Mendel found that some traits could not easily be unambiguously classified as either this or that; they were not included in his study. For example, size and form of the leaves; length of the flower stalk; size of the flowers, pods, and seeds—none of these traits possessed constant differentiating characters demanded by Mendel's criteria for further study. In other words, Mendel's studies were highly restricted to certain kinds of traits. Thus, he realized that his "laws of hybrids" could not be applied to the inheritance patterns of all traits. We now know that sizes and shapes of body parts usually have a complex heredity involving the pleiotropic interaction of many genes. These traits must be expressed by measurement (inches, pounds, etc.) and hence are called "quantitative traits." Quantitative traits are usually influenced by environmental factors that are difficult to control in breeding experiments (the latter being a fact that Mendel fully realized).

Mendel did enough experimenting on the flowering time of peas to recognize that varietal hybrids were phenotypically intermediate between their pollen and seed parents. In other words, neither long nor short flowering time could be considered to be a dominating trait. No data was presented in his paper because, "the experiments are not yet concluded."[1] However, he proposed that, "the constitution of the hybrids with respect to this character probably follows the rule ascertained in the case of the other characters."[2] Unfortunately he did not elaborate on this. My interpretation of this statement is that he assumed that the hybrid character would segregate in the next generation into three classes (two pure breeding and one segregating again), just like the seven traits for which he presented data. Flowering time in most plants is a quantitative trait (governed by multiple genes), and hybrids from pure-line parents usually exhibit a complicated segregation pattern in their offspring. My guess is that when Mendel saw the intermediate character of the hybrids, he realized that flowering time was not an "either/or" situation and so he did not bother investigating the progeny of these hybrids. In other words, he must have known that there was more to the laws of heredity than those he had postulated to explain the behavior of his seven selected traits.

My major criticism of Mendel's results stems from my own experience in determining such phenotypes as kernel color or seed shape in corn. Peas are probably no different in this respect. For example, I sometimes have difficulty in determining if a corn kernel is yellow or white, wrinkled or smooth. A few kernels will be light yellow. Into which phenotypic group should these be placed? Does one small dent qualify a kernel as wrinkled? Sometimes I just can't tell. I ask my genetics class what I should do with such data. All too often I get the response, "Throw it out." Then I explain that a scientist should report what she finds. She does not tamper with the data. You do not throw out data because it fails to conform to your own biased way of viewing how the world should behave. If you cannot unambiguously determine what the phenotype is, then start a new column in your tally sheet to record ambiguous data; and be sure you report all of the data. That apparently was not done by Mendel. He reports no ambiguous data. All of his peas were either wrinkled or smooth, either yellow or green. Real data, especially large numbers (in the hundreds he recorded) do not always come in these either-or phenotypes. I have no doubt that such ambiguous observations were encountered in recording the data, but they somehow disappeared in his published report. According to Mendel,

> in some of the seeds of many plants the green colour of the albumen (endosperm) is less developed and at first may be easily overlooked. The cause of this partial disappearance of the green colouring has no connection with the hybrid-character of the plants, as it likewise occurs in the parental variety. This peculiarity [bleaching] is also confined to the individual and is not inherited by the offspring. . . . Seeds which are damaged by insects during their development often vary in colour and form, but, with a little practice in sorting, errors are easily avoided.[3]

As much as I admire Mendel, and knowing what I now do about unintentional bias in experimental work, I believe that he should have known that the phenotype of every pea seed cannot be unambiguously identified, and he should have made some provision for recording such data. Hindsight is always better than foresight! Unfortunately, we have nothing but two of Mendel's published papers and some of his letters as actual documentation. After Mendel died, the new abbot of the monastery burned all of Mendel's experimental records.

Sir Ronald Fisher, the great English mathematician, published a critique of Mendel's pea experiments in 1936.[4] One of the things revealed by Fisher's analysis is that "Mendel left uncounted, or at least unpublished, far more material than appears in his paper."[5] This is not a serious objection as long as there was no selection of certain plants or seeds from the superabundance of material that in most cases he had available to him. Statistical tools to test the "significance of deviations from expectation in a binomial series had been familiar to mathematicians at least since the middle of the eighteenth century."[6] These tools give us, in an infinite number of trials of the same size, the proportion that would have larger deviations of observed from expected values than the trial being tested. By modern convention, scientists would reject a hypothesis if less than 5 percent of an infinite number of replicate experiments of the same size would have as large as (or greater) deviations from expected values than the observed values under test. Such a result is said to be *statistically significant*. The expected values, of course, are generated by the hypothesis under test (e.g., 3:1 ratio).

Suppose, for example, that we toss a coin ten times. Assuming that heads and tails are equally frequent expectations on any one toss, we expect in our experiment to observe five heads and five tails. This is the most likely outcome in an infinity of similar trials of size ten. Most folks would likely not be concerned about an observation of 6:4 or even 3:7; these deviations from the expected values are simply attributed to chance. Of course the 6:4 outcome has a lower probability of occurring by chance than the 5:5; and the 3:7 outcome has a lower probability than 6:4. Somewhere along the line, however, most folks have a point at which they begin to doubt the validity of their hypothesis (the assumption that heads and tails are equally frequent occurrences). Certainly if ten heads appeared in a row, I would want to inspect the coin to be sure it wasn't a double-headed coin. Yes, ten heads can occur by chance alone, but with very low probability; $(1/2)^{10} = 1/1024$. As a biological example, suppose that our genetic hypothesis for a set of F_2 data is a 3:1 ratio. In a sample of size forty, this theory would lead us to expect thirty with dominant trait and ten with recessive trait. Small deviations from this expected value (e.g., 28:12) would not be significant. That is to say, the probability of such an observed deviation from the expected numbers occurs very often by chance alone. Better than 90 percent of an infinite number of trials of size forty should have as large as (or larger) deviation than what we observed (28:12 versus 30:10). If chance is the only factor causing deviation of observations from expectations, as the sample size increases, the effects of sampling error should be proportionately less. That is, as the sample size increases toward infinity, the plus and minus effects of chance should progressively come to balance each other, so that at infinity, the ratio expected is the ratio observed. We do not deal with infinity in the real world of experimental data, but we do in our probability theories by which we analyze these observations.

We know that Mendel studied mathematics at Vienna, but it may not have been probabilities. Nevertheless, the tools were available in his time by which he could have statistically analyzed his data, but they were not used in support of his hypotheses. Fisher thought that since Mendel "was engaged in other researches of a statistical character, in meteorology, and in connection with sun-spots, . . . that it is scarcely conceivable, had the matter caused him any anxiety, that he knew of no book or friend that would enable him to examine objectively whether or not the observed deviations from expectations conformed with the laws of chance."[7] Fisher apparently believed that Mendel thought his data was so good that it didn't need the support of a statistical test. Fisher finds it remarkable that there is "only one case in the record in which Mendel was moved to verify a ratio by repeating the trial."[8] That instance occurred when he was trying to establish the numerical ratios of *AA* to *Aa* genotypes among F_2 progeny that showed the dominant phenotype. Fisher believes that Mendel's genetic theory led him to expect a 1:2 ratio, respectively. One of his trials of size 100 gave 40 homozygotes (*AA*) and 60 heterozygotes (*Aa*). This is not a significant deviation (at the 5 percent level of rejection of 1:2 hypothesis), but poor enough to cause Mendel to repeat the experiment; the second trial gave 65:35, and according to Fisher, "as near to expectation as could be desired."[9] If Mendel was trying to fudge his results, why would he bother reporting such data and then repeat the experiment?

When *Aa* hybrids are self-pollinated, three-fourths of their progeny ais expected to show the dominant trait. Mendel chose to test his F_2 progeny by allowing the dominant phenotypes to self-pollinate. He selected ten seeds at random from the progeny, grew them out, and reportedly thereby confirmed the 1:2 ratio. That is, for every plant of dominant phenotype that failed to segregate recessive types there were two that did so. However, as Fisher points out, the probability that ten progeny from hybrid plants would fail to segregate out any recessive types is $(0.75)^{10}$ or 0.0563. "Consequently, between five and six percent of the heterozygous parents will be classified as homozygotes, and the expected ratio of segregating to nonsegregating families is not 2:1 but 1.8874:1.1126."[10] Mendel's results (201 classified as homozygotes and 399 as heterozygotes) is closer to the uncorrected 2:1 expectation than to Fisher's corrected ratio. Fisher calculates from combined tests of the 2:1 ratio that "A deviation as fortunate as Mendel's is to be expected once in twenty-nine trials. Unfortunately the same thing occurs again with the trifactorial [three pairs of segregating alleles] data."[11]

Assuming that a valid hypothesis is being tested, a random sample of statistical tests of experimental data should display a broad range of probabilities (some experiments giving good agreement with expectations and sometimes not so good). However, one trial in twenty (5 percent) is expected, by chance alone, to have such a large deviation of observed from expected values that it calls for a rejection of the hypothesis (significant). Fisher found that a statistical test (called the chi-square test)[12] on the sums of various kinds of experiments (e.g., 3:1 phenotypic ratios, 1:1 gametic ratios, etc.) generally gave probabilities greater than 0.95 (the one notable exception was a probability of 0.74 for the sum of the tests for 2:1 ratios). Collectively, all of Mendel's results gave a probability of 0.99993. Fisher seems to have interpreted Mendel's data as being too good to represent a random sample of experimental results. "There can be no doubt that the data from the later years (after 1862) of the experiment have been biased strongly in the direction of agreement with expectation."[13]

Fisher suggests that Mendel may have had to rely on inexperienced help in recording his data. After all, Mendel was in a teaching position. He therefore may have had some of his students helping him to record the results of his numerous experiments. Or perhaps the monastery had a gardener who helped him with this task. In any event, whoever supposedly helped him was undoubtedly a novice in scientific work and therefore was unaware of the biasing pitfalls inherent in observation techniques. It may well be true that Mendel had an idea about the outcome of his experiments before he did them. Suppose that he told some of his novice help what kind of ratios he expected to see. Unless he carefully explained to them the necessity for accurate reporting (including recording of ambiguous phenotypes) it would not be surprising that there were no unambiguous results reported. If a helper ran into a dented pea, and if he suspected that the wrinkled tally was deficient with regard to the expected ratio, he might subconsciously be tempted to believe that such a pea belonged to the truly wrinkled class, and record it as such. It had to go into one of the two columns (if he was presented with only two phenotypic columns in which to record the data). So the results that Mendel obtained were usually very close to what he expected. Thus Fisher tried to exonerate Mendel from some guilt in the fabrication or adulteration of his seed data. "Such an explanation, however, gives no assistance in the case of the tests of gametic ratios and of other tests based on the classification of whole plants. . . . The bias seems to pervade the whole of the data."[14]

The eminent geneticist Sewall Wright suggests that Mendel must have known that not all of the ten seeds per plant that constituted the test of heterozygosity would germinate and grow to maturity.[15] So Mendel "may well have planted somewhat more than 10 in each case to be sure of having at least 10 to examine for segregants. If the average was 12, the probability of misclassifying Aa falls from.056 to.031."[16] What Wright does not say is that if Mendel did plant only ten seeds and one or more did not germinate, then the probability of misclassification increases markedly. It is highly unlikely that every seed planted grew to maturity. Mendel made no mention of germination success. If we take him at his word, he planted ten seeds. He did not report how many of them reached maturity to constitute the test. This part of his experimental design or reporting thereof could certainly have been improved, but it does not necessarily imply intellectual dishonesty. Wright goes on to say,

In the 91 percent of the data concerned with one factor ratios and their testing, the amount of bias toward expectation is slight and on a rough estimate would need less than two such misentries in one thousand to reduce [chi-square value] below expectations as much as observed. . . . Taking everything into account, I am confident, however, that there was no deliberate effort at falsification.[17]

Almost twenty years later, Monaghan and Corcos[18] performed chi-square tests on each of Mendel's experiments individually, not in groups of experiments as Fisher did. Individually the probabilities formed a rather broad range, with one as low as 10 to 20 percent. None were significant at less than 5 percent, however. Remember, 1 time in 20 by chance alone, a random sample of an infinite number of experiments should theoretically give a significant result. This is disconcerting to me, and I'm sure it is to most people who have tried to come to their own conclusions regarding Mendel's work because many more than twenty chi-square tests are possible of Mendel's reported data. Monaghan and Corcos propose that the earliest that Mendel could have been led by his experimental results to formulate a hypothesis of hybrid development was at the end of the second year. In comparing the chi-square probabilities from data derived in the first two years, with those in the remaining six years, they found no increase (i.e., no trend toward higher chi-square probabilities). Thus, they conclude that there is no bias. The idea being that if Mendel did not have a theory in mind until after the second year, he could not have planted the seeds of bias (pun intended!) in the minds of his helpers. All of this of course depends on when Mendel developed the model for how hybrids segregate. If he developed the model before he began his experiments, the bias may have existed from the inception of the project. Monaghan and Corcos also compared Mendel's results with several other workers and came to the conclusion that "if Mendel was biased, so were all the other workers."[19]

A much more sophisticated analysis by Franz Welling[20] may have finally pinpointed the problem. He proposes that "all chi-square statistics of genetic segregations fall short because the variance of genetic segregations is smaller and not of a binomial type as assumed (by Fisher and other critics). Furthermore, this variance and the corresponding chi-square statistics are not homogeneous in different segregation types. Consequently, it is not possible to summarize the different chi-square statistics as Fisher did."[21] Welling concludes his paper: "However, since Fisher's two decisive statements are, as we have seen, incorrect, it is to be hoped that this defamatory questioning of Mendel's accuracy henceforth will stand corrected."[22] To which I say, Amen!

It is ironic that none of the multitude who studied Mendel's paper before 1936 came to the same conclusion as Fisher did. But after Fisher's paper appeared, his conclusions apparently were unquestioned until the

1980s. Is it possible that no one felt competent enough to argue with one of the greatest (if not the greatest) statisticians of all time? There is nothing wrong with the mathematics of Fisher's chi-square tests. But there is definitely something wrong with the interpretation that he put on them. Fisher was using this statistical tool to answer the question: "What is the probability that the data represent a truly random sampling? Chi-square does not provide an answer to that question."[23]

EOANTHROPUS DAWSONI

According to Stephen Jay Gould, "The most famous and spectacular fraud of twentieth-century science" was the Piltdown forgery.[24] Piltdown is a town in England. In 1908 Charles Dawson claimed to have found some primate skull fragments in a gravel pit there along with fragments of other fossil mammals. Dawson was a successful lawyer and amateur archeologist. He had collected local fossils from boyhood. This collection became so impressive that by the age of twenty-one he was elected a fellow of the Geological Society of London, thereby bringing him into contact with such distinguished scientists as Sir Arthur Keith and Sir Arthur Smith Woodward. Dawson had visited the pit prior to his discovery, and found two farmhands digging in the iron-stained gravel used for road construction. Dawson learned that they had not come across any fossil bones, but he asked them to be on the lookout for them. On one of his return visits to the gravel pit, a laborer handed him an unusually thick portion of a human cranium. As soon as he found time, Dawson returned to the pit and spent an entire day digging in what appeared to him to be an unfossiliferous bed. Several years later, Dawson returned to the pit and found a second and larger piece of the same skull, and a portion of a hippopotamus tooth. The next successful dig at the pit occurred on June 3, 1912, when he brought with him Father Teilhard de Chardin and professor Smith Woodward. Dawson found another skull fragment and Teilhard discovered an elephant's tooth. Other visits were made, until the winter rains flooded the pit, resulting in finding three pieces of the right parietal bone of the skull that fitted together perfectly. Woodward found a fragment that fitted the broken edge of the occipital bone and gave a line of contact with the left parietal bone. Woodward described their most remarkable discovery: "Mr. Dawson was exploring some untouched remnants of the original gravel at the bottom of the pit, when we both saw the human lower jaw fly out in front of the pick-shaped end of the hammer which he was using."[25] The jaw was definitely apelike. So here they had a modern human skull and the jaw of an ape. If they were both parts of the same individual, the implications for evolutionary theory would be astounding.

The trio did not rush into publication of their findings. Woodward thought that any premature publicity would likely result in damaging public intrusion at the excavation site.[26] According to Ronald Millar in his book *The Piltdown Men*, the discoverers feared the public far less than their professional colleagues. They realized that any claim for such a discovery would be highly controversial and they feared unrehearsed rebuttals. In other words, they felt that the fewer people who knew of their discovery, the better the defense they could present for their position in the debate that was certain to follow. The first published report of Piltdown man appeared in the December 5, 1912, issue of *Nature*. Piltdown man was formally presented for public inspection at the office of the Geological Society in London on December 18, 1912. Woodward, with the help of F. O. Barlow (an expert plaster-cast maker at the British Museum), had reconstructed the skull from fragments. So it fell to Woodward to interpret the findings of those who had worked on the project. He described the general features of the skull as having the steep forehead of a modern human, with hardly any brow ridges. It seemed to him that the only external appearance of primitiveness was the position of the occiput, indicating that the attitude of the neck was like that of an ape. Woodward pointed out that another striking feature of the skull was its unusual thickness, about twice that of modern humans. Then turning to the jaw, he claimed that it agreed exactly with that of a young chimpanzee with the exception of the teeth. The jaw had only two molar teeth intact, and their crowns had "a marked regular flattening such as has never been observed among apes, though it is occasionally met with in lower types of men."[27]

Before I conclude the story, let me digress to set the historical stage on which Woodward's team made its discoveries and interpretation of the Piltdown fossils. It is always important, in the analysis of any scientific discovery, to understand the historical events that preceded it, and what the "Zeitgeist" (as the Germans call it) or "the spirit of the times" was. It is difficult for many people to accept that the idea of Earth being millions of years old was first supported scientifically in 1833 with the publication of the third and final volume of Sir Charles Lyell's *Principles of Geology*. (The best current estimates of the age of Earth are more like 4.5 billion years.) In 1593 James Ussher, the archbishop of Amagh, calculated from the life spans of the Old Testament patriarchs that Adam and Eve, along with the rest of the universe, were created in the year 4004 B.C.E. Of course the Church had always believed in a recent creation, but Ussher gave it the exact date. All of history and prehistory had to fit into this small time span. It was generally accepted that the great empires of Egypt, Assyria, Persia, Greece, and Rome had flourished before the rise of Christianity. But if this were true, these events would take up most if not all of the available four millennia. Where did, for example, Britain and her prehistory fit in?

The first recorded suggestion that primitive humans might be older than

Ussher's date was made by John Frere, F.R.S. (Fellow of the Royal Society), in a paper read to the Geological Society on June 22, 1797. He proposed that the flint instruments which had been recovered twelve feet below the earth's surface were "fabricated and used by a people who had not the use of metals. . . . The situation at which these weapons were found may tempt us to refer them to a very remote period indeed, even beyond that of the present world."[28] Fossils of extinct life-forms had been recognized as such before 1812, but in that year French naturalist Georges Cuvier published his manifesto that established the science of paleontology. Cuvier proposed that Earth had experienced a series of global catastrophes that destroyed all life-forms, only to be repopulated each time by divine creation with new forms. Noah's flood was thought to be the latest catastrophe. Cuvier argued from the fossil record, with its apparent episodes of extinctions and creations, that the age of Earth must be ancient. Even before Cuvier's contribution, Scottish geologist James Hutton had proposed a "doctrine of uniformitarianism" in his book *Theory of the Earth,* published in 1785. This doctrine proposed that the geological processes that were then seen at work in the world (volcanism, earthquakes, glaciation, erosion, etc.) have probably always been in operation since the world began with about the same tempo. Cuvier's worldwide catastrophes simply did not fit in with the uniformitarian concept. Gradually, however, the tide of opinion began to turn away from catastrophism and toward uniformitarianism. In his publication of 1833, Lyell made a strong scientific case for the acceptance of uniformitarianism in geology, and proposed that Earth must be millions of years old rather than only a few thousand as calculated by Ussher.

The first discovery of a prehistoric human fossil was made by Josseph Neander in 1856 in a valley (now called Neander Valley) seven miles east of Dusseldorf in what was then Rhenish Prussia (now Germany). Neanderthal "man" (it could have been a woman) consisted of a skull without facial bones, a clavicle, a scapula, two ulnae, five ribs, and a pelvis. This discovery was ignored by both the French and the English, as if they were not interested in "foreign" fossil men. Then the bombshell hit! Charles Darwin published *The Origin of Species* in 1859. His theory of evolution by means of natural selection provided new hope of finding fossil human forms quite different than anything then known. Neanderthal man fit into an evolutionary scenario very nicely, but not into the creation story. On April 9, 1863, a French newspaper in Abbevillois announced that a local amateur, Boucher de Parthes, had found an ancient human jaw together with a number of flint instruments at the Moulin Quignon gravel pit near Abbeville in northern France. Flint tools had been discovered widely in Europe prior to this announcement, but de Parthes claimed that his jaw belonged to the primitive manufacturer of the flint artifacts. Two well-known geologists/paleontologists, Sir John Evans and John Prestwich, had visited the Moulin Quignon gravel pit on several occasions, but they had found neither human remains nor flint artifacts. After learning of de Parthes' discovery, they hurried to France and received a cordial welcome from de Parthes. Ronald Millar describes the event:

> A pilgrimage was made to the Moulin Quignon gravel pit. . . . But the famous black bed from which the human remains had been recovered was now obscured by a fall of gravel. As the pair were about to leave, a workman who had offered his services as guide took two highly coloured flint implements from his pocket and handed them to Prestwich. He said that he had found the flints while working in the Moulin Quignon pit. They were crude, badly-shaped and smeared over with a deeply iron-stained clay. With intense excitement the Englishmen realized that the dark soil must have come from the black bed. They eagerly gave the workman the two francs he demanded. It was while washing the dark earth from the flints at the first cottage they reached that Evans and Prestwich realized that they had been hoodwinked. Not only the earth came away but most of the color as well.[29]

A month later, a Geological Society congress opened in Paris on May 9, 1863, and closed at Abbeville on May 13. A large number of flints and the Moulin Quignon jaw with its single molar were on display. Demonstrations on the possible origins of the flint instruments were given. Two fossils of undoubted antiquity were offered to serve as controls for the jaw; one was a molar of some unknown animal, and the other was a molar of an extinct cave-dwelling hyena. Chemical tests were performed that yielded ambiguous results. Then the jaw was sawed in half. A distinctive odor was detected, similar to that of limb amputations performed on live subjects, a product not to be expected of an ancient bone. In addition, the coloring could be scrubbed off, revealing a surface with little of the erosion common in old bones. It was concluded that the jawbone was probably no older than any other old bones that could be dug from a cemetery. According to Millar, "The weight of opinion at the time, however, seems in favour of the proposition that Boucher de Parthes was the innocent dupe of some unscrupulous workman who took advantage of the amateur's gullibility."[30]

By 1868, Neanderthal fossils were being discovered quite regularly throughout much of Europe. In that year Edward Lartet, chair of paleontology at the Paris National Museum of Natural History, made a journey to Cro-Magnon, near Les Eyzies in the Dardogne, to identify some skeletons found in a rock-shelter uncovered by a railway excavation. He concluded that the skeletons were of three men, one woman, and an unborn infant. Furthermore, they were neither fully modern, nor as deviant as Neanderthals. The best-preserved skull was estimated to have a cranial capacity of 1,500 cubic centimeters, well above the modern average of 1,350 cc. The consensus of anthropologists was that these fossils should be considered as a new race of *Homo sapiens*, to be known as Cro-Magnon man. Later Cro-Magnon fossils were found to be associated with the earliest cave paintings, a very marked cultural advance over the Neanderthals.

In *The Origin of Species*, Darwin had little to say about human evolution. He merely suggested that his theory of evolution would shed some light on the origin of humans. In 1870 *The Descent of Man and Selection in Relation to Sex* was published. In this book Darwin proposed that man had a long evolutionary history like that of any other living creature. Although the brain of modern man has allowed him to develop a sophisticated culture, his body still bears the "indelible stamp of his lowly origin."[31] His work was not well received because it implied that we had a brute origin like the rest of the animal world. It implied that we were not created specially in the image of God. It implied that the biblical story of creation was in error. The implications explode into areas that are sacred and cherished, because we want to believe that we are the supreme form of life on Earth, that everything was created for us to enjoy and subdue. Those beliefs give us license to treat other forms of life and the environment as we see fit to suit our needs alone. Who would want to give up this exalted view of our importance? At the time that Darwin wrote *The Descent of Man*, there was still very little fossil evidence to support his theories. He proposed modern humans had descended from the Old World (catarrhine) monkeys, giving hope to convinced scientists that diligence and patience would eventually result in discovery of a succession of evolving forms, each more primitive than the last. Ultimately, this chain of discoveries would lead to the most primitive form of all, the so-called missing link, an intermediate between humans and apes.

Another paleontological hoax occurred in Calivaras, California, in 1886. The gravels of this locality contained some fossils, primarily stumps of ancient palm trees. A recent flood from a nearby stream had washed a number of skeletons out of an Indian burial site. These bones were found by some local miners who decided to play a trick on the town physician who was an avid collector of fossils. The miners claimed to have dug a very ancient skull out of a mine shaft. The trusting physician, however, did not keep it to himself. He submitted it as authentic to J. D. Whitney, chief of the California Geological Survey. It was soon discovered that the matrix inside the skull did not resemble the Sierran gravels of the mining pit. The perpetrators of the hoax stuck to their story in an attempt to save face for all concerned. So it was not until 1911, after both Whitney and their physician friend had died, that the true story of the Digger Indian skull could be published.

Dutch paleontologist Eugene Dubois traveled to the Dutch East Indies looking for ancient humans in 1866. He was convinced that if humans had evolved from apes, and apes live in tropical climates, the most likely place to look for fossils of ancient man was in tropical regions. It wasn't until November of 1880 that he found anything of importance. At Kedung Brubas, forty kilometers east of Trinial in Java, he found a fragment of a lower jaw. By 1892 he had found a right upper third molar, a skull cap, a left upper second molar, and a left femur. He published his findings in 1894 and concluded that his fossils belonged to an ancient extinct giant gibbon which he named *Pithecanthropus erectus*. His conclusion was hotly debated because the cranial capacity of the skull was low enough to be that of a gibbon, but the femur (upper thigh bone) bore a close resemblance to that of modern man. His critics claimed that Dubois had mixed the bones of an animal with those of a human. By 1896 the experts seemed to be about equally divided between those who thought *Pithecanthropus* was a kind of anthropoid ape, those who claimed that it was human, and those who saw it as an intermediate form. Dubois was deeply affected by the controversy. He became progressively more eccentric and finally withdrew himself and his fossils from the debate entirely to the privacy of his house at Haarlem. It was not until 1920 that he could be persuaded to allow the fossils to be transferred to the Leiden Museum. Later studies of Dubois's fossils led anthropologists to the conclusion that their affinities were so close to that of modern humans that they belonged in the same genus. Hence the name was changed from *Pithecanthropus* to *Homo erectus* (man who walks upright).

A massive, chinless jaw had been found in a sandpit in Mauer near Heidelberg, Germany, that was so unlike anything yet discovered that it was named *Homo heidelbergensis*. The jaw, likened to that of an orangutan, was suggested to be an intermediate form between ape and man. Some paleoanthropologists, such as Professor Henry Osborn, considered Heidelberg man to be "a Neanderthaler in the making," that is, a more primitive, more powerful, and more apelike ancestral form. Some participants in the debate, who wanted to give the fossil a distinct generic ranking, referred to it as *Paleoanthropus heidelbergensis*. These and other finds of debatable hominid ancestors had led to the conclusion that human ancestry had been very complex, and that Neanderthal man, like modern man, had been divided into a number of races. At first it was thought that the Neanderthals might have evolved into *Homo sapiens*. Later opinion swung to favor the idea that Cro-Magnon man had migrated into western Europe, and by his superior intelligence drove the Neanderthals to extinction. Cro-Magnon man was the 'brainy' species *Homo sapiens*, whereas Neanderthals were assigned to a closely related species *Homo neanderthalensis*. Thus, the search was on to find the true "missing link" between the most primitive human species *Homo erectus* (thought to be about a million years old) and its possible ape ancestors (thought to be between ten and twenty million years old). The most important question was which evolved first, increased cranial capacity or the jaw. In other words, was the "missing link" an intelligent ape or a man with an ape-sized brain? The general opinion of the time strongly favored the former, that man's most distinguishing character, his intelligence, evolved first, and this had in some way served to guide the rest of his evolution.

This is the historical background on which the Piltdown hoax was played. Woodward had described the major features of the Piltdown man. He concluded that Piltdown man was the missing ancestral link between humans and apes. It had a high vaulted forehead and an apelike jaw that fit in very nicely with the then popular theory about human origins. He assigned Piltdown man to the species *Eoanthropus dawsoni* or the "Dawn man of Dawson." Many disagreed with the conclusion of Woodward and Dawson, suggesting that the jaw had been washed into the Piltdown gravel from an older geological deposit. The argument could not be resolved because the knob or condyle of the mandible (lower jaw) that articulates with the skull was missing. Charles Darwin had proposed in *The Descent of Man* that the progenitors of humans would likely have great canine teeth. The canine teeth were missing from the Piltdown specimen, so it could not confirm Darwin's prediction. The intact molars, however, were remarkably flat, as would be expected if during mastication the jaw was as free to move as that in modern man and not restricted to the ape's limited sideways motion. On August 30, 1913, Teilhard de Chardin had the unbelievably "good fortune" to find the missing canine tooth of the Piltdown man very close to where its jaw had been found. Its size, shape, and nature of wear was just what Woodward and Dawson had hoped for. Then in 1915 Dawson found a second Piltdown man at Sheffield Park, about two miles from the initial discovery. There were parts of a brain case and a molar tooth. It was not until February 17, 1917, six months after Dawson died, that Woodward publicly announced Dawson's second discovery. Most of the doubters of the first Piltdown man now became believers, because no coincidence of nature could have brought the bones of a man and an ape together at two different sites. Woodward continued to dig for fossils after Dawson's death in August of 1916, but without much success.

Meanwhile, other anthropological finds were being made that seemed to isolate *Eoanthropus dawsoni* without any certain affinities in the broad stream of human evolution. In 1921 a discovery was made in Rhodesia consisting of part of a hip bone, two ends of a femur, and a tibia. The skull was Neanderthal-like, but the limb bones lacked the curvature normally associated with the European Neanderthal specimens. The foramen magnum (the hole at the base of the skull through which the spinal cord enters the brain) was so far back that a slouching attitude was almost inescapable. Some critics thought that Rhodesian man was a composite of bones from two species (*Homo sapiens* and *Homo neanderthalensis*). Professor Raymond Dart discovered, in a limestone quarry of Taung in South Africa, the first specimens of what he named *Australopithecus africanus*. Dart thought he might have found the "missing link," but Woodward, Grafton Elliot Smith, and Sir Arthur Keith (among others) placed the fossils squarely in the same sub-family as the chimpanzee and the gorilla. Dart was disappointed at his failure to sway his critics, and eventually deposited the fossil with F. O. Barlow at the British Museum. Dart's interpretations were not confirmed until 1936 when other Australopithecines were independently found at Sterkfontein, Krondraai, and Swartkraans in South Africa.

Meanwhile, the skull of "Peking man" had been discovered in 1929 by W. G. Pei, near Choukoutien, China. It was considered to belong to an entirely different genus than any other hominid fossil, and was given the name *Sinanthropus pekinensis*. Its relationship to the genera *Homo* and *Eoanthropus* was fiercely debated. But before a unanimous consensus could be reached, all of the *Sinanthropus* fossils were lost in 1941 while being transferred from Peking to the SS *President Harrison*. Many more discov-eries of fossils from Java were made during 1936-1939 that were also assigned to the genus *Pithecanthropus*. Collectively, this mass of fossil mate-rial painted a picture of hominids evolving humanlike jaws and teeth early, and only gradually evolving larger brains later on. This was exactly opposite to the conclusions drawn from Piltdown man. The controversy stagnated. A French mineralogist, A. Carnot, reported in 1892 that the amount of fluorine in fossil bones tends to accumulate with their geological age. This informa-tion was mysteriously ignored until 1949, when Dr. Kenneth Oakley of the British Museum rediscovered it and realized that it might be used to deter-mine whether two bones found in a single deposit were of the same or dif-ferent age. The fluorine content of the Piltdown jaw and cranium was so low that they were probably less than 50,000 years old. A "dawn man" could not possibly be as recent as this, so the Piltdown man was soon out of the picture as an ancestor of modern man. Sir Arthur Smith Woodward died at the age of eighty, just a year before this bad news broke, which is just as well because according to Millar the results of the fluorine test would have killed him.

The recent age of the Piltdown fossils suggested that perhaps they were not authentic. In 1953 three anthologists—J. S. Weiner, W. E. le Gros Clark, and K. P. Oakley—decided to reexamine the Piltdown casts in Oxford Uni-versity's anatomy department with this possibility in mind. Because the jaw was claimed to be human, based primarily on the flat wear of the molar teeth, particularly close inspection was made of the teeth casts. It was then that they discovered the unmistakable marks of artificial abrasion. Le Gros Clark later wrote, "Indeed so obvious did they (the criss-cross scratches) seem it may well be asked—how was it that they had escaped notice before?"[32] Somebody had filed the teeth down to give them the appearance of wear seen typically in old human teeth. He then answered his own ques-tion thus, "they had never been looked for. . . . nobody previously had examined the Piltdown jaw with the idea of a possible forgery in mind, a deliberate fabrication."[33] Clark's conclusion is a reminder to us that the observations that scientists make are not necessarily completely objective or theory-free. Sometimes one sees what one expects or wants to see (as discussed in chapter 4, "Grand Illusions").

A full-scale examination of the Piltdown fossils themselves was then launched. What A. S. Underwood had identified in 1916 as an area of secondary dentine in a Piltdown tooth was, on reexamination, identified as an overabraded patch which had been plugged with some kind of plastic material "remarkably like chewing gum." New fluorine tests were performed because improvements in the technique were giving more accurate results than Oakley's original method. The jaw proved to be significantly younger than the cranium. Both the skull and the jaw had been artificially stained with potassium dichromate to give them the appearance of antiquity. The jaw was identified as belonging to an orangutan, an idea that was suggested as long ago as 1927. Thus the fraud was exposed in all its facets in 1953.

Two major questions need to be answered. First, why was it that such a misfit fossil was ever seriously considered suitable for scientific investigation by all of the greatest English paleontologists? And second, who was responsible for perpetrating this elaborate hoax?

Stephen Jay Gould suggests four possible reasons for Piltdown's acceptance as a genuine fossil.[34] First, international rivalry may have played a big part. England had no fossils of primitive man, whereas France had an abundance of Neanderthals and Cro-Magnons with their tools and art. One doesn't expect nationality to make any difference in the attitude of scientists toward the material that they study. Unfortunately, sometimes chauvinism can get in the way of the desired objectivity. British workers thought that the French were flaunting their wealth of scientific evidence and winced at the thought that English ancestors might have originated on French soil. Piltdown man was the great English hope for regaining the lead in anthropology. Smith Woodward proposed that "The Neanderthal race was a degenerate offshoot of early man while surviving modern man may have arisen directly from the primitive source of which the Piltdown skull provides the first discovered evidence."[35]

A second contributory factor was the a priori preference (largely cultural) of leading paleontologists for a "brainy ape" ancestry. Because the large brain of modern man and the opposable thumb are the main structures that make us anatomically superior to other primates, it was logical to expect that the brain must have evolved first and then the rest of the body followed later. Grafton Elliot Smith wrote in 1924, "Man at first . . . was merely an Ape with an overgrown brain. The importance of the Piltdown skull lies in the fact that it affords tangible confirmation of these inferences."[36] Piltdown fossils were found in gravels of an age equivalent to the strata in which Peking man was found. Peking man had a brain only about two-thirds the size of Piltdown man's brain. Thus, the modern white European descendants of Piltdown man had evolved a fully modern brain while other races of humans would have attained a fully modern brain only recently. The white races, having lived longer in the exalted state of mental superiority, would naturally be expected to excel in the arts of civilization.

A third reason for Piltdown's acceptance was that the anomaly of a fully modern cranium and a simian jaw had been reduced by matching facts to fit expectations. Those who favored a "brain first" evolutionary scenario never expected that the cranium could have developed completely before any changes had been made in the jaw. One thing that has never been explained to my satisfaction is the reported extraordinary thickness of the Piltdown skull; that certainly is not a modern attribute. Perhaps in the zeal of proving that a hoax had been perpetrated, we chose to ignore some facts that didn't really jibe with the result we wanted to obtain. Instead of modeling their theories to fit the facts, Piltdown's champions modeled the facts to fit their a priori theory that the modern cranium would show some simian features and the primitive jaw would show some modern features. For example, Smith Woodward originally estimated the cranial capacity of Piltdown man at 1,070 cubic centimeters. Estimating cranial capacity from fragmentary material may be subject to considerable error. Sir Arthur Keith later convinced him to revise his estimate nearer to the lower end of the modern cranial capacity range of 1,400-1,500 cc. This revision had the effect of making the skull seem more modern while still varying in the direction of simian features. Grafton Elliot Smith asserted that "We must regard this as being the most primitive and most simian human brain so far recorded; one, moreover, such as might reasonably have been expected to be associated in one and the same individual with the mandible which so definitely indicates the zoological rank of its original possessor."[37] Similarly, Keith tended to see human attributes in the orangutan jaw; e.g., he thought that the teeth were inserted into the jaw in a human fashion.

Finally, the keepers of the Piltdown fossils treated them more like art objects rather than subjects for hands-on scientific investigation. For example, when the famous anthropologist Louis Leakey went to see the fossils in 1972, he was allowed to look at them but was prohibited from handling them. Only the plaster casts could be touched. The casts were acknowledged to be very good, but the detection of fraud could not have been made without access to the original fossils. Artificial staining of the bones and filing of the teeth cannot be detected on plaster casts.

Now, turning to the question of who was responsible for the hoax, it is natural to try to first establish a motive. Who would benefit by seeding the Piltdown gravels with artifacts? Who would have the knowledge and cleverness to put together a forgery that would fool most of the best English paleontologists for decades? Who would have access to the bones that were altered and seeded into the Piltdown gravels? These are a few of the questions that, if answered, could help pinpoint the perpetrator of the hoax. Several suspects have been suggested with little or no substantial evidence. Of course Dawson has always been the chief suspect because he was the "discoverer" of the fossils, but a motive for Dawson's involvement is elusive. Dawson was a highly respected amateur anthropologist, not a professional one. It has been suggested that some of his professional counterparts were extremely jealous of his standing in the scientific community and wanted to humble him by a hoax that, if he bit on it, would send his reputation into the dumpster. Others have suggested that Dawson might have wanted to enhance his reputation by finding the "missing link" that everyone had been looking for. After years of searching for and not finding it, he might have decided to manufacture it. A third possibility is that Piltdown man was planned as a joke that got out of hand, rather than a malicious forgery. Gould suspects that Dawson and Teilhard de Chardin were coconspirators in such a hoax for the following reasons: "Dawson to expose the gullibility of pompous professionals; Teilhard to rub English noses once again with the taunt that their nation had no legitimate human fossils, while France reveled in a superabundance that made her the queen of anthropology."[38] Perhaps they had planned on revealing the hoax themselves one day, but the joke became a nightmare. Arthur Smith Woodward, Grafton Elliot Smith, and Arthur Keith, among other top British anthropologists, had "staked their careers on the reality of Piltdown."[39] By this time, it was too late to confess the truth. Dawson suddenly became ill and died in 1916, leaving Teilhard alone to carry the burden of the hoax.

Gould thinks that he has found a fatal slip in one of Teilhard's letters to Kenneth Oakley after Oakley had exposed the fraud. Teilhard tried to divert suspicion away from Dawson. But in describing the second Piltdown find, Teilhard writes: "He [Dawson] just brought me to the site of Locality 2 and explained me [sic] that he had found the isolated molar and the small pieces of skull in the heaps of rubble and pebbles raked at the surface of the field."[40] Other records indicate that Dawson did take Teilhard to the second site in 1913, and took Woodward there in 1914, but no fossils were found on either occasion. Woodward received a letter from Dawson dated January 20, 1915, containing the good news of discovery of two cranial fragments. In July 1915 Dawson wrote again to relate the discovery of a molar tooth. Dawson became sick later in 1915 and died the next year. Teilhard was called into military service as a stretcher-bearer in December 1914 and was at the front lines by January 22, 1915. If Dawson did not discover the molar until July 1915, how could Teilhard have known about it unless he was also involved in the hoax? Gould finds it difficult to believe that if Dawson had discovered it back in 1913, when Teilhard last was at the second site, he would have withheld that information from his old friend Woodward for two years. Teilhard had met Dawson in the spring or summer of 1909. Dawson found his first piece of skull at Piltdown in autumn 1909. In his letter to Oakley, Teilhard claims that he did not meet Dawson until 1911. Gould suggests that "the later (and incorrect) date, right upon the heels of Dawson's first 'find', certainly averts suspicion from Teilhard."[41]

In the closing paragraphs of his article "The Piltdown Conspiracy," Gould reminds us that the Piltdown hoax was not a trivial episode in the history of science.

> We cannot simply laugh and forget. Piltdown absorbed the professional attention of many fine scientists. It led millions of people astray for forty years. It cast a false light upon the basic process of human evolution. Careers are too short and time too precious to view so much waste with equanimity.[42]

INSTRUMENTS DON'T LIE—OR DO THEY?

The extent to which fraud exists in the scientific community remains unresolved. One of the most fundamental assumptions underlying all scientific work is that of intellectual honesty. Scientific discoveries of great importance are bound to be tested in the crucible of repeatability by independent investigators. If an experiment cannot be verified in this way, it will never be acknowledged as a valuable contribution to the advancement of knowledge. The self-monitoring system that is built into scientific methodology should deter anyone from attempting to perpetrate a fraud in the realm of science. But time and time again a few mavericks attempt to beat the system.

In chapter 10, "Déjà Vu," the scandal of the "patchwork mouse" is discussed. It became news on April 15, 1974. By mid-August 1974 another scandal was made public, this time at the Institute for Parapsychology in Durham, North Carolina. Dr. Joseph B. Rhine, the institute's founder, had pioneered in extrasensory perception (ESP) research in the 1930s. He had long been concerned with attempts to tighten up parapsychology's procedures and to eliminate ways by which experimental subjects could cheat or have any statistical advantage over random guessing.

As with all scientific work involving human subjects, every attempt should be made to prevent the conscious or unconscious biases of the investigator from influencing the experimental results. It is for this reason that "double-blind" trials are often part of the experimental design. For example, in the case of testing a drug for its effects in a defined group of people (either ill with a disease or healthy), doses of the drug would receive random code numbers (assigned by a source independent of the researcher or research team) and interspersed with randomly numbered placebos (a solution or pill—comparable in volume, color, taste, and in all other respects to the treatment, but without the drug). Neither the human subjects nor the investigator(s) would know what was in a given sample until after the results had been gathered. Then the seal on the coded information would be broken and the effectiveness of the drug could be ascertained statistically without bias.

Dr. Rhine advocated using instrumentation for gathering and recording data wherever possible in order to remove as much interaction as possible between the subjects and the investigator. Rhine hired a young medical student named Walter J. Levy to work at the institute during his summer vacation in 1969 because he seemed to exhibit an unusual talent for running experiments and recording data by instruments without human intervention. After Levy graduated from the Medical College of Georgia in 1973, he was hired by Rhine as a full-time researcher. It was reported that Levy had been selected to become the director of the institute upon Rhine's retirement.

Levy was attempting to determine if rats could anticipate random events by ESP and/or if they could effect physical changes by sheer willpower (psychokinesis). To this end, he implanted electrodes in the brains of rats in a zone where electrical stimulation would produce intensely pleasurable sensations. The stimuli were generated at random intervals by a computer that interfaced with a device that monitored the random decay of radioactive atoms in a sample of strontium-90. The system was designed to stimulate the rat's pleasure zone 50 percent of the time without outside influence. If the rats could anticipate the computer by ESP or influence the decay of the radioactive source by psychokinesis, then the pleasure score would exceed 50 percent.

I'm not a believer in ESP, but I am willing to be shown that it exists through controlled experiments. I don't know of any scientists who believe in ESP, but I suppose there must be some somewhere. What kind of grant proposal would be funded for ESP research when so many more worthwhile projects go begging? I find it difficult to believe that the government would get involved in a project with such a low status among scientists. However, ESP research was amply funded in the USSR during the cold war for its potential military/political uses. Former president Ronald Reagan's wife Nancy believes in astrology, along with millions of other people, but the vast majority of them are not scientists. Private funding for astrology, ESP, and other "pseudosciences" would most likely come from this segment of the public. If psychokinesis exists at all, it seems much more likely that it would be in humans or other primates than in the minds of rats. If I were serving on a committee to screen grant proposals, this kind of an experiment would not get my vote. And yet, if ESP exists it would be of the greatest importance to know how to tap into its potentialities. But it will never be discovered as long as people with funding priorities, like myself, sit on such committees. The private sector may be the only hope for such projects.

Early in May 1974, Levy's rats were reportedly obtaining pleasure scores markedly in excess of 50 percent, indicating that they had psychic powers. One of Levy's assistants became suspicious, however, when he observed Levy to be loitering needlessly around the recording equipment. The assistant enlisted the aid of two colleagues; one watched from concealment while the other two helped Levy run a test. Levy was observed to tamper with the recorder, causing a higher pleasure score to be recorded. By installing another set of instruments without Levy's knowledge, they confirmed their suspicions by recording the expected 50 percent score. The three "watchdogs" reported their observations to Rhine on June 12, 1974. When Levy was confronted with the evidence, he confessed and submitted his resignation. Levy insisted that the reported incident was the only time he had falsified the data, but Rhine was not going to take anything he said at face value. Other staff members were immediately assigned by Rhine to check on the more critical aspects of Levy's earlier data, and he cautioned other psychic researchers not to rely on these data until they could be independently verified. Attempts to repeat some of Levy's data were unsuccessful; the results all fell back to the level of chance. Other researchers were reluctant to continue with this line of investigation if it seemed unlikely to yield any positive results. Therefore, the test of independent repeatability is unlikely to be even attempted on the rest of Levy's work in the future. Unfortunately the stench of mendacity tends to become associated with his mentors and the institution in which it was perpetrated, and it raises suspicions about all of science and scientists in the public mind.

Levy claimed that he had been overburdened by administrative duties and had been under great pressure to produce positive results. Does this sound familiar? If not, it will after you read "The Patchwork Mouse" in chapter 10. An unscientific bias existed in Levy's mind concerning what he wanted to prove or what he thought the experiment would yield. When the results would not conform to his preconceived ideas, he felt compelled to force the data to reflect his expectations. Negative results usually don't qualify for extension grants to continue the project. No grant often means no job. Talk about pressure to fudge!

Earlier that same year, Rhine had published a soul-searching article in the March 1974 *Journal of Parapsychology* in which he addressed the need for eternal vigilance against fraud in this field. He even commented that "apparatus can sometimes be used as a screen to conceal the trickery it was intended to prevent."[43] Ironically, with his article barely in print, a classical example of fraud turned up in his own laboratory.

WHY BOTHER GATHERING DATA?

In the February 24, 1977, issue of *Nature*, Robert J. Gullis admitted that the data in four published papers bearing his name as either principal author or coauthor were simply "figments of my imagination."[44] Three of the papers dealt with levels of a chemical called cyclic guanosine monophosphate (cGMP), and the fourth with a related chemical called cyclic adenosine monophosphate (cAMP). These chemicals are proving to be extremely important in regulation of many diverse cellular activities. Hence, research involving these substances is of great concern to those who hope to learn how the cell goes about its normal functions and why some cells (e.g., cancerous cells) behave abnormally. The February 1977 report differs from the earlier cases of fraud discussed in this chapter in that both the accusation and the admission of guilt were published simultaneously. Gullis was a postdoctoral scientist working at the Max Plank Institute for Biochemistry in West Germany. He left the institute in September 1976, having spent two postdoctoral years there engaged mainly in ascertaining the concentration of cGMP in neuroblastoma cells and neuroblastoma/glioma hybrid cells. Bernd Hamprecht, a colleague at the same institute and a coauthor with Gullis on all four papers, said of him that

> Several of my colleagues (Brandt, Traber and van Calker) repeated this work but were unable to reproduce it. Dr. Gullis was therefore asked to return to our laboratory and repeat his essential experiments under our supervision. . . . In none of the experiments was Dr. Gullis able to obtain his previous results. Neither morphine nor levorphanol nor the enkephalins nor cholinergic antagonists changed the level of cyclic GMP in hybrid cells.[45]

Gullis confessed that "the curves and values published are mere figments of my imagination, and during my short research career I published my hypothesis rather than experimentally determined results. The reason was that I was so convinced of my idea that I simply put them down on paper; it was not because of the tremendous importance of published papers to the career of a scientist."[46] He concluded with an apology to the scientific community. His reference to the value of published papers to a scientist's career is certainly valid. A scientist "lives or dies" by what he submits in writing to the scrutiny of his peers. His entire reputation goes on the line each time he takes pen in hand for publication. Journals are the primary vehicles by which scientists communicate their observations and interpretations to their colleagues. It is also the primary means by which one's proficiency and expertise as a scientist can be demonstrated. One of the major motivations for many people to follow a scientific career is the potential recognition and prestige one can receive through publication in reputable refereed journals. Advancements in most academic careers are dependent upon the *number* of papers published because it is often difficult to evaluate the relative merits of different research projects in terms of sophistication of design, amount of time required, difficulty of obtaining data, pragmatic values, and the like. In these instances (as explained in "Publications" in chapter 7), some academicians use the adaptive strategy of fragmenting a research project for publication into two or more subtopics, thereby inflating the list of papers bearing their names as authors. Whenever advancement in rank (prestige) and/or monetary returns (salary, grants) is dependent upon producing "positive results" (be they immediately technologically useful discoveries or papers in areas of pure research) the researcher is subject to some very powerful pressures that, if succumbed to, could tend to compromise one's integrity as a scientist. As long as the environment in which scientists work contains these pressures, there will probably continue to be a small percentage of them that yield to the temptation to forge, cheat, steal, exaggerate, plagiarize, hoard, and in other ways behave in a spirit counterproductive to the advancement of scientific knowledge.

SUMMARY

In this all-too-brief survey, we have sampled the entire range of fraud from the case of Mendel, where it is not even sure that fraud was committed, to the Piltdown fiasco, where we are sure that a hoax was committed but the perpetrator(s) has not been unequivocally identified, to the Levi case where he admitted tampering with some of the data, to Gullis who confessed to fabrication of all his data. Daniel Koshland Jr., then editor of *Science*, says, "Sloppy experimentation and poor scholarship are condemned. Outright fraud is intolerable."[47] He further gives us an admonition: "You may escape detection by falsifying an insignificant finding, but there will be no reward. You may falsify an important finding, but then it will surely form the basis for subsequent experiments and become exposed."[48] Although "the scientific enterprise has grown a thousand-fold since the 1800s," Koshland does not see any proportionate increase in scientific fraud. Nonetheless, he does see some dangers in modern science that were not factors in earlier times. One is the nature of interdisciplinary research where experts in different fields pool their resources to solve a complex problem. When no one person has expertise in all of the fields, there can be dangers. A second problem arises when a project leader is saddled with too many projects and too little time to adequately monitor the work done by the members of his team. Finally, "A principal investigator must not only devise critical tests for his findings, but must also generate an atmosphere that encourages coworkers to report the bad news as well as the good news."[49]

According to Richard S. Nicholson,[50] a booklet published by the National Academy Press, *On Being A Scientist*,[51] "covers such topics as the treatment of data (Is fabrication a more serious offense than 'cooking' or 'trimming' data?), values in science, the risk of self-deception, and the primacy of discovery."[52] Nicholson suggests that this booklet could be used most effectively "as a basis for discussion and debate among graduate students and postdocs."[53] I believe, however, that the attitude of "honesty is the best policy" is learned at a very early age in the home. Attempting to change these attitudes at the late stage of graduate school may be difficult, so it must be vigorously pursued at all levels as one of the highest goals of a scientific education.

NOTES

1. James A. Peters, ed., *Classic Papers in Genetics* (Englewood Cliffs, N.J.: Prentice Hall, 1959), 14.
2. Ibid.
3. Ibid., 8.
4. R. A. Fisher, "Has Mendel's Work Been Rediscovered?" in *The Origin of Genetics: A Mendel Source Book*, ed. Curt Stem and Eva R. Sherwood (San Francisco: W. H. Freeman and Company, 1966), 139-72.
5. Ibid., 146.
6. Ibid., 150.
7. Ibid.
8. Ibid., 154.
9. Ibid.
10. Ibid., 154-55.
11. Ibid., 155.
12. The chi-square test is "a statistical procedure that enables the investigator to determine how closely an experimentally obtained set of values fits a given theoretical expectation." Robert C. King and William D. Stansfield, *A Dictionary of Genetics*, 5th ed. (New York: Oxford University Press, 1997), 59.
13. Fisher, "Has Mendel's Work Been Rediscovered?" 162.
14. Ibid.
15. Sewall Wright, "Mendel's Ratios," in Stem and Sherwood, *The Origin of Genetics*, 173-75.
16. Ibid., 174-75.
17. Ibid., 175.
18. Floyd Monaghan and Alan Corcos, "Chi-square and Mendel's Experiments: Where's the Bias?" *Journal of Heredity* 76 (July-August 1985): 307-309.
19. Floyd Monaghan and Alan Corcos, "Correction: Chi-square and Mendel's Experiments," *Journal of Heredity* 77 (July-August 1986): 283.
20. Franz Welling, "What About R. A. Fisher's Statement of the 'Too Good' Data of J. G. Mendel's Pisum Paper?" *Journal of Heredity* 77 (July-August 1986): 281-83.

21. Ibid., 281.

22. Ibid., 283.

23. Ira Pilgrim, "A Solution to the Too-Good-to-Be-True Paradox and Gregor Mendel," *Journal of Heredity* 77 (May–June 1986): 218.

24. Stephen Jay Gould, *Hen's Teeth and Horse's Toes* (New York: W. W. Norton & Company, 1983), 202.

25. Ronald Millar, *The Piltdown Men* (London: Victor Gollancz, 1972), 123.

26. Ibid., 124.

27. Ibid., 129.

28. Ibid., 19.

29. Ibid., 69.

30. Ibid., 74.

31. Ibid., 45.

32. Ibid., 207.

33. Ibid.

34. Stephen Jay Gould, *The Panda's Thumb* (New York: W. W. Norton & Company, 1980), 115–19.

35. Ibid., 116.

36. Ibid., 117.

37. Ibid., 118.

38. Ibid., 114.

39. Ibid.

40. Gould, *Hen's Teeth and Horse's Toes*, 209.

41. Ibid., 212.

42. Ibid., 225–26.

43. J. B. Rhine, "Security Versus Deception in Parapsychology," *Journal of Parapsychology* 38 (March 1974): 110.

44. R. J. Gullis, "Statement," *Nature* 764 (February 24, 1977), 764.

45. B. Hamprecht, "Statement," *Nature* 764 (February 24, 1977), 764.

46. Ibid.

47. Daniel E. Koshland Jr., "Fraud in Science," *Science* 235 (January 9, 1987): 141.

48. Ibid.

49. Ibid.

50. Richard S. Nicholson, "On Being a Scientist," *Science* 246 (October 20, 1989): 305.

51. Committee on the Conduct of Science, *On Being a Scientist* (Washington, D.C.: National Academy of Sciences Press, 1989).

52. Richard S. Nicholson, "On Being a Scientist," 305.

53. Ibid.

CHAPTER 12

WHAT A TRIP!

In this romp through the history of science, we have gone back to the first great scientific revolution—the proposition that Earth goes around the Sun and not vice versa (chapter 5). We have also investigated the origin of the polymerase chain reaction that catalyzed the current biotechnology revolution. Between these two chronological extremes, we have freely jumped around like a time traveler in our pursuit of at least a partial answer (through analysis of selected historical case studies) to some questions of fundamental importance for the development of a well-rounded understanding and appreciation of science.

CONNECTIONS

No attempt has previously been made to tie together events or concepts from one chapter to another. Now, however, we can reflect on a few of the possible connections that can be made. For example, ethics and morality were prime concerns in chapter 1 (human and animal experimentation) and chapter 11 (fraud). Although the subject of animal rights was not an issue at the time Galvani (chapter 6) did his frog experiments, it would be today. Morality and religious beliefs are inextricably intertwined. At any given time, actions that might be considered ethical by one social community might be considered unethical by another. Furthermore, even within the same community, what is considered moral at one time may be considered amoral at another time (or vice versa). So the grave robbing/disturbance by some modern physical anthropologists/archaeologists (chapter 7, "Bones of Contention") also become moral and religious issues for Native Americans, Australian Aborigines, and other native groups. Religious beliefs (chapter 5) are also at the heart of current debates between "creation scientists" and evolutionists. Darwin's theory of evolution was incomplete, mainly because an understanding of the biological basis of heredity was then unknown. Mendel (chapter 11) is often credited as having proposed a corpuscular theory of heredity that eventually became the basis of a new scientific discipline called genetics. Mendel experimented with peas, and so did Lysenko (chapter 5). The Lysenko affair involved basic philosophical differences between Mendelians and anti-Mendelians. Different philosophies of science were discussed at length in Wenner's *The Anatomy of a Scientific Discovery* (chapter 4). The mindset adopted by a system of belief may predispose a person (scientist or not) to "see" things that do not exist (chapter 4). The Lysenko affair occurred primarily because of ideological belief in dialectical materialism—at the heart of Communism in the former USSR. Lysenko's success in getting his scientifically unorthodox theories accepted was due to the dictatorial power wielded by Stalin. Politics cannot be separated from science in any modern society, even our own, largely because so much scientific research is dependent upon government financing. Availability of monetary support is just one "stone in the road" (chapter 7) that prevents science from advancing as rapidly as we might wish it would. Vast amounts of money from several nations have been poured into the conduct of ever more powerful nuclear-fusion experiments in the hope of producing a virtually limitless supply of energy for future generations. That is why the announcement of a successful inexpensive electrochemical fusion experiment by Pons and Fleishman in 1989 was such a shock to the scientific world (chapter 4). All nuclear power plants currently in use produce energy by controlled nuclear fission (not fusion). The first insights into the radioactivity (gamma rays) produced by uncontrolled (natural) fission of unstable elements was discovered by accident (chapter 6). Artificially produced X rays are the energy equivalents of gamma rays, and were used by Franklin to analyze the structure of DNA crystals. It was her X-ray diffraction photographs that Watson and Crick used to decipher and confirm their model of DNA structure (chapter 2). Knowledge of this structure stimulated innumerable research projects that eventually led to recombinant DNA techniques whereby genes can be transplanted from one organism to another, even between members of different species. This led to questions concerning the safety and ethics of such experiments (back to chapter 1 again). It should be obvious by now that the chapters of *Death of a Rat* are interconnected in many different and interesting ways, some more directly than others. If the contents of each chapter are conceptualized as pieces of a jigsaw puzzle, then, as each piece is put in place, a picture emerges. That picture is a deep and abiding understanding and appreciation of science.

THE FUTURE

It is indeed presumptuous of anyone to try and predict what is likely to occur in rapidly evolving fields like science and technology, even for the near future. However, at the end of the twentieth century (the twenty-first century does not begin until January 1, 2001) in a book such as this, it is expected that the author will at least take a stab at prognostication, knowing full well that he is climbing out on a limb that can be sawed off at any time. So, here goes!

Any accurate history of science will not only relate the positive things that have come from science, but must also expose its faults, errors, and frailties. Only this approach can lead to a genuine understanding and appreciation of science. The take-home lesson of Anthony Standen's book *Science Is a Sacred Cow*[1] is that science should be respected for the good it can do, but should not be worshiped as if it were faultless. "To a not inconsiderable segment of the public, the word 'science' conjures up images of Chernobyl, Bhophal, Thalidomide, Challenger, and the atomic bomb. Too often, science is perceived as the cause of problems rather than the solution. . . . For every scientific (or engineering) action, there is an equal and opposite social reaction."[2] The knee-jerk reactions of antiscience/technology activists has probably been more pesky than a major problem for most of science. However, in the latter half of the twentieth century, fundamental antievolutionists have been a persistent and growing threat to science education. As pointed out in chapter 5, their attacks on evolution are also attacks on all branches of science. Their failure to receive federal or state judicial support of equal time for "creation science" in science classrooms has not deterred them. They have infiltrated local school boards at the grass-roots level and influenced curriculum development and textbook selection that fosters their anti-science agenda. This strategy is likely to continue to be a constant threat to quality science education in the twenty-first century. Concerned parents and educators will need to use every means available to combat this menace. Publications of the National Center for Science Education (NCSE) will continue to be a major source of information to parents and teachers in their fight against anti-science in all of its guises.

It seems obvious to me that the molecular biology revolution, begun in 1953 with the double helical model of DNA structure, will continue expanding in the twenty-first century. Sequencing of the entire human genome (structural genomics) is expected to be completed by the year 2005 or even earlier. It is likely to take decades more to discover all of (or even a significantly large number of) the proteins specified by the human genome, to identify the function(s) of each protein, and to correlate mutant proteins with their phenotypic/pathological effects (functional genomics). As this knowledge becomes available, however, much human misery is expected to be alleviated or prevented. At the same time, similar long-range research programs will decipher the functional genomics of livestock and agriculturally important crops (especially rice, wheat, corn, soybeans, and potatoes). Genetic engineering based on this knowledge is expected to lead to greater agricultural plant productivity through such mechanisms as increased resistance to disease, pests, and herbicides. Fear of consuming genetically engineered foods likely will gradually wane as did the fears that sparked the recombinant DNA debate in the 1970s. As the functional genomics of pathogenic microorganisms are deciphered, science will discover new targets of vulnerability for treatments by both natural substances and synthetic ("designer") drugs.

Despite all of the scientific advances that may occur in the first half of the twenty-first century, growth in world population size will more likely lead to increased epidemics (due to overcrowding, ineffectiveness of most current antibiotics, and emerging diseases), increased environmental pollution, and increased shortages of food, water, fuel, lumber, and other natural resources. Until this population problem abates, I doubt that scientific advances on all fronts are likely to significantly improve the human condition in most parts of the world. Unless some unforeseeable technological breakthroughs occur, "Few physicists expect fusion to be a viable energy source before the middle of the [twenty-first] century."[3] Many (perhaps most) scientists and nonscientists do not share this "gloomy Gus" outlook (at least not publicly). But skeptics who thought that a moon voyage was impossible eventually had to eat humble pie. Let us all hope that the dire predictions of *The Population Bomb*[4] will also be wrong. Global sociological changes will be needed to halt population growth. This, in my opinion, is not likely to happen voluntarily. Population growth certainly will be checked if the four horsemen of the apocalypse (famine, disease, pestilence, and war) are allowed full gallop. Meanwhile, support for science should be given top priority, for it is the only realistic alternative to delay the horsemen from riding with free reins.

NOTES

1. Anthony Standen, *Science Is a Sacred Cow* (New York: E. P Dutton & Co., 1958).

2. Norman Augstine, "What We Don't Know Does Hurt Us. How Scientific Illiteracy Hobbles Society," *Science* 279 (March 13, 1998): 1640–41.

3. James Riordon, "Common Ground for Fusion," *Science* 285 (August 6, 1999): 821.

4. Paul R. Ehrlich, *The Population Bomb* (New York: Ballantine Books, 1968).

BIBLIOGRAPHY

Abelson, Philip H. "Excessive Zeal to Publish." *Science* 218 (December 3, 1982).

Augstine, Norman. "What We Don't Know Does Hurt Us: How Scientific Illiteracy Hobbles Society." *Science* 279 (March 13, 1998).

Baringa, Marcia. "Biotech Nightmare: Does Cetus Own PCR?" *Science* 251 (1991).

Baskin, Yvonne. "DNA Unlimited." *Discover* 11 (July 1990).

Beecher, Henry K. *Research and the Individual.* Boston: Little, Brown and Company, 1970.

Beil, L. "Dilutions or Delusions?" *Science News* 134 (July 2, 1988).

———. "Nature Douses Dilution Experiment." *Science News* 134 (July 30, 1988).

Berg, Paul, et al. Letter to the editor, "Potential Biohazards of Recombinant DNA Molecules." *Science* 185 (July 26, 1974).

Berra, Tim M. *Evolution and the Myth of Creationism.* Stanford, Calif.: Stanford University Press, 1990.

Blaser, Martin J. "The Bacteria Behind Ulcers." *Scientific American* 274, no. 2 (February 1996).

Burke, James. *The Day the Universe Changed.* Boston: Little, Brown and Company, 1985.

Carlson, Peter. "A Seattle Man Vents His Spleen Against Those Who Would Use It for Profit." *People* 24 (September 23, 1985).

Carpenter, Kenneth J. *The History of Scurvy and Vitamin C.* Cambridge, Mass.: Harvard University Press, 1986.

Cherfas, Jeremy. "Two Bomb Attacks on Scientists in the U.K." *Science* 248 (June 22, 1990).

Cohen, I. Bernard, ed. *The Career of William Beaumont and the Reception of His Discovery*. New York: Arno Press, 1980.

Committee on the Conduct of Science. *On Being a Scientist*. Washington, D.C.: National Academy of Sciences Press, 1989.

Corcoran, Elizabeth. "Patent Medicine: New Patients Challenge Congress and Courts." *Scientific American* 259 (September 1988).

Crawford, Mark. "Court Rules Cells Are the Patient's Property." *Science* 241 (August 5, 1988).

Crease, R. P. "Righting the Antibiotic Record." *Science* 246 (November 17, 1989).

Crow, James F. "Eighty Years Ago: The Beginnings of Population Genetics." *Genetics* 119 (July 1988).

Culliton, Barbara J. "Harvard Tackles the Rush to Publication." *Science* 241 (July 29, 1988).

———. "Inside the Gallo Probe." *Science* 248 (June 22, 1990).

———. "Mo Cell Case Has Its First Court Hearing." *Science* 226 (November 16, 1984).

———. "White House Wants Fetal Research Ban." *Science* 242 (September 16, 1988).

del Regato, J. A. "Letters: Yellow Fever Research." *Science* 224 (June 8, 1984).

Dickerson, David. "Europe Split on Embryo Research." *Science* 242 (November 25, 1988).

Ehrlich, Paul R. *The Population Bomb*. New York: Ballantine Books, 1968.

Epstein, Sam, and Beryl Epstein. *Dr. Beaumont and the Man with the Hole in His Stomach*. New York: Arno Press, 1980.

Fisher, R. A. "Has Mendel's Work Been Rediscovered?" In *The Origin of Genetics: A Mendel Source Book,* ed. Curt Stern and Eva R. Sherwood. San Francisco: W. H. Freeman and Company, 1966.

Franklin, Jon, and John Sutherland. *Guinea Pig Doctors: The Drama of Medical Research Through Self-Experimentation*. New York: William Morrow and Company, Inc., 1984.

———. "Letters: Yellow Fever Research." Science 224 (June 8, 1984).

Fricke, Hans. "Coelacanths: The Fish That Time Forgot." *National Geographic* 1173 (June 1988).

Gallo, Robert C., et al. "Frequent Detection and Isolation of Cytopathic Retroviruses (HTLV-III) from Patients with AIDS and at Risk for AIDS." *Science* 224 (May 4, 1984).

———. "Isolation of Human T-Cell Leukemia Virus in Acquired Immune Deficiency Syndrome (AIDS)." *Science* 220 (May 20, 1993).

Gardner, Eldon J. *History of Biology*, 2d ed. Logan, Utah: Burgess Publishing Company, 1965.

Gelmann, Edward P., et al. "Proviral DNA of a Retrovirus, Human T-Cell Leukemia Virus, in Two Patients with AIDS." Science 220 (May 20, 1993).

Gingerich, Owen. "The Galileo Affair." *Scientific American* 247 (August 1982).

Gold, Michael A. *A Conspiracy of Cells.* Albany: State University of New York Press, 1986.

Gould, James L. "The Dance-Language Controversy." *Quarterly Review of Biology* 51, no. 2 (June 1976).

———. *Hen's Teeth and Horse's Toes.* New York: W. W. Norton & Company, 1983.

———. "Honey Bee Communication." *Nature* 252 (November 22, 1974).

———. *The Panda's Thumb.* New York: W. W. Norton & Company, 1980.

Grobstein, Clifford. *A Double Image of the Double Helix: The Recombinant DNA Debate.* San Francisco: W. H. Freeman and Company, 1979.

Gullis, R. J. "Statement." *Nature* 764 (February 24, 1977).

Hallock, Grace T., and C. E. Turner. *Health Heroes: Walter Reed.* New York: Metropolitan Life Insurance Company, n.d.

Hamprecht, B. "Statement." *Nature* 764 (February 24, 1977).

Hellman, A., M. N. Oxman, and R. Pollack, eds. *Biohazards in Biological Research: Proceedings of a Conference Held January 22-24, 1973.* Cold Spring Harbor, N.Y.: Cold Spring Harbor Laboratory, 1973.

Hopson, Janet L. "Fins to Feet to Fanclubs: An (Old) Fish Story." *Science News* 109 (January 10, 1976).

Huber, Peter. *Liability: The Legal Revolution and Its Consequences.* New York: Basic Books, 1988.

———. "Litigation Thwarts Innovation in the U.S." *Scientific American* 260 (March 1989).

Huffman, John C. "Scientific Suppression?—letter #2." *Science News* 134 (July 23, 1988).

Johanson, Donald C., and Maitland Edey. *Lucy: The Beginnings of Humankind.* New York: Simon and Schuster, 1981.

Kaplan, John. "The Use of Animals in Research." *Science* 242 (November 11, 1988).

King, Robert C., and William D. Stansfield. *A Dictionary of Genetics*, 5th ed. New York: Oxford University Press, 1997.

Kleppe, K., et al. "Studies on Polynucleotides XCVI. Repair Replication of Short Synthetic DNAs as Catalyzed by DNA Polymerases." *Journal of Molecular Biology* 56 (1971).

Klotz, Irving M. "The N-Ray Affair." *Scientific American* 242 (May 1980).

Koestler, Arthur. *The Case of the Midwife Toad.* New York: Random House, 1971.

Koshland, Daniel E., Jr. "Fraud in Science." *Science* 235 (January 9, 1987).

Kuhn, Thomas S. "Logic of Discovery or Psychology of Research." In *Criti-*

cism and the Growth of Knowledge, ed. I. Lakatos and Musgrave. Cambridge: Cambridge University Press, 1970.

——. *The Structure of Scientific Revolutions.* Chicago: University of Chicago Press, 1962.

Landman, Otto E. "The Inheritance of Acquired Characteristics." *Annual Reviews of Genetics* 25 (1991).

Latour, Bruno. *Science in Action: How to Follow Scientist and Engineers Through Society.* Cambridge: Harvard University Press, 1987.

Lawler, Andrew. "Key NASA Lab Under Fire for Animal Care Practices." *Science* 268 (June 23, 1995).

Lewin, Roger. "Ethiopia Halts Prehistory Research." *Science* 219 (January 14, 1983).

——. "Extinction Threatens Australian Anthropology." *Science* 225 (July 27, 1984).

——. "Genome Projects Ready to Go." *Science* 244 (April 29, 1989).

Luria, S. E. *A Slot Machine, A Broken Test Tube: An Autobiography.* New York: Harper & Row, 1984.

Maddox, John. "Maddox on the 'Benveniste Affair.'" *Science* 241 (September 23, 1988).

Marshall, Eliot. "Smithsonian, Indian Leaders Call a Truce." *Science* 245 (September 15, 1989).

Medawar, Peter B. *Advice to a Young Scientist.* New York: Harper and Row, 1981.

Medvedev, Zhores A. *The Rise and Fall of T. D. Lysenko.* New York: Doubleday & Company, 1971.

Millar, Ronald. *The Piltdown Men.* London: Victor Gollancz, 1972.

Monaghan, Floyd V., and Alan F. Corcos. "Chi-Square and Mendel's Experiments: Where's the Bias?" *Journal of Heredity* 76 (July–August 1985).

——. "Correction: Chi-Square and Mendel's Experiments." *Journal of Heredity* 77 (July–August 1985).

——. "Mendel the Empiricist." *Journal of Heredity* 76 (January–February 1985).

——. "Possible Influences of Some Nineteenth-Century Chemical Concepts on Mendel's Ideas About Heredity." *Journal of Heredity* 74 (July–August 1983).

Monath, T. P. "Glad Tidings from Yellow Fever Research." *Science* 229 (August 23, 1985).

Mullis, Kary B., "The Unusual Origin of the Polymerase Chain Reaction." *Scientific American* 262 (April 1990).

Murray, Mary. "A Long-Disputed Paper Goes to Press." *Science News* 131 (January 24, 1987).

Nicholson, Richard S. "On Being a Scientist." *Science* 246 (October 20, 1989).

Norman, Colin. "Letters: Yellow Fever Research." *Science* 224 (June 8, 1984).

———. "The Unsung Hero of Yellow Fever." *Science* 223 (March 20, 1984).

Palca, Joseph. "The True Source of HIV." *Science* 252 (May 10, 1991).

Papovic, M., M. G. Sarngadharan, and Elizabeth Reed. "Detection, Isolation, and Continuous Production of Cytopathic Retroviruses from Patients with AIDS and Pre-AIDS." *Science* 224 (May 4, 1984).

Pappworth, Maurice H. *Human Guinea Pigs: Experimentation on Man.* London: Routledge & Kegan Paul, 1967.

Park, Davis & Company. "The Era of Antibiotics." Narrative accompanying *A Pictorial History of Medicine,* 1964.

———. "James Lind: Conquest of Scurvy." *A History of Medicine in Pictures* (1959).

Peat, F. David. *Cold Fusion: The Making of a Scientific Controversy.* Chicago: Contemporary Books, 1989.

Peters, James A. *Classic Papers in Genetics.* Englewood Cliffs, N.J.: Prentice-Hall, 1959.

Peterson, I. "Publication Bias: Looking for Missing Data." *Science News* 135 (January 7, 1989).

Pilgrim, Ira. "A Solution to the Too-Good-to-Be-True Paradox and Gregor Mendel." *Journal of Heredity* 77 (May–June 1986).

Placa, Joseph. "Fetal Tissue Transplants Remain Off Limits." *Science* 246 (November 10, 1989).

Pool, R. "More Squabbling Over Unbelievable Result." *Science* 241 (August 5, 1988).

———. "Unbelievable Results Spark a Controversy." *Science* 241 (July 22, 1988).

Revkin, Andrew C. "Dilutions of Grandeur." *Discover* 10 (January 1989).

Rhine, J. B. "Security Versus Deception in Parapsychology." *Journal of Parapsychology* 38 (March 1974).

Rider, Caroline V. "Scientific Suppression?—letter #1." *Science News* 134 (July 23, 1988).

Riordon, James. "Common Ground for Fusion." *Science* 285 (August 6, 1999).

Roberts, Royston M. *Serendipity: Accidental Discoveries in Science.* New York: John Wiley & Sons, 1989.

Rogers, Michael. "The Follies of Science." *Newsweek* 113 (May 8, 1989).

Rubinstein, Ellis. "The Untold Story of HUT-78." *Science* 248 (June 22, 1990).

Ruggere, Christine A. Letter to the editor. *Science* 286, no. 901 (October 29, 1999).

Sarngadharan, M. G., et al. "Antibodies Reactive with Human T-Lympotropic Retroviruses (HTLV-III) in the Serum of Patients with AIDS." *Science* 224 (May 4, 1984).

Schüpbach, Jörg, et al. "Serological Analysis of a Subgroup of Human T-Lymphotropic Retroviruses (HTLV-III) Associated with AIDS." *Science* 224 (May 4, 1984).

Singer, Maxine, and Dieter Soll. Letter to the editor. *Science* 181 (September 23, 1973).

Sinoussi, F. Barré, et al. "Isolation of a T-Lymphotropic Retrovirus from a Patient at Risk for Acquired Immune Deficiency Syndrome (AIDS)." *Science* 220 (May 20, 1993).

Standen, Anthony. *Science Is a Sacred Cow*. New York: E. P. Dutton & Co., 1958.

Stansfield, William D. *The Science of Evolution*. New York: Macmillan Publishing Company, 1977.

Stem, Curt, and E. R. Sherwood, eds. *The Origin of Genetics; A Mendel Source Book*. San Francisco: W. H. Freeeman and Company, 1966.

Stewart, Walter, and Ned Feder. "The Integrity of the Scientific Literature." *Nature* 325 (January 1, 1987).

Sun, Marjorie. "UC Told to Review Impact of Research." *Science* 238 (November 27, 1987).

———. "Weighing the Social Costs of Innovation." *Science* 223 (March 30, 1984).

Swan, H. T. "The Antibiotic Record." *Science* 247 (March 23, 1990).

Tangley, I. "Mouse Mixup May Alter Research Results." *Science News* 122 (July 24, 1982).

Templeton, Nancy Smyth. "The Polymerase Chain Reaction: History, Methods, and Applications." *Diagnostic Molecular Pathology* 1, no. 1 (1992).

Thomson, Keith S. "Marginalia: A Fishy Story." *American Scientist* 74 (March–April 1986).

———. "Marginalia: The Second Coelacanth." *American Scientist* 77 (November–December 1989).

Thrower, Norman J., ed. *Standing on the Shoulders of Giants*. Berkeley: University of California Press, 1990.

Valdes, Jose J. "The Kammerer Case." *Science News* 107 (June 21, 1975).

Veldink, Connie. "The Honey Bee Language Controversy." *Interdisciplinary Science Reviews* 14, no. 2 (1989).

Villiers, Alan. "That Extraordinary Sea Genius, Captain James Cook." *Nutrition Today* 4, no. 3 (Autumn 1969).

von Frisch, Karl. *The Dance Language and Orientation of Bees*. Cambridge: Harvard University Press, 1967.

Waugh, Dexter. "Sickle Cell Contributions Wane; AIDS Focus Blamed." *San Francisco Examiner*, December 3, 1988.

Weiss, Rick. "Forbidding Fruits of Fetal-Cell Research." *Science News* 134 (November 5, 1988).

Welling, Franz. "What About R. A. Fisher's Statement of the 'Too Good' Data of J. G. Mendel's Pisum Paper?" *Journal of Heredity* 77 (July–August 1986).

Wells, Patrick H., and Adrian M. Wenner. "Do Honeybees Have a Language?" *Nature* 241 (January 19, 1973).

Wenner, Adrian M. *The Bee Language Controversy: An Experience in Science*. Boulder, Colo.: Educational Programs Improvement Corporation, 1971.

———. "Sound Production in the Waggle Dance of the Honey Bee." *Animal Behavior* 10, nos. 1 and 2 (January–April, 1962).

Wenner, Adrian M., and Patrick H. Wells. *Anatomy of a Controversy: The Question of a "Language" Among Bees*. New York: Columbia University Press, 1990.

Winkler, Karen J. "Histories Fail to Explain Science to the Laymen, Scholar Says." *Chronicle of Higher Education* 30, no. 23 (August 7, 1985).

Wright, Sewall. "Mendel's Ratios." In *The Origin of Genetics,* ed. Curt Stem and Eva R. Sherwood. San Francisco: W. H. Freeman and Company, 1966.

Wu, C. H. "Electric Fish and the Discovery of Animal Electricity." *American Scientist* 72 (September–December 1984).

INDEX

Leyden jar, 176–78
life expectancy, 12
Lysenko, Trofim Denisovich, 138, 142–52, 209

Marx, Karl, 139
medical treatments, experimental, 16
Mendel, Gregor, 141–42, 255–66, 318–26
mice. *See* animal experimentation, mice
midwife toad, 300–304
molecular genetics, 43–48
morality and science, 12, 13, 16
  self-experimentation, 28
Mullis, Kary, 249–55

National Aeronautics and Space Administration (NASA), 58
National Institutes of Health (NIH), 44–48, 204, 235
  guidelines, 47–48
Nazi Germany, 18, 48
Neanderthal fossils, 328–30
Nelson-Rees, Walter A., 197–203
nitrous oxide, 179–80
Nobel, Alfred Bernhard, 183–86
Nobel Prizes, 185–186
N rays, 111–16
nuclear fission, 189–90
nuclear power, 47, 118
nucleic acids, structure of, 71, 83

Origin of Species, The (Darwin), 328, 330

Pappworth, H. Maurice, 15–20
Pauling, Linus, 75, 82, 84
penicillin, 168–74
People for the Ethical Treatment of Animals (PETA), 50, 56
philosophies of science, 346

Piltdown forgery, 326–27, 332–37
polymerase chain reaction (PCR), 248–55
population control, 42–43, 238–39, 348
pox. *See* syphilis
predecessors, debt to, 271–72
principle of equity, 19
prisoners
  of war, abuse of, 18
  use in research, 17
publication of research, 221–22, 226–27, 232–36

quality of life, 12–13
quantitative genetics, 222–24

rabbits. *See* animal experimentation, rabbits
radioactivity
  artificial, 189
  natural, 188–89
rats. *See* animal experimentation, rats
recombinant DNA, 43–48, 241, 248–54
Reed, Walter, 280–87
religion as impediment to science, 129–38, 155–63
  Catholic Church, 129–38
  fundamentalists, 155–63, 193
research, scientific
  ends of, 15
rhesus monkeys. *See* animal experimentation, rhesus monkeys
Röentgen, W. K., 111, 186–88
RU486, 238–39

Sagan, Carl, 53
St. Martin, Alexis, 29–39
Sanger sequencing technique, 250
science education in the United States, 267–69